Technology and Applied Principles of Heat Transfer

Technology and Applied Principles of Heat Transfer

Edited by **Nathan Rice**

$\mathcal{C}$LANRYE
INTERNATIONAL

New Jersey

Published by Clanrye International,
55 Van Reypen Street,
Jersey City, NJ 07306, USA
www.clanryeinternational.com

Technology and Applied Principles of Heat Transfer
Edited by Nathan Rice

International Standard Book Number: 978-1-63240-481-7 (Hardback)

Contents

Preface

Heat transfer is fundamentally described as the energy in transit due to temperature difference. Heat transfer calculations in various facets of engineering applications are significant to aid engineering design of heat exchanging apparatus. Reducing computational time is a demanding task faced by researchers and users. The book discusses the calculation procedure in some application fields, such as differential evaluation of heat recoveries with CFD in a tube bank, heating and ventilation of equipment, and methods for analytical solution of nonlinear problems. Numerical evaluation is the prerequisite for design and manufacture of heat exchanging tools. Numerical and experimental knowledge, as well as the analytical solution of heat transfer is also discussed. Furthermore, the book elaborates on the study of heat transfer phenomenon and its applications.

This book has been the outcome of endless efforts put in by authors and researchers on various issues and topics within the field. The book is a comprehensive collection of significant researches that are addressed in a variety of chapters. It will surely enhance the knowledge of the field among readers across the globe.

It is indeed an immense pleasure to thank our researchers and authors for their efforts to submit their piece of writing before the deadlines. Finally in the end, I would like to thank my family and colleagues who have been a great source of inspiration and support.

Editor

Heat Transfer Calculations

Calculation Methods for Heating and Ventilation System of Electrical Machines

Otilia Nedelcu and Corneliu Ioan Sălişteanu

Additional information is available at the end of the chapter

1. Introduction

To design as accurately electrical machines it is necessary as the cold machine to be efficient and to determine values and distribution of temperature in the machine (along the rotor). For this, must do a complete and accurate analysis of temperature field in steady state (or in transient regime) which requires the knowledge and location of the temperature loss amounts, the knowledge of the velocity distribution of the cooling fluid in the cooling channels different of the machine, the determination of thermal flux values and the distribution of these values in some component parts of the electric machine.

Considering the analogy between a ventilation network of cooling and a nonlinear resistive circuit, the flows of the cooling agent of an electric machine can be determined.

The aim of thermal computation of electric machines is necessary to establish heating, temperature overrunning or overheating in some parts of the machine from the environment temperature at a given operating condition of the machine.

The service life of the insulations used in electric machines is limited heating which greatly decreases in the same time with temperature increasing. Heating determined is stabilized, because the electrical machine which will be discussed has a long-lasting operating regime. Inside the electric machine a heat quantity is produced which has to be exhausted to outside for limiting the excessive heating, heating due to the losses which occur in the active materials.

2. Heat transfer

The heat transfer is the science of spontaneous and irreversible processes of heat propagation in space and is the exchange of thermal energy between two objects, two regions of the same part and respectively two fluids, as a result of a difference of temperature between two fluids.

In technical, the heat transfer must be:

- more intense (e.g. steam boilers, the optimal exchange in heat exchangers);
- Stopped (e.g. steam by thermal insulation).

The heat is the transfer of energy between phisical-chemical systems or between different parts of the same systems, if there is a transformation where is not performed any mechanical work.

The heat transfer deals with the processes where the thermical energy at higher parameters is changed in thermal energy to lower parameters, usually, the parameter which appreciates the heat quality is the temperature, defined as a global measure of processes' intensity which determines the internal energy of an object (thermal agitation of molecules in liquid and gases, atoms vibration and free electrons motion in metals, etc.).

The heat exchange follows thermodynamics principles: **the first principle of thermodynamics** expresses the energy conservation rule and it establishes the quantitative relationship between heat and mechanical work, and allows integral transformation of heat in mechanical work and the reverse; **the second principle of thermodynamics** states the necessary conditions for transforming heat in mechanical work and establishes natural meaning of heat propagation, always from the source with higher temperature to the source with lower temperature.

The main objectives of the heat transfer are: first, the determination or ensuring the amount of heat exchanged per time unit in fixed temperature conditions, and second, checking the compatibility of materials used with the regime of temperature faced by, in determining the temperature field. We can say that an exchange of heat device is an optimal solution from thermic, hydraulic, mechanic, economic and reliability points of view, mainly, the heat transfer being the determining factor. We should add to these, finding the enhancing methods and procedures or, in certain cases, heat transfer breaking. [5]

Conduction, radiation and convection are distinct forms of heat transfer. As defined above, only **conduction** and **radiation** are heat exchange processes due solely to the difference in temperature, **convection** is a more complex process, necessarily involving the mass transfer. Because, the convection makes energy transfer from regions with higher temperature in regions with lower temperature, it was generally accepted „heat transfer by convection" as being the third mode of heat exchange.

Basic sizes in heat transfer

The temperature field. The status scalar parameter, the temperature, in any point of space M(x, y, z) depends on position and time, namely it is a function of the following form.

$$t = f(x,y,z,\tau) \tag{1}$$

The temperature field is such, the total of the temperature t values throughout all space, at some time τ, the expression (1) representing the equation of this field.

The temperature field may be constant (stationary or permanent) and transient (non-stationary or variable) as time τ is explicitly or nonexplicitly in equation (1) and namely:

-   the temperature constant field has the equation:

$$t = f_1(x,y,z), \quad \frac{\partial t}{\partial \tau} = 0 \tag{2}$$

-   the temperature transient field is expressed by equation (1)

Regarding the number of coordinates, the temperature field can be one, two or tridirectional. So, in equations (1) and (2), the temperature field is tridirectional in transient regime, respectively, constant. If the temperature is depending on two coordinates and time, the field is bidirectional transient, with the equation:

$$t = f_2(x,y,\tau), \quad \frac{\partial t}{\partial z} = 0 \tag{3}$$

and if it is expressed as a function of a coordinate and time, the field is onedirectional transient, with equation:

$$t = f_3(x,\tau), \quad \frac{\partial t}{\partial y} = \frac{\partial t}{\partial z} = 0. \tag{4}$$

The equation of temperature in the onedirectional constant field is the most simple:

$$t = f_4(x), \quad \frac{\partial t}{\partial y} = \frac{\partial t}{\partial z} = \frac{\partial t}{\partial \tau} = 0. \tag{5}$$

All points of space considered which have the same temperature t at a moment in time τ form **isothermal surface**. Because, a point in an object can not concomitant has two different temperature values, the result is that isothermal surfaces are continuous surfaces which do not intersect one another. [5]

The temperature gradient is a measurement which expresses elementary increase of temperature to a point of a temperature field, at a given time τ.

The temperature gradient is a normal vector to isothermal surface and it is numerically equal with the limit of the report between the temperature variation Δt between two isothermal surfaces and the distance Δn of these two, measured on normal to surface, when Δn tends to zero, namely:

$$grad\, t = \lim_{\Delta n \to 0} \frac{\Delta t}{\Delta n} = \frac{\partial t}{\partial n} \quad [°C/m] \tag{6}$$

The heat amount ΔQ passing through an object or from an object to another, by an isothermal surface S, in time unit $\Delta\tau$, is called **heat flow (thermal flow)** Q:

$$\dot{Q} = \frac{\Delta Q}{\Delta\tau} \quad [W], \tag{7}$$

The unitary heat flow q_s is the heat flow which crosses the surface unit in the time unit:

$$\dot{q}_S = \frac{\dot{Q}}{S} = \frac{\Delta Q}{S\Delta\tau} \quad [W/m^2] \tag{8}$$

where S is the area of heat exchange surface, in m^2.

Electrical analogy of heat transfer

Two systems are analogous when they have different natures, but respect similar equations with similar boundary conditions. This implies that equations describing the action of a system can be transformed into the equations of the other system by simply exchanging the variable symbols. So, Ohm's law for Electrotechnics expressing the link between DC I, voltage (potential difference) ΔU and the electrical resistance R_e, has an analog form in heat transfer, by the relationship between unitary thermal flow q, temperature difference (thermal potential) Δt and a size called **thermal resistance** R_t, namely:

$$I = \frac{\Delta U}{R_e} \quad ; \quad \dot{q}_S = \frac{\Delta t}{R_t} \tag{9}$$

In equation (9), when q is measured in W/m^2 and Δt is measured in °C, the thermal resistance R_t is expressed in $(m^2 \cdot °C)/W$.

Based on this analogy, a series of concepts of DC circuits can be applied to the heat transfer problems (for example, electrical circuit has an equivalent thermal circuit and reverse) and in dynamic regime (for example, electrical modeling of some transient thermal processes). The electrical analogy of heat transfer can be used as a calculation tool and as viewing heat transfer equations by linking them to the electrical engineering domain.

Corresponding to the three fundamental types of heat transfer, we can establish the expressions for calculating the thermal resistance to conduction, convection and respectively, radiation, which can take place in the complex processes of heat exchange, equivalent electrical diagrams serially or mixed connected.

Inverse thermal resistance is called **thermal conduction.** [5]

3. The calculation methods for heating system of electrical machines

The heat flow Q (the amount of thermical energy which pass through a given surface S in time unit) expressed in J/s is numerically equal with losses of power P, expressed in W, which determine heat in any part of the machine. So, instead of heat flow, in relations used, will use losses P, directly. The object characteristics and the difference between object temperature to the environment that transfers heat determines heat transfer on the mass and on the surface of that object. For thermal calculation, the analogy is made between electrical resistance from the power of a circuit and thermal resistance R from heat flow Q.

Based on this analogy, the temperature drop in mass or surface of an object corresponding to heat flow Q, (or losses P) is determined by multiplying between thermical resistance of the object opposed to heat flow in its mass, respective its surface, R, and losses P, relation (10):

$$\theta = PR\left[°C\right], \tag{10}$$

Where: $\theta = v_1-v_2-$ is the temperature drop, over temperature or object heating, in °C.

On the relation (10) we may work with equivalent *thermal diagrams* with means which, for certain values of thermal resistance and losses, the heat θ is resulting.

The way to solve the thermal schemes, depending on the thermal resistances connection (series, parallel or mixed) is identical to the electrical schemes. [2, 3]

Next we will present the thermal resistances expressions depending on the types of heat transmission and the structure encountered in building electrical machines.

Heat transmission by radiation and convection in the case of heat transmission in axial channels and radial channels of electrical machines

In heat transmission by radiation and convection in the case of heat transmission in axial channels and radial channels of electrical machines, the transmission coefficients of heat depend on the air velocity which blows on the surface. For certain lengths of the channel, transmission coefficients of heat will be taken from the diagrams shown in „figure 1" for axial channels and „figure 2" for radial channels.

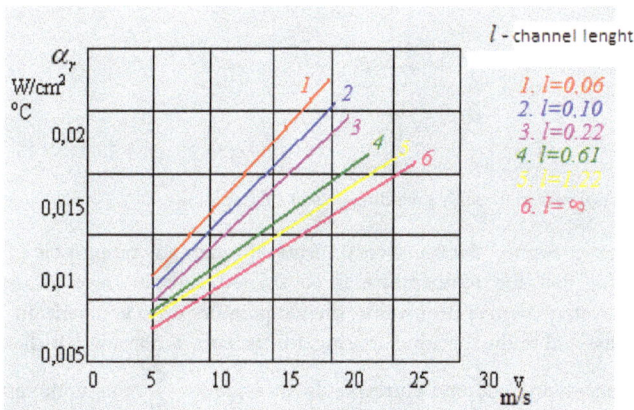

Figure 1. Transmission coefficient of heat in axial channels

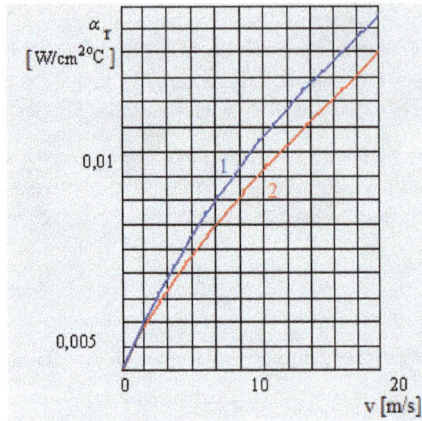

Figure 2. Transmission coefficient of heat in radial channels

4. Common thermal schemes used by electrical machines

Operating winding of electrical machines

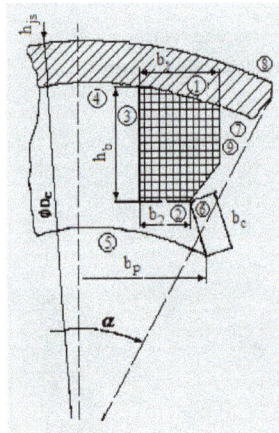

Figure 3. Transversal section through a winding pole of a DC machine.

In "figure 3", it is presented the transversal section through a winding pole of an electrical DC machine, for that the temperature is considered constant around the coil mass, calculating the average temperature value, the temperature drop is practically null, as well as in the coil mass and in the machine housing, for the part of housing which contributes to cooling of a magnetizing coil and corresponds an angle $\alpha = \dfrac{\pi}{2p}$, this is the angle between symmetry axes from a main pole and the auxiliary neighboring poles.

The real thermal diagramme of heat transmission to a coil from a pole of DC machine is indicated in „figure 4", where there are indicated losses by Joule effect of one coil P_{cub}, thermal resistance from coil to housing R_1 corresponding the domain ① of „figure 3", thermal resistance from coil to pole by pole piece R_2 corresponding the domain ② of „figure 3", thermal resistance from coil to pole by the pole body R_3 corresponding the domain ③ to figure 3, thermal resistance from pole to housing R_4 considering the pole-housing jonction $\approx 0,1$ mm corresponding the domain ④, thermal resistance from pole to air by air-gap R_5 corresponding the domain ⑤, thermal resistance from pole to the space air between poles R_6 corresponding the domain ⑥, thermal resistance from housing to air inside machine R_7 corresponding the domain ⑦, thermal resistance from housing to environmental medium R_8 corresponding the domain ⑧, thermal resistance from coil to air R_9 corresponding the domain ⑨. All resistances are measured in $^0C/W$.

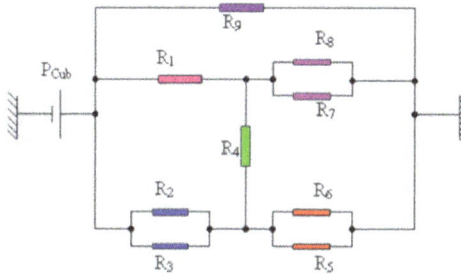

Figure 4. Main diagram of heat transmission to a coil from a pole of DC machines

For learning the total equivalent resistance we need to reproduce intermediate thermal diagrams, „figure 5.a, b".

Thermal resistances for determining the heating of magnetizing coil from „figure 5.a" correspond to relations:

$$R_A = \frac{R_2 R_3}{R_2 + R_3}; \ R_B = R_1; \ R_C = R_4 ;$$
$$R_D = \frac{R_5 R_6}{R_5 + R_6}; \ R_E = \frac{R_7 R_8}{R_7 + R_8}$$

(11)

And for „figure 5.b" after transforming the triangle R_D, R_C and R_E from „figure 5.a" into star R_x, R_y and R_z the relations for thermal resistances are:

$$R_x = \frac{R_C R_D}{R_C + R_D + R_E}; \ R_y = \frac{R_C R_E}{R_C + R_D + R_E}; \ R_z = \frac{R_D R_E}{R_C + R_D + R_E};$$
$$R_I = R_A + R_x; \ R_{II} = R_B + R_y$$

(12)

According to (12) and the equivalent diagram from „figure 5.b" we can determine the total equivalent resistance:

$$R_T = \frac{\left(\dfrac{R_I R_{II}}{R_I + R_{II}} + R_z\right) R_9}{\dfrac{R_I R_{II}}{R_I + R_{II}} + R_z + R_9} \tag{13}$$

After having known the total equivalent resistance R_T , we can determine the heating of magnetizing coil towards the cooling air in the machine:

$$\theta_{cub} = R_T P_{cub} \tag{14}$$

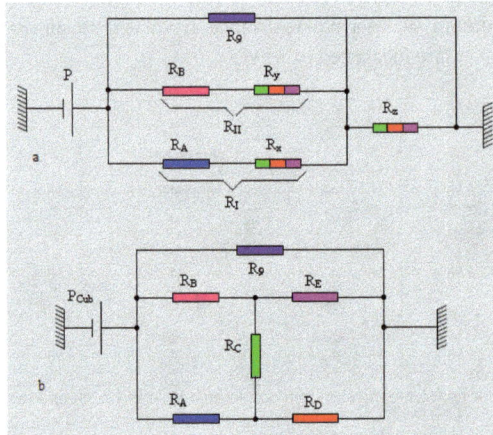

Figure 5. Intermediate thermal diagrams for determining the heating of magnetizing coil: a – real diagram; b – equivalent diagram.

The calculation of thermal resistances from the real diagram (figure 4)

The thermal resistance from coil to machine housing

Thermal resistance from coil to machine housing R_1 is calculated according to total thickness β_{t1} (thickness coil insulation β_b, insulation thickness to the housing β_c, thickness of protecting lac from impregnation $\beta_l \approx 0{,}025$ cm), of equivalent thermal conductivity λ_1 (thermal conductivity when coil is separately isolated λ_b, thermal conductibility for insulation from housing λ_c, thermal conductivity for protecting lac λ_l) and from the corresponding surface S_1 by which the heat is transmitted, considering only the side parts. [2, 3]

$$\beta_{t1} = \beta_b + \beta_c + \beta_l \quad \left[cm\right];$$

$$\lambda_1 = \frac{\beta_{t1}}{\dfrac{\beta_b}{\lambda_b} + \dfrac{\beta_c}{\lambda_c} + \dfrac{\beta_l}{\lambda_l}} \tag{15}$$

$$S_1 = 2b_1\left(l_m + b_m + 2b_1\right), \quad \left[cm^2\right]$$

Where:
b_1 - the thickness of the coil to the upper part, in cm;
b_m – the polar core width, in cm;
l_m – the polar core length, in cm.

These are obtained by:

$$R_1 = \frac{\beta_{t1}}{\lambda_1 S_1} \quad \left[°C/W\right].\tag{16}$$

The thermal resistance from the coil to the polar piece.

The thermal resistance from the coil to the polar piece, R_2, is calculated as the thermal resistance from coil to machine housing for same insulation, but get calculating only the isolated frame bottom and the transmission surface of heat S_2.

$$R_2 = \frac{\beta_{t2}}{\lambda_2 S_2} \quad \left[°C/W\right]$$
$$\beta_{t2} = \beta_{t1} \quad and \quad \lambda_2 = \lambda_1 \tag{17}$$
$$S_2 = 2b_2\left(l_m + b_m + 2b_2\right), \quad \left[cm^2\right]$$

where b_2 is the thickness the coil to the below part, in cm.

The thermal resistance from the coil to the pole object

The thermal resistance from the coil to pole object R_3, is calculated according to the thickness of the insulated coil β_b only when the coil is isolated separately, β_m, the thickness of the insulation toward the body of the pole, the impregnation layer $\beta_l \approx 0{,}025$ cm, the thermal conductivity of the insulated coil λ_b, thermal conductivity of insulation from the pole body λ_m, the thermal conduction of the protecting lac λ_l and of the corresponding surface S_3.

$$\beta_{t3} = \beta_b + \beta_m + \beta_l \quad \left[cm\right]$$
$$\lambda_3 = \frac{\beta_{t3}}{\dfrac{\beta_b}{\lambda_b} + \dfrac{\beta_m}{\lambda_m} + \dfrac{\beta_l}{\lambda_l}} \tag{18}$$
$$S_3 = 2b_3\left(l_m + b_m + 2b_3\right), \quad \left[cm^2\right]$$

$$R_3 = \frac{\beta_{t3}}{\lambda_3 S_3} \quad \left[°C/W\right]\tag{19}$$

The thermal resistance from the pole core to the housing

The thermal resistance from the pole core to the housing R_4 is calculated according to the thickness of the parasite airgap to the pole-housing jonction at about 0,1 mm, for thermal conductivity in electrical machines whose stator is not impregnated after winding $\lambda_4 =$

0,00025 W/cm°C, or for electrical machines with the impregnated stator $\lambda_4 = \lambda_t$, and, also according to the corresponding calculated surface with b_m the polar core width and l_m the polar core length.

$$R_4 = \frac{0,01}{\lambda_4 S_4} \left[°C/W \right] \qquad (20)$$

The thermal resistance to the heat transmission from the core pole to airgap

$\alpha_0 = 1,42 \times 10^{-3}$ W/cm² °C	Iron and steel surfaces, grouted (housing surfaces)
$\alpha_0 = 1,67 \times 10^{-3}$ W/cm² °C	Iron and steel surfaces, nongrouted
$\alpha_0 = 1,33 \times 10^{-3}$ W/cm² °C	Cooper or brass surfaces, and coil impregnated or varnished surfaces

Table 1. The coefficient values of heat transmission α_0, on the non-blast surfaces

Considering the corresponding surface S_5 calculated according to the length and width of the polar piece l_p respectively b_p, the thermal resistance to the heat transmission from the core pole to airgap is determinate by relation:

$$R_5 = \frac{1}{\alpha_{t5} S_5} \left[°C/W \right] \qquad (21)$$

The coefficient of heat transmission on the object surface α_{t5} is determined admitting that $\theta'_{aer} = 20°C$ is half of the total heating of the air in the machine, $k = 0,8$ chosen from „table 2″, the coefficient of heat transmission chosen from „table 1″ is $\alpha_0 = 1,67 \times 10^{-3}$W/cm² °C and air velocity in the airgap $v_\delta = \frac{1}{2} v_r$.

$$\alpha_{t5} = \alpha_0 \left(1 + k\sqrt{v_\delta}\right)\left(1 - 0,5a\right) \left[W / cm^2 °C \right]$$

$$a = \frac{\theta'_{aer}}{\theta_a}$$

$$a = \frac{20°}{70°} = 0,28 \text{ class of insulation B, } a = \frac{20°}{90°} = 0,22 \text{ class of insulation F}$$

$$S_5 = b_p l_p \ [cm^2].$$

The name of the surface and the type of blast	k
The blast surface in the most perfect way	1,3
The surface of the front linkage of induced winding	1,0

The name of the surface and the type of blast	k
The surface of the active part of the induced	0,8
The surface of the magnetizing coils	0,8
The collector surface	0,6
The outer surface of the housing	0,5

Table 2. The coefficient values of blast intensities k

The thermal resistance to heat transmission from polar piece to the space between poles

In the heat transmission to space between poles, by polar pieces, the thermal resistance is determined regarding the coefficient of heat transmission α_{t6} and the surface located on the polar width of the object b_c l_m polar core length:

$$R_6 = \frac{1}{\alpha_{t6}S_6} \quad \left[^\circ C/W\right]$$

$$S_6 = 2b_c l_m \left[cm^2\right] \tag{22}$$

$$\alpha_{t6} = 1{,}6.7 \times 10^{-3}\left(1 + 0{,}8\sqrt{v_6}\right)\left(1 - 0{,}5a\right) \left[W/cm^2{}^\circ C\right], \ cuv_6 = v_p$$

The thermal resistance to heat transmission from the housing to interior

When heat transmission is done from housing to interior, the thermal resistance R_7 is determined the same as in heat transmission to the space between poles, by polar pieces with $\alpha_{t7} \approx \alpha_{t6}$, $v_7 = v_6$, and the corresponding surface S_7 is calculated regarding the interior diameter of the housing D_c, the number of pole pairs p and the housing length l_c.

$$R_7 = \frac{1}{\alpha_{t7}S_7} \quad \left[^\circ C/W\right]$$

$$S_7 \approx \frac{\pi D_c}{4p}l_c - \left(b_m + 2b_1\right)\left(l_m + 2b_1\right) \quad \left[cm^2\right] \tag{23}$$

The thermal resistance to heat transmission from housing to exterior (quiet air)

The heat transfer from the housing to outer where the air is quiet, thermal resistance is calculated taking into account the heat transmission coefficient $\alpha_0 = 1{,}42 \times 10^{-3}$ W/cm²⁰C determined in „table 1", by the appropriate surface S_8, at which involved, the height of stator yoke h_{js}, the inner diameter of the housing D_c, the number of pole pairs p and the housing length l_c.

$$R_8 = \frac{1}{\alpha_0 S_8} \quad \left[^\circ C/W\right]$$

$$S_8 = \frac{\pi\left(D_c + 2h_{js}\right)}{4p}l_c \tag{24}$$

The thermal resistance to heat transmission from coil directly to cooling air in the machine

When the heat transmission is done from coil directly to cooling air in the machine, the thermal resistance depends on the thickness and the thermal conductivity of coil insulation when the coil is isolated separately β_b and λ_b, by the coefficient of heat transmission α_{t9}, and by the corresponding surface S_9.

$$R_9 = \frac{\beta_b}{\lambda_9 S_9} + \frac{1}{\alpha_{t9} S_9} \quad \left[{}^\circ C / W \right]$$

$$\alpha_{t9} = \alpha_0 \left(1 + 0{,}8\sqrt{v_9} \right)\left(1 - 0{,}5a \right) \quad \left[W / cm^2 {}^\circ C \right], \tag{25}$$

$$\alpha_0 = 1{,}33 \bullet 10^{-3}. \, W / cm^2 {}^\circ C, \text{ from table 1; } v_9 = v_6;$$

$$S_9 \approx 2\left(b_m + l_m + 4b_1 \right)1{,}1h_b \quad \left[cm^2 \right]$$

5. The losses by Joule effect, a single coil

The losses by Joule effect P_{cub} are the total losses in the magnetizing reeling P_{cuc} reported to the number of poles of the electrical machine $2p$.

$$P_{cub} = \frac{P_{cuc}}{2p} \quad \left[W \right] \tag{26}$$

6. The total heat of magnetizing reeling

Heating the magnetizing coil to the air in the machine, is determined by the relation (14) together with heating the air to the contact with the coil ($\theta_{aer} \approx 20^\circ C$ – from rotor to stator) determining heating of the magnetizing reeling towards the external environment.

$$\theta_{cue} = \theta_{cub} + \theta'_{aer} < \theta_{admis} \tag{27}$$

7. The diagrams of ventilation and calculation of aerodynamic resistances

In the construction of rotating electrical machines, the cooling air is most frequently used. Depending on the construction type and the type of protection of the electrical machine the best ventilation system is chosen.[1]

Calculating ventilation of rotating electrical machine is done as follows:

-    establishing the ventilation diagram and calculating aerodynamic resistances;
-    calculating the pressure determined by means of air entrainment;
-    calculating air flow at different sections.

Preparing the ventilation diagram for a rotating electrical machine requires prior execution of the diagram to the scale of general assembly (longitudinal section and cross section) and establishing the routes where cooling air circulates. For calculation aerodynamic resistances we need to know longitudinal and transversal dimensions of all pipes that form the routes for air circulation.

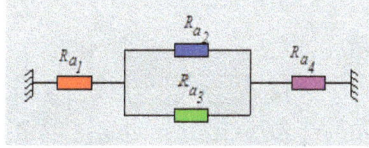

Figure 6. The diagram of mixed connection of aerodynamic resistances

When we solve an aerodynamic diagram, we aim to find the equivalent resistance R_{ae}.. Between aerodynamic diagrams used for air circuits and electrical diagrams may be an analogy, the aerodynamic resistances act as passive elements.

Determining the equivalent resistance depends on how aerodynamic resistances are connected:

-   to series arrangement R_{ae} is the sum of the aerodynamic resistances and it is determined by relation:

$$R_{ae} = \sum_{i=1}^{n} R_{ai}$$

(28)

-   to shunt connection R_{ae} is determined by relation:

$$\frac{1}{\sqrt{R_{ae}}} = \sum_{i=1}^{n} \frac{1}{\sqrt{R_{ai}}}$$

(29)

-   to mixed connection, „figure 6", R_{ae} is calculated according to relations of calculation for the series arrangement and shunt connection and we obtain:

$$R_{ae} = R_{a1} + \frac{R_{a2} R_{a3}}{\left(\sqrt{R_{a2}} + \sqrt{R_{a3}}\right)} + R_{a4}$$

(30)

Bilateral axial-radial ventilation for an electrical machine in protected construction

In „figure 7.a" are represented the longitudinal section and the frontal view for an induction motor with the rotor in short-circuit, in protected construction, and with arrows we indicate routes that circulate the cooling air.

Typically, in asynchronous machine the role of inductor is accomplished by the stator, while the rotor is the induced machine. The machine excitation is done in AC. The collector body is of „collector rings" type.

The asynchronous machine can not develop a couple unless the speed „n" of the rotor is different by the speed „n_1" of the stator rotating field. [2, 3]

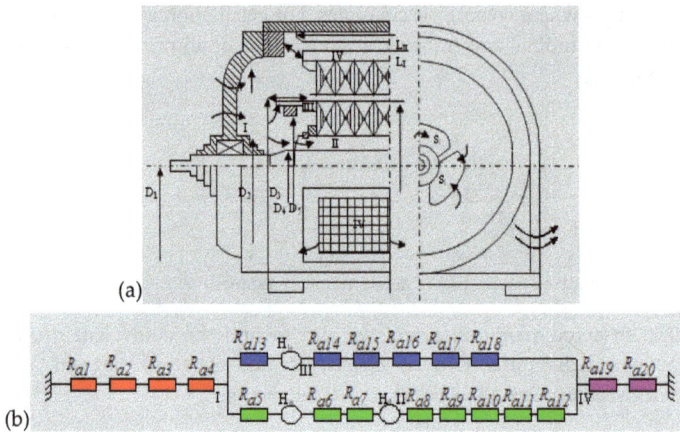

Figure 7. Ventilation diagram of asynchronous electrical machine with bilateral radial ventilation: a) physical diagram; b) main diagram.

Because depression created in the rotor bars ends and the rotor ribs (the electrical machine has not its own ventilator), the air is absorbed through the windows of the two fore-shields, washes the active elements and it is thrown out through the windows of its middle, by the bottom of the housing.

Air flow is symmetrical to the middle of the machine length, and is, therefore, sufficient to calculate ventilation only on one side, which is half of the machine.

The sources of pressure are denoted by „H", the aerodynamic resistance by „R_a" and are divided in sections. [3, 11, 12]

8. Section I

We established I section, which contains aerodynamic resistances R_{a1}, $R_{a\,2}$, $R_{a\,3}$, and $R_{a\,4}$, which are connected in series and is calculated the total aerodynamic resistance for section I, according the relation:

$$R_{atrl} = \sum_{i=1}^{4} R_{ai} \qquad (31)$$

For determining total aerodynamic resistance of the I section we calculate, in turns, each aerodynamic resistance which is part of the I section regarding the coefficients of aerodynamic resistances and the sections of the channels by which the air circulates.

The air enters through the shield windows under sharp edges (blinds) by narrowing from the external environment to the input section. From „table 3", for sharp edges $\zeta_{as} = 62 \cdot 10^{-2}$, and for narrowing $\dfrac{S_1}{S_{ext}} \approx 0$, as in „figure 8" $\zeta_{ing} \approx 32 \cdot 10^{-2}$.

Conditions of entry	ζ
Entry with sharp edges	62×10^{-2}
Entry with edges at angle of 90°	30×10^{-2}
Entry with rounded edges	$(12,5...0) \times 10^{-2}$

Table 3. The values of the resistance coefficient to the air admission pipes

The angle of deviation	ζ
135°	32×10^{-2}
90°	70×10^{-2}

Table 4. The values of the resistance coefficient to deviation in different angles

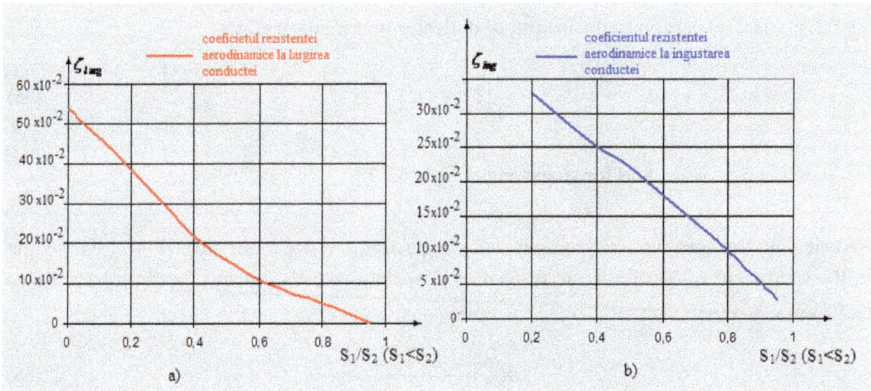

Figure 8. The aerodynamic coefficient resistance to widening and narrowing core

Aerodynamic resistance R_{a1} is calculated according to the net input surface in the shield windows S_1, to the coefficient of aerodynamic resistance to narrowing ζ_{ing} and to the coefficient for sharp edges ζ_{as}.

$$R_{a1} = \frac{\zeta_{as} + \zeta_{ing}}{S_1^2} = \frac{(62 + 32) \cdot 10^{-2}}{S_1^2} \qquad (32)$$

Where: $- S_1 = k_i S_i \; \left[m^2\right]$

- S_i is the total surface of the shield windows, in m²;
- k_i is the coefficient of reduction in section because the coatings required by the machine protection: $k_i = 0,6 \div 0,7$

There is a relaxation of the air coming out of the shield and to the input in the machine and a deviation to 135° by deflector (figure 7.a)

$$S_2 = \frac{\pi}{4}\left(D_1^2 - D_2^2\right) \ \left[m^2\right] \tag{33}$$

For $\frac{S_1}{S_2} = ... \Rightarrow \zeta_{larg} = ...$, and for the deviation to 135° from „table 4" $\Rightarrow \zeta_r = 32 \cdot 10^{-2}$

The aerodynamic resistance R_{a2} is calculates according to the net output surface of windows shield S_1, to the coefficient of aerodynamic resistance to enlargement ζ_{larg} and to the coefficient to the air relaxation of shield output ζ_{as}.

$$R_{a2} = \frac{\zeta_{larg} + \zeta_r}{S_1^2} \tag{34}$$

Next, there is a relaxation to the output of deflector under sharp edges.

$$S_3 = \frac{\pi}{4}\left(D_3^2 - D_4^2\right) \ \left[m^2\right] \tag{35}$$

For $\frac{S_2}{S_3} = ... \Rightarrow \zeta_{larg} = ...$, and for sharp edges $\zeta_{as} = 62 \cdot 10^{-2}$

We calculate the aerodynamic resistance R_{a3} regarding the net input surface in the machine S_2, the coefficient of aerodynamic resistance to widening ζ_{larg} and the coefficient for sharp edges ζ_{as}.

$$R_{a3} = \frac{\zeta_{larg} + \zeta_{as}}{S_2^2} \tag{36}$$

At the rotor winding input occurs a narrowing and there are sharp edges for the short-circuited rotor.

$$S_4 = \frac{\pi}{4}\left(D_5^2 - D_4^2\right) \ \left[m^2\right] \tag{37}$$

For $\frac{S_4}{S_3} = ... \Rightarrow \zeta_{ing} = ...$, and for sharp edges $\zeta_{as} = 62 \cdot 10^{-2}$.

We calculate the aerodynamic resistance R_{a4} regarding the net input surface under rotor winding S_4, the coefficient of aerodynamic resistance to narrowing ζ_{ing} and the coefficient for sharp edges ζ_{as}.

$$R_{a4} = \frac{\zeta_{ing} + \zeta_{as}}{S_4^2} \tag{38}$$

After we calculated the aerodynamic resistances R_{a1}, R_{a2}, R_{a3}, R_{a4}, regarding the narrowing or widening ventilation channels by relation (31) we calculate the *total aerodynamic resistance of the section I.*

9. Section II

The II section contains aerodynamic resistances R_{a5}, R_{a6}, R_{a7}, R_{a8}, R_{a9}, R_{a10}, R_{a11}, R_{a12}, connected in series and we calculate the total aerodynamic resistance to the section II with the relation:

$$R_{atrII} = \sum_{i=5}^{12} R_{ai} \tag{39}$$

To calculate the aerodynamic total resistance by section II, we calculate, as for section I, each aerodynamic resistance, in turns, considering the widening and the narrowing of the ventilation channels.

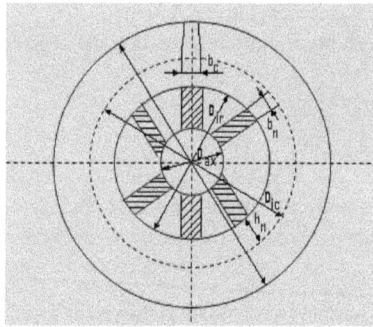

Figure 9. Cross section in the rotor with ribbed shaft

When the air begins to cross the section II, it enters under the rotor core which occurs in narrowing, and the input is done under sharp edges (figure 9).

$$S_5 = \frac{\pi}{4}\left(D_{ir}^2 - D_{ax}^2\right) - n_n b_n h_n \quad \left[m^2\right] \tag{40}$$

Where:
- D_{ir}, is the interior diameter of the armature core disc, in m;
- D_{ax} is the diameter of the shaft, in m;
- b_n is the ridge width, in m;
- $h_n = \dfrac{D_{tr} - D_{ax}}{2}$ is the ridge height, in m;
- n_n is the ridges number.

For $\dfrac{S_5}{S_4} = ... \Rightarrow \zeta_{ing} = ...$ and for sharp edges $\zeta_{as} = 62 \cdot 10^{-2}$.

We calculate the aerodynamic resistance R_{a5}, regarding the net surface to the output under the rotor core S_5, the coefficient of aerodynamic resistance to the narrowing ζ_{ing} and the coefficient for sharp edges ζ_{as}.

$$R_{a5} = \frac{\zeta_{ing} + \zeta_{as}}{S_5^2} \qquad (41)$$

For the input of radial ventilation channels of the rotor under sharp edges occurs a deviation to 90°, $\zeta_c = 70 \cdot 10^{-2}$, from „table 3" $\left(\zeta_{as} = 62 \cdot 10^{-2}\right)$ and a narrowing.

$$S_6 = 0,9\pi D_{ir} b_v \frac{n_v}{2} \; \left[m^2\right] \qquad (42)$$

Where:
-    b_v is the width of the radial channel for ventilation, in m;
-    n_v is the number of radial channels for ventilation;
-    0,9 is the coefficient which takes into consideration the presence of spacers.

For $\dfrac{S_6}{S_5} = ... \Rightarrow \zeta_{ing} = ...$

The aerodynamic resistance R_{a6} is calculated, using the relation (42) which takes into consideration the net surface to the input of radial channels of the ventilator S_6, the coefficient of aerodynamic resistance to narrowing ζ_{ing} and the coefficient for sharp edges ζ_{as}.

After the air entered in the channels of ventilation the relaxation occurs (figure 9)

$$S_7 = 0,9\pi D_{ic} b_v \frac{n_v}{2} \; \left[m^2\right] \qquad (43)$$

Where:

-    D_{ic} is the diameter of the rotor slots base, in m.

For $\dfrac{S_6}{S_7} = ... \Rightarrow \zeta_{larg} = ...$

The aerodynamic resistance R_{a7} is calculated, taking into consideration the relaxation in the channels of ventilation:

$$R_{a7} = \frac{\zeta_{larg}}{S_6^2} \qquad (44)$$

A new narrowing occurs to the entry in the region of rotor coils under sharp edges, for the short-circuited rotor or the rounded edges of rotors and widening to the outside rotor.

$$S_8' = S_7 - z_2 b_{c2} b_v \frac{n_v}{2} \quad \left[m^2 \right] \tag{45}$$

$$S_8'' = 0{,}9\pi D_r b_v \frac{n_v}{2} - Z_2 b_{c2} b_v \frac{n_v}{2} \quad \left[m^2 \right] \tag{46}$$

Where:

- D_r is the exterior diameter of the rotor, in m;
- z_2 is the number of rotor slots;
- b_{c2} is the width of the rotor slot, in m.

For $\dfrac{S_8'}{S_7} = ... \Rightarrow \zeta_{ing} = ...$ and for sharp edges $\zeta_{as} = 62 \cdot 10^{-2}$.

For $\dfrac{S_8'}{S_8''} = ... \Rightarrow \zeta_{larg} = ...$

We obtained the aerodynamic resistance R_{a8}, calculated to the input of the coils region by narrowing the ventilation channel and the output of rotor by widening the ventilation channels for short-circuited rotor where the edges are sharp:

$$R_{a8} = \frac{\zeta_{ing} + \zeta_{as} + \zeta_{larg}}{S_8'^2} \tag{47}$$

And R_{a8}, for phase-wound rotor where the resistance of coils rounded edges are neglected is:

$$R_{a8} = \frac{\zeta_{ing} + \zeta_{larg}}{S_8'^2} \tag{48}$$

At the transition from the rotor coils zone to the stator coils zone and the input under rounded edges a widening (or a narrowing) occurs $\left(\zeta_r = 12{,}5 \cdot 10^{-2} \right)$.

$$S_9' = 0{,}9\pi D b_v \frac{n_v}{2} - z_1 b_{c1} b_v \frac{n_v}{2} \quad \left[m^2 \right] \tag{49}$$

Where:

- D is interior diameter of the stator, in m;
- z_1 is the number of slots of the stator;
- b_{c1} is the width of stator slots, in m.

For the rapport $\dfrac{S_8''}{S_9'}$ or $\dfrac{S_9'}{S_8''}$ (the lower section is used to the numerator) we determine ζ_{larg} or ζ_{ing}. For determining the resistance we put at the denominator the smallest section.

The resistance R_{a9} is obtained, taking into account the net surface in the stator coils zone S_9, the aerodynamic resistance coefficient to narrowing/widening ζ_{ing} /ζ_{larg} and the coefficient for the input under sharp edges ζ_r.

$$R_{a9} = \frac{\zeta_{larg(ing)} + \zeta_r}{S_{8(9)}^{"2}} \tag{50}$$

In the stator coils zone and to the output between coils, the net surface increases and a relaxation occur.

$$S_{10} = 0,9\pi\left(D + 2h_{cl}\right)b_v\frac{n_v}{2} \quad \left[m^2\right] \tag{51}$$

$$S_9' = S_{10} - z_1 b_{c1} b_v\frac{n_v}{2} \quad \left[m^2\right] \tag{52}$$

Where:

-   h_{cl} is the height of the stator slot, in m.

For $\dfrac{S_9'}{S_9''} = ... \Rightarrow \zeta'_{larg} = ... $; For $\dfrac{S_9''}{S_{10}} = ... \Rightarrow \zeta''_{larg} = ... $

The resistance R_{a10}, is calculated taking into account the net surface in the stator coils zone S_9 and the net surface to the output between coils S_{10}, the aerodynamic resistance coefficient to widening ζ_{larg} in both zones.

$$R_{a10} = \frac{\zeta'_{larg}}{S_9'^2} + \frac{\zeta''_{larg}}{S_9''^2} \tag{53}$$

To the exterior armature core disks in the radial channels of the stator a widening is produced, again, the surface S_{11} is bigger than the surface S_{10}.

$$S_{11} = 0,9\pi D_e b_v\frac{n_v}{2} \quad \left[m^2\right] \tag{54}$$

Where:

-   D_e is the exterior diameter of the stator core disk, in m.

For $\dfrac{S_{10}}{S_{11}} = ... \Rightarrow \zeta'_{larg} = ... $

The result is R_{a11}:

$$R_{a11} = \frac{\zeta_{larg}}{S_{10}^2} \tag{55}$$

To the output of radial channels of the stator, the air suffers a relaxation again.

$$S_{12} = \pi D_e \frac{L_1}{2} \quad [m^2]$$ (56)

Where:

- L_1 is the geometric length of the machine plus the thickness pressure rings, in m.

For $\dfrac{S_{11}}{S_{12}} = ... \Rightarrow \zeta_{larg} = ...$

Because there is a widening of ventilation channels to the output of the stator radial channels R_{a12} is calculated regarding the aerodynamic resistance coefficient to widening and S_{11}:

$$R_{a12} = \frac{\zeta_{larg}}{S_{11}^2}$$ (57)

10. Section III

The aerodynamic resistances R_{a13}, R_{a14}, R_{a15}, R_{a16}, R_{a17}, R_{a18} constitute the section III, they are serially connected. Taking into account the equation (28) the total aerodynamic resistance is calculated by the section III with equation:

$$R_{atrIII} = \sum_{i=13}^{18} R_{ai}$$ (58)

After the air got out of the stator radial channels, it will cross the rotor ends coils by deviation to 135° for which $\zeta_c = 32 \cdot 10^{-2}$ (figure 7 and figure 10).

$$S_{13} = \pi D_3 l_{br} \quad [m^2]$$ (59)

Where:

- l_{br} is the axial length of the rotor coil end, in m.

Because there is no widening or narrowing, the resistance R_{a13} is calculated regarding the crossing surface to the rotor coil ends and the coefficient for rounded edges:

$$R_{a13} = \frac{\zeta_c}{S_{13}^2}$$ (60)

After that a narrowing to input follows between rotor bars ends and sharp edges. ζ_{as}= 62 x 10^{-2} for short-circuited rotors from bars with sharp edges. For the rotor with bars with rounded edges ζ_{as}=0.

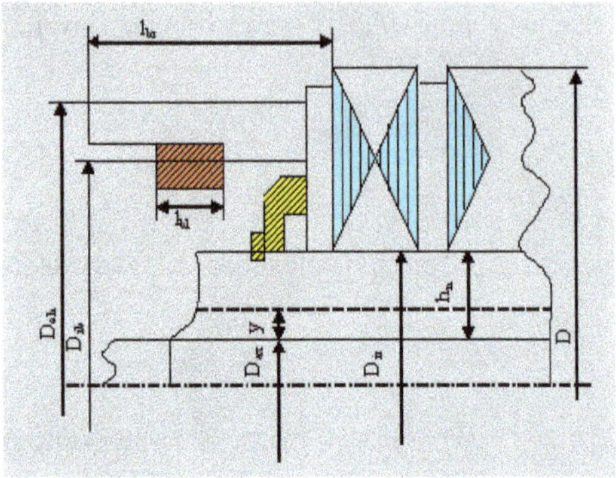

Figure 10. Partial longitudinal section through a rotor cage induction motor with ribbed shafts and bars ends having a ventilator effect (the rotor from figure 7)

$$S_{14} = \pi D_3 \left(l_{br} - l_{bi} \right) - z_2 b_{r2} \left(l_{br} - l_{bi} \right) \quad \left[m^2 \right] \tag{61}$$

Where:

- l_{bi} is the shorting ring width, in m (figure 10)

For $\dfrac{S_{14}}{S_{13}} = ... \Rightarrow \zeta_{ing} = ...$

Resulting the resistance R_{a13} calculated using the net surface to the input between the rotor bars ends and the coefficients of aerodynamic resistances to narrowing under sharp edges:

$$R_{a14} = \frac{\zeta_{as} + \zeta_{ing}}{S_{14}^2} \quad or \quad R_{a14} = \frac{\zeta_{ing}}{S_{14}^2} \tag{62}$$

The isolated support of the bandages does not allow air to enter through the coils and as a result in the right front end, to winding motors only a narrowing occurs.

To the ends output (the bars) of the rotor coils there is a relaxation of the air, a relaxation caused by widening the channel (figure 7)

$$S_{15} = \pi D l_d \quad \left[m^2 \right] \tag{63}$$

For $\dfrac{S_{14}}{S_{15}} = ... \Rightarrow \zeta_{larg} = ...$

Is obtained:

$$R_{a15} = \frac{\zeta_{ing}}{S_{14}^2} \tag{64}$$

There is a passing through the spaces between the stator coils ends which can be of two types.

- prefabricated coils
- winding to the round conductor.

The prefabricated coils (from profiled conductor or bars) are represented in „figure 11".

We determine the surface S_{16} under the front part to which we consider the narrowing of the lower layer.

$$S_{16} = \pi Dl_f \quad \left[m^2\right] \tag{65}$$

We admit then, that the distances d_1 and k between coil ends, to lower part, are equal to those at the upper part (d_1 and k are constant on front ends height).

Narrowing at the lower layer (figure 11)

$$S'_{16} = \left[l_{bs}d_1 + \left(l_{mi} - b_c\right)a + kc - b_d d_1 n_d\right]z \quad \left[m^2\right] \tag{66}$$

Where:

- n_d is the number of distances between coil adjacent ends;
- l_{mi} is the dental corresponding to the mean diameter of the coil lower side, in m.

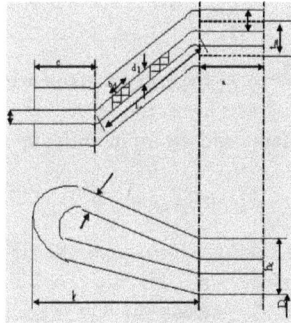

Figure 11. End of prefabricated stator coil

For $\dfrac{S'_{16}}{S_{16}} = ... \Rightarrow \zeta_{ing} = ...$, and for rounded edges $\zeta_r = 12,5 \times 10^{-2}$

The resistance R'_{a16} is:

$$R'_{a16} = \frac{\zeta_r + \zeta_{ing}}{S'^2_{16}} \tag{67}$$

For the output relaxation from lower layer S''_{16} is:

$$S_{16}^{''} \approx \pi\left(D+h_{c1}\right)l_f \quad \left[m^2\right] \tag{68}$$

Where:

h_{c1} is height of stator slot, in m.

For $\dfrac{S_{16}'}{S_{16}''} = ... \Rightarrow \zeta_{larg} = ...$

We obtained:

$$R_{a16}^{''} = \frac{\zeta_{larg}}{S_{16}'^{2}} \tag{69}$$

Narrowing in the upper layer and the input under the slightly rounded edges.

$$S_{16}''' \approx S_{16}'; \quad \text{for} \quad \frac{S_{16}'}{S_{16}''} = ... \Rightarrow \zeta_{ing} = ...$$

For the slightly rounded edges from „table 3" $\zeta_r = 12{,}5 \times 10^{-2}$

The resistance R''_{a16} is:

$$R_{a16}^{'''} = \frac{\zeta_r + \zeta_{ing}}{S_{16}'^{2}} \tag{70}$$

The total resistance to airflow between the ends of stator winding is the sum of calculated resistances for narrowing in the lower layer, the relaxation to the output from lower layer and the narrowing in the upper layer and the input under the slightly rounded edges:

$$R_{a16} = R_{a16}' + R_{a16}'' + R_{a16}''' \tag{71}$$

In the case of winding from round conductor (from wire), the air can not penetrate through the coil ends, because spaces between coils are obstructed of isolated strips between phases, so the air avoid them. The distance X between the end of stator coil and deflector, must be large enough not to exclude the air (figure 7.a)

The narrowing in the section X.

$$S_{16} \cong \pi\left(D+h_{c1}\right)X \left[m^2\right] \tag{72}$$

For $\dfrac{S_{16}}{S_{15}} = ... \Rightarrow \zeta_{ing} = ...$ and the resistance R_{a16} is:

$$R_{a16} = \frac{\zeta_{larg}}{S_{16}^2} \tag{73}$$

After passing by front ends of stator winding there is a relaxation of air after widening the ventilation channel.

$$S_{17} \approx \pi\left(D + 2h_{c1}\right)\left(l_f + X\right) \ \left[m^2\right] \tag{74}$$

For $\dfrac{S_{16}}{S_{17}} = ... \Rightarrow \zeta_{ing} = ...$ and aerodynamic resistance R_{a17} is:

$$R_{a17} = \frac{\zeta_{larg}}{S_{16}^2} \tag{75}$$

The air flows in a narrow channel between housing ribs to a deviation of 135° (ζ_c= 62 x 10⁻²).

$$S_{18} = \pi D_c p - b_{nc} p n_{nc} \ \left[m^2\right] \tag{76}$$

Where:
- b_{nc} is the housing rib width, in m;
- n_{nc} is the number of the housing ribs;
- D_c, p (from figure 7.a) in m.

For $\dfrac{S_{18}}{S_{17}} = ... \Rightarrow \zeta_{ing} = ...$ and so:

$$R_{a18} = \frac{\zeta_c + \zeta_{as} + \zeta_{ing}}{S_{18}^2} \tag{77}$$

11. Section IV

The total resistance of section IV is calculated as in the other sections with the equation (26) for the series connection:

$$R_{atrIV} = \sum_{i=19}^{20} R_{ai} \tag{78}$$

A deviation at 90° occurs (ζ_c= 70 x 10⁻²) to the output sided windows (figure 7.a) without the narrowing or widening of ventilation channels:

$$S_{19} = \frac{L_n B_n}{2} \ \left[m^2\right] \tag{79}$$

And the result is:

$$R_{a19} = \frac{\zeta_c}{S_{19}^2} \tag{80}$$

The output of the air from the housing is done to the deviation at $90°(\zeta_c= 70 \times 10^{-2})$ for the relaxation of the outside air (figure 7.a).

$$S_{20} = \frac{1}{2}L_p h_p k_i \quad \left[m^2\right]$$

(81)

For this k_i we must take into account the calculation of R_{a1}.

For $\dfrac{S_{20}}{S_{ex}} \approx 0 \Rightarrow \zeta_{larg} = 60 \cdot 10^{-2}$ and so:

$$R_{a20} = \frac{\zeta_c + \zeta_{larg}}{S_{20}^2}$$

(82)

In the scheme from „figure 7.b", H_b is the pressure produced by the ends of moving rotor bars, H_n is the pressure produced by rotor ribs, and H_d is the pressure produced by rotor spacers.

12. The calculation air flow

The calculation of air flow is done after we solved the ventilation scheme regarding the air flows of different sections and after we calculated the aerodynamic resistances values and pressures values created by sources.

Bilateral radial-axial ventilation for an electrical machine in protected construction

We calculate the flows for the asynchronous electrical machine with bilateral radial ventilation from „figure 7.a", the cooling air is driven only by the effect of ventilator of ends of rotor bars and of ribs and spacers on rotor.

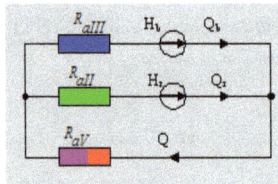

Figure 12. Variation for the representation of scheme from figure 7.b

The flows Q_b, Q_r and Q by sections III, II, and V, respectively I and IV from „figure 7.b", are determined by equations:

$$Q_b = \sqrt{y} \quad \left[m^3/s\right]$$

(83)

$$Q_r = \frac{1}{\sqrt{R_{all}}}\sqrt{R_{aIII}Q_b^2 - H_b + H_r} \quad \left[m^3/s\right]$$

(84)

$$Q = Q_b + Q_r \tag{85}$$

When the flow which enters and exit from the machine is less than the flow calculated for half of the machine it is necessary to adopt a solution by decreasing the equivalent resistances on the sections by sections increasing, or giving up the constructive solution adopted and finding other solutions.

The example of calculation for asynchronous electrical machine with bilateral radial ventilation:

- We calculate the aerodynamic resistance on every section, starting from the nominal data.

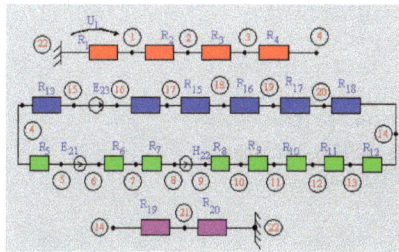

Figure 13. The principal ventilation scheme of asynchronous electrical machine with bilateral radial ventilation

- Equivalent resistance by section I is: $R_{atrI} = \sum_{i=1}^{4} R_{ai}$

- Equivalent resistance by section II is: $R_{atrII} = \sum_{i=5}^{12} R_{ai}$

- Equivalent resistance by section III is: $R_{atrIII} = \sum_{i=13}^{18} R_{ai}$

- Equivalent resistance by section IV is: $R_{atrIV} = \sum_{i=19}^{20} R_{ai}$

By means of the computer program PANCIA (Program Analysis of Analog Circuits) we obtain the system solution. [7, 8, 9, 10]

• **The unknowns of the system**								
	V1	V2	V3	V4	V5	V6	V7	V8
V9	V10	V11	V12	V13	V14	V15	V16	V17
V18	V19	V20	V21	I21	I22	I23		

• **The system of equations**
+ (+G1+G2)*V1 + (-G2)*V2 = 0
+ (-G2)*V1 + (+G2+G3)*V2 + (-G3)*V3 = 0

$+ (-G3)*V2 + (+G3+G4)*V3 + (-G4)*V4 = 0$
$+ (-G4)*V3 + (+G4+G5+G13)*V4 + (-G5)*V5 + (-G13)*V15 = 0$
$+ (-G5)*V4 + (+G5)*V5 + (+1)*I21 = 0$
$+ (+G6)*V6 + (-G6)*V7 + (-1)*I21 = 0$
$+ (-G6)*V6 + (+G6+G7)*V7 + (-G7)*V8 = 0$
$+ (-G7)*V7 + (+G7)*V8 + (+1)*I22 = 0$
$+ (+G8)*V9 + (-G8)*V10 + (-1)*I22 = 0$
$+ (-G8)*V9 + (+G8+G9)*V10 + (-G9)*V11 = 0$
$+ (-G9)*V10 + (+G9+G10)*V11 + (-G10)*V12 = 0$
$+ (-G10)*V11 + (+G10+G11)*V12 + (-G11)*V13 = 0$
$+ (-G11)*V12 + (+G11+G12)*V13 + (-G12)*V14 = 0$
$+ (-G12)*V13 + (+G12+G18+G19)*V14 + (-G18)*V20 + (-G19)*V21 = 0$
$+ (-G13)*V4 + (+G13)*V15 + (+1)*I23 = 0$
$+ (+G14)*V16 + (-G14)*V17 + (-1)*I23 = 0$
$+ (-G14)*V16 + (+G14+G15)*V17 + (-G15)*V18 = 0$
$+ (-G15)*V17 + (+G15+G16)*V18 + (-G16)*V19 = 0$
$+ (-G16)*V18 + (+G16+G17)*V19 + (-G17)*V20 = 0$
$+ (-G18)*V14 + (-G17)*V19 + (+G17+G18)*V20 = 0$
$+ (-G19)*V14 + (+G19+G20)*V21 = 0$
$+ (+1)*V5 + (-1)*V6 = -E21$
$+ (+1)*V8 + (-1)*V9 = -E22$
$+ (+1)*V15 + (-1)*V16 = -E23$

• Potential at nodes	• Currents and voltages sides		
V1=-31.752799	U1=31.752799	I1=0.365521	
V2=-43.259396	U2=11.506597	I2=0.365521	
V3=-57.299053	U3=14.039657	I3=0.365521	
V4=-58.125131	U4=0.826077	I4=0.365521	
V5=-335.922113	U5=277.796983	I5=0.108940	
V6=864.077887	U6=533.914906	I6=0.108940	
V7=330.162981	U7=36.309700	I7=0.108940	
V8=293.853281	U8=36.309700	I8=0.108940	
V9=2393.853281	U9=164.468886	I9=0.108940	
V10=2357.543581	U10=721.801534	I10=0.108940	
V11=2193.074695	U11=1.873768	I11=0.108940	
V12=1471.273161	U12=43.867956	I12=0.108940	
V13=1469.399393	U13=0.726124	I13=0.256581	
V14=1425.531437	U14=1.424024	I14=0.256581	
V15=-58.851254	U15=0.251449	I15=0.256581	
V16=1441.148746	U16=11.463264	I16=0.256581	
V17=1439.724722	U17=0.130856	I17=0.256581	
V18=1439.473272	U18=2.347715	I18=0.256581	
V19=1428.010008	U19=511.729234	I19=0.365521	
V20=1427.879152	U20=913.802203	I20=0.365521	
V21=913.802203	U21=-1200.000000	I21=0.108940	
V22=0	U22=-2100.000000	I22=0.108940	
	U23=-1500.000000	I23=0.256581	

• Balance of powers	• Nodal analysis results
- The power generated: 744.373309	V1 = -31.752799
- The power consumption:	V2 = -43.259396
744.373309	V3 = -57.299053
	V4 = -58.125131
	V5 = -335.922113
	V6 = 864.077887
	V7 = 330.162981
	V8 = 293.853281
	V9 = 2393.853281
	V10 = 2357.543581
	V11 = 2193.074695
	V12 = 1471.273161
	V13 = 1469.399393
	V14 = 1425.531437
	V15 = -58.851254
	V16 = 1441.148746
	V17 = 1439.724722
	V18 = 1439.473272
	V19 = 1428.010008
	V20 = 1427.879152
	V21 = 913.802203
	I21 = 0.108940
	I22 = 0.108940
	I23 = 0.256581

PANCIA is based on the modified nodal method (MNM) which is, due to its flexibility, the most used and the most efficient method of simulating the electrical circuits on computer. [7, 8] The system of the equations that describes the equivalent circuit as ventilation scheme is a nonlinear algebraic system because the constants of pressure have the expressions:

$$\Delta H_K = R_{aK} Q_K^2 \quad \left(U_K = R_K I_K^2 \right)$$

For every nonlinear resistance, we considered the linear section of the characteristic U-I, so the circuit obtained is a linear resistive circuit.

In the example given, we found that the method used is good for calculating the ventilation systems of different types of electrical machines.

13. The thermal calculation. The heat transfer

Our thermal model is based on the analyses of the characteristic circuit providing almost instant calculation speeds. This allows the user to perform any calculation on time. Alternative numerical methods, for example, the analysis of finite elements [FEA] and calculable dynamic fluids [CFD], normally require some days/weeks to put together a three-dimensional model. The numeric mode may then take some hours (days for a complex problem of transit) to calculate a solution. [13]

The characteristic diagram (the characteristic circuit) is like an electrical scheme:

- The thermal resistance instead the electrical resistance
- Sources of power instead of current sources
- Thermal capacity instead electrical capacity
- Nodal temperatures instead voltages
- Power flow through the resistance instead the current.

All thermal resistances and thermal capacity from the model Motor-CAD are automatically calculated from geometric dimensions and materials properties. The user must be familiar with complex phenomena of heat transfer such as the dimensional analysis of conversion. Motor-CAD automatically selects and solves most appropriate formulation for a given surface and selected cooling time. Motor-CAD configures efficient, precise and robust mathematical algorithms for forced and natural convection, for cooling liquid, for radiation and conduction. A vast library of lamination correlations and demonstrated turbulent convections is used to give precise models for all internal and external surfaces. The air gap model includes laminar, vortex and turbulent convection.

The calculation in stationary regime

Motor-CAD simplifies the analyses of output dates of the stationary regime by using schematic diagrams, simple and easy to read. [13]

In "figure 14" we observe the solved main scheme. This scheme is used very much by post processing calculation in stationary regime. The resistances and the power sources indicated in scheme are color coded as machine components shown in editors of cross sections. The vertically oriented resistances represent radial heat transfer and horizontally oriented resistances represent axial heat transfer. The resistances which contain the letter "C" are resistances of convection and resistances which contain the letter "R" are resistances of radiation. Two colors resistances represent interface resistances between two components.

The diagram also shows labels for each of its components (resistances and nodes). With a single click of a button the user can change the scheme to indicate the nodal temperature and the thermal resistance or power values, so will avoid elaborate images. The main scheme is very useful for visualizing the place where there may be restrictions on the heat transfer and what can be done to improve cooling.

The calculation in transient regime

The Motor-CAD software allows the thermal transition of the motor to be calculated and the outputs to be viewed in graphical format. The opposite graph shows how the test data is important to emphasize on the precision of calculation. Also, Motor-CAD includes facility analyses of the operation cycle. A major advantage of this is that it allows the motor to be directed to its full potential without overheating.

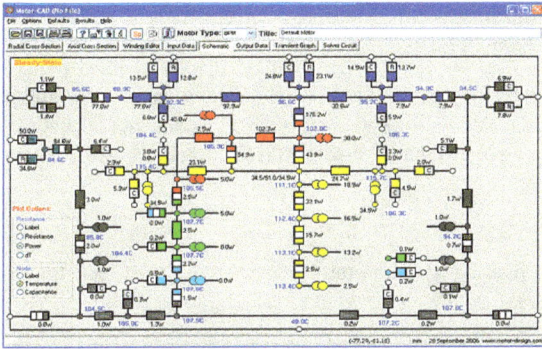

Figure 14. The main scheme (modeling by the characteristic circuit)

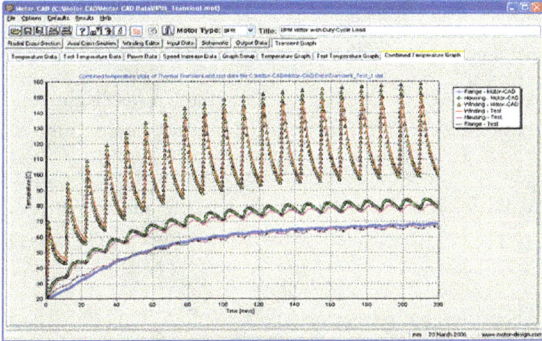

Figure 15. The variation in transient regime

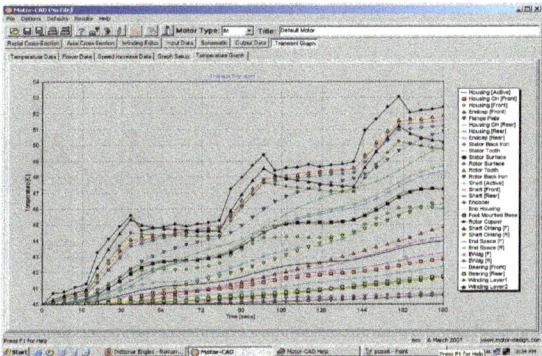

Figure 16. The variation in transient regime on temperature in diverse points in time of the machine

The transient regime can be modeled, so in any case the thermal capacities are automatically added to circuit. It is essential to consider the transient thermal analysis when using operational complex loads if the motor will be used to its full potential. The graph below shows a typical transient regime. In this case we have a repetitive duty cycle (three overloads followed by half loads).

The Motor-CAD software incorporates mathematical efficient and suitable algorithms for the cooling methods. All these, and facility of prescribe miscalculation, if is necessary, provides customization calculations so as to suit the unique construction methods.

14. Conclusions

Determining or providing the amount of exchanged heat per unit in time in given conditions of temperature is a problem to be taken into account for calculating heating and ventilation for the electrical machines. Were presented the basic sizes of the heat transfer studied in the literature and the distinct forms of heat transfer, the conduction, the radiation and the convection, the basic equations and rules of conduction, radiation and convection. Between the thermal sizes and the electrical sizes we showed a correspondence using the method of thermal resistances and concentrated losses.

By analogy between a network of cooling ventilation and a nonlinear resistive circuits were determined the values of the cooling fluid flows of an electrical machine.

Also by analogy, the temperature drop in the mass or on the surface of an object, corresponding the caloric flow (or loses), was determined by the product between the thermal resistance of the object opposed to the caloric flow in the mass or on surface and loses.

Solving the thermal schemes, were done analogous with solving the electrical schemes, as in electrical circuits of conexions between thermal resistances are series, parallel or mixed.

As we said the exhaust heat from the electric machine is as important as and phenomena related to the proper functioning of the machine. Choosing the most appropriate ventilation is done even by the assignation of electromagnetic stresses.

Were showed the equivalent resistances,and the aerodynamic schemes used were solved according to the conexions between aerodynamic resistances (series, parallel or mixed).

The discussed example for axial-radial bilateral ventilation in an electrical machine under protected construction for which the air circulation is symmetrical towards the mid of machine and it is sufficient to calculate ventilation only for one part, which is half of the machine.

After calculating the aerodynamic resistances, were calculated the pressures and air flows using the computer program PACEN (Program of Analysis of Nonlinear Electrical Circuits), and were obtained the system solutions.

Motor-CAD is an example for the analytical analysis of the characteristic circuit, is a program dedicated to electrical motors and generators and has all the indications on limiting design effort.

Motor-CAD has some advantages over CFD:

- The problem of defining time (minutes to hours/days/weeks)
- The calculating rate (instantly to hours/days)
- The postprocesing time (instantly to hours)
- Easy to use.

Motor-CAD allows optimization of the cooling system and identifies vulnerable points of design and/or problems of fabrication; it checks if a provided motor is suitable for a particular application.

The final conclusion for using Motor-CAD is that the program is a quick method designing for motors and electrical generators of thermal point of view, taking into account the fact that the thermal analogous circuit of the electrical machines is as important as designing the electromagnetical, mechanical.

Author details

Otilia Nedelcu and Corneliu Ioan Sălişteanu
Department of Electronics, Telecommunications and Energy Engineering,
Valahia University of Targoviste, Romania

15. References

[1] Bâlă C. (1982) – Electrical machines – Didactic and Pedagogical Publishing, Bucharest, Romania.

[2] Cioc I., Bichir N., Cristea N. (1981) – Electrical machines. Design guideline. Vol. I, II, III. –Romanian Writing Publishing, Craiova, Romania.

[3] Nedelcu O. (2010) – Modeling of heating and ventilation of electrical machines, Bibliotheca Publishing, Targoviste, Romania.

[4] Chua L. O., Lin P. M.,(1975) - Computer-Aided Analysis of Electronic Circuits: Algorithms and Computational Techniques, Prentice Hall, Englewood Cliffs, New Jersey.

[5] Leca,A., Prisecaru,I. (1994) – Thermophysical and thermodynamical properties. – Technical Publishing, Bucharest, Romania.

[6] Dumitriu L., Iordache M.,(1998) –Modern theory of electric circuits- vol.1- Theoretical foundation, Applications, Algorithms and computer programs, All Educational Publishing, Bucharest, Romania.

[7] Iordache M., Dumitriu L., (2004) – Computer assisted analysis of nonlinear analog circuits, Politechnica Press publishing, Bucharest, Romania.

[8] Iordache M., Dumitriu L., (1999) - PANCIA – Program Analysis of Analog Circuits, User manual, Politechnica Press publishing, Bucharest, Romania.

[9] McCalla W. J., (1988) - Fundamentals of computer-aided circuit simulation, Kluwer Academic Publishers, Boston.

[10] Hănţilă Fl.,(1979) -A Method for Solving Nonlinear Resistive Networks, Rev. Roum. Sci. Techn. Électrotechn. et Énerg., 24, 2, pp. 21-30

[11] Nedelcu O, Iordache M, Enescu D (2006) "An efficient method for computing of the electric machine ventilation"- The Sixth WESC TORINO, ITALY, July 10-12, ISBN10: 88-87380-51-1 , p 251.

[12] Nedelcu O., Enescu D., Sălisteanu C.I., (2009) – Determination of temperature field distribution and rate of heat transfer in DC machine – WESAS Genova, ISBN- 978-960-474-130-4.

[13] Motor-Cad help

Methodology to Calculate Boundary Conditions in a Single Isolated Helically Segmented Finned Tube Module

E. Martínez, W. Vicente, G. Soto, A. Campo and M. Salinas

Additional information is available at the end of the chapter

1. Introduction

Helically segmented finned tubes are used in compact heat recoveries in order to save energy in industrial applications. These equipments are small because the gas phase turbulence and the heat transfer surface are increased by the presence of fins; both are relevant in heat transfer. However, the gas phase pressure drop is elevated and consequently, operational problems such as backpressure can emerge. Therefore, a study focusing on finned tubes is important in order to understand the fluid dynamics and heat transfer phenomena. There are two main methods for the analysis; the first uses integral analysis (gross effects) and the second uses Computational Fluid Dynamics (CFD) techniques. Integral analyses allow a quick evaluation of thermo-physical phenomena with minimum computational infrastructure but only gross effects can be examined. These analyses are primarily used in the design of equipment because only inlet and outlet fluid properties are important. The CFD technique requires good computational support and long calculation times, but it provides complete and detailed information on the intricate thermo-physical phenomena. This modern analysis requires a correct implementation of boundary conditions in order to adequately represent the flow hydrodynamics and heat transfer phenomena.

The implementation of boundary conditions is relevant in differential analyses because predictions depend from it. The differential analyses can be developed by means of analytical solutions or by means of numerical methods. In the case of helically segmented finned tube bank analyses, the analytical solutions are not possible because the geometry is complex. Then, a numerical simulation of helically segmented finned tube bank is the best option. The numerical simulations can be carried out by means of three different CFD alternatives such as Direct Numerical Simulation (DNS), Large Eddy Simulation (LES), and Reynolds Average Navier-Stokes Equations (RANS). The DNS technique is limited to low Reynolds flows with simple configurations. The LES technique is less demanding than the DNS, but it takes considerable computing resources and computing time because the required calculations are always three-dimensional and unsteady. Finally, the RANS technique, which is widely used in industrial applications, considers average spatial and temporal scales of turbulent fluctuations and solves the transport equations as a function of these average variables. However, these equations are not closed and additional models (turbulence models) are indispensable to close the system. So, the numerical analysis on helically segmented finned tube bank (complex geometry) is proposed with the RANS technique.

In the open literature there are several papers have been focused on numerical analysis of small finned tube banks, the majority of them are restricted to numerical simulations on bare tube layout (symmetric tube layout). For example, Beale and Spalding [1], Comini and Croce [2] and Beale [3, 4] have performed simulations on laminar flow regimes exclusively. Other authors like Benhamadouche and Laurennce [5] and Salinas-Vazquez et al. [6] have conducted studies on turbulent flow regimes. These simulations have analyzed symmetric tube layout with periodic boundary conditions. On the other hand, there are few numerical simulations focused on helically segmented finned tubes (asymmetric finned tubes). The papers are focused on laminar flows with Dirichlet boundary conditions. For example, Hofmann [7] and Mcilwain [8, 9] conducted two-dimensional simulations on a single helically segmented finned tube. Afterwards, Lemouedda et al. [10] developed a three-dimensional numerical simulation in a small finned tube bank. There are no reports of numerical simulations of asymmetric finned tube layout (helically segmented finned tube bank) under periodic boundary conditions and the effect of inside fluid temperature has not been considered. Therefore, a methodology to calculate flow properties in different zones of finned tube bank is required in order to implement boundary conditions on a single isolated finned tube module. This methodology considers calculations in entire and partial finned tube layout and it is applied to calculate boundary conditions in a numerical simulation. Then, a compact heat recovery in staggered layout is represented as some single isolated finned tube modules in order to save computational resources. The single isolated finned tube module is simulated and predictions are compared with results from correlations available in the open literature.

2. Methodology

The differential analysis of compact heat recoveries with CFD techniques requires high calculation times because a full finned tube bank needs to be simulated. The dimensions of computational domain are high due to size of finned tubes and number of finned tubes used in the equipment. So, the necessity to reduce calculation times is relevant in numerical simulations because these times can be excessively-high. The only way to reduce calculation times is by reduction of computational domain but it requires a correct implementation of boundary conditions for representing adequately physical phenomena. A complete finned tube bank can be represented by a single isolated finned tube module in the fully developed flow region as shown in figure 1. The single isolated finned tube module consists of an arrangement of entire and partial finned tubes. This finned tube layout can allow a computational domain reduction of 99% but requires values of boundary conditions in intermediate regions of finned tube bank. The values of boundary conditions for velocity, pressure, and temperature should be calculated for entire and partial finned tubes in intermediate regions of finned tube bank. Then, a method to calculate those boundary conditions must be developed in order to represent the finned tube bank as a single isolated finned tube module (figure 1). This methodology is based on integral models which have been validated experimentally [22] with precision higher than 90% for pressure and 95% for temperature. So, a numerical simulation of single isolated finned tube module in the fully developed flow with periodic boundary conditions is done. Numerical predictions are compared with results obtained from best correlations available in the open literature, which are presented in section 2.2. So, the mean pressure drop, mean temperature difference, mean Nusselt number and mean friction factor are compared.

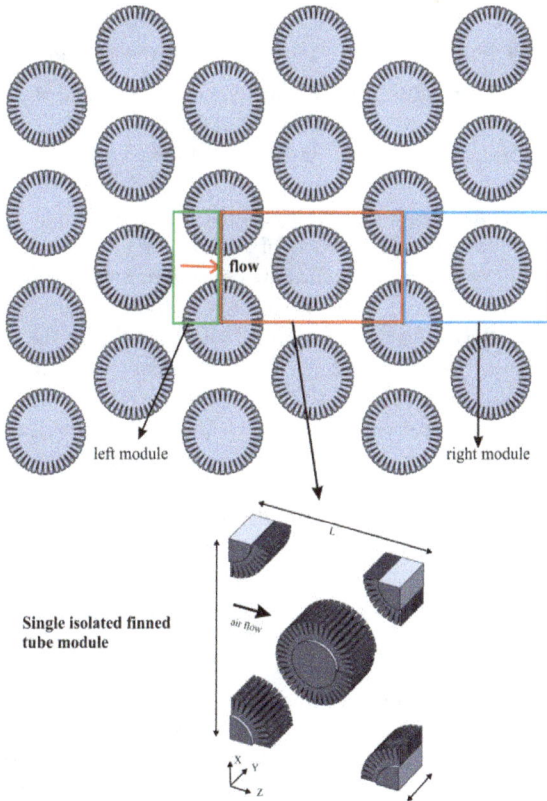

Figure 1. Finned tube bank in staggered layout.

2.1. Finned tube bank analysis

The complete finned tube bank may be represented as a single isolated fined tube module in the fully developed flow as shown in figure 1. This finned tube layout requires values for velocity, pressure, and temperature in intermediate regions of finned tube layout for both outside and inside flows. These values are obtained by means of an integral analysis, which considers the Logarithmic Mean Temperature Difference method (LMTD). However, this method only can be applied for entire finned tubes in arrangement of two or more finned tube rows and two or more finned tubes per row. Then, the complete finned tube bank and the single isolated finned tube module should be analyzed in order to apply LMTD method, which is described in section 2.2. The finned tube bank can be divided in single isolated finned tube modules as shown in figure 2a. In this figure, a full finned tube bank (6 finned tube rows with 4 finned tubes per row) is composed by 12 single isolated finned tube modules (black and red boxes). Every finned tube module contains 2 finned tubes because there are one entire finned tube and 4 quarters of finned tube. Then, two single isolated finned tube modules like (red boxes) can be represented as an equivalent small finned tube bank (blue box) as shown in figure 2b. Therefore, the finned tube bank showed in figure 1 is represented as a single isolated finned tube modules arrangement as presented in figure 2a.

Figure 2b shows a small finned tube bank (blue box) composed of 2 finned tubes per row and 2 rows of finned tubes, which is arranged in order to obtain 2 single isolated finned tube modules (red box). Then, the analysis of single isolated finned tube modules must consider minimum arrangements of 2 finned tubes per row and 2 rows of finned tubes. This consideration does not affect predictions of friction factor and Nusselt number because these dimensionless parameters are not function of number of finned tubes involved in the arrangement if mass flow is corrected to the new finned tube layout, which is demonstrated in the sensitive analysis (section 4.1). So, models for evaluating heat transfer and pressure drop can be applied to the equivalent small finned tube bank (figure 2b). The pressure drop depends mainly of flow hydrodynamics, which shows similar velocity fields for every single isolated finned tube module as is discussed in results analysis (section 4.2). Therefore, the models for evaluating pressure drop can be applied directly in the equivalence small finned tube bank (figure 2b). However, the pressure drop cannot be considered as a boundary condition because only represents the pressure difference at the inlet and exit of single isolated finned tube modules. The pressure in the boundaries of single isolated finned tube module is obtained with the analysis of finned tube bank from figure 1. In this figure, the single isolated finned tube in the fully developed region (red box) is located near to the exit of module. This finned tube module has to the right a part of finned tube module (blue box) while at left has one and a part finned tube module (green box). In the case of an atmospheric discharge of flow gases (zero relative pressure), which is correct because flue gases of compact heat recoveries cannot be used in additional industrial process, the relative pressure drop in the last part of single isolated finned tube module (blue box, figure 1) is calculated as an proportional arithmetic mean pressure drop to the part of this finned tube module. The proportional part of this module corresponds to a value of 0.75. Therefore, the relative pressure at the exit of the single isolated finned tube module in the fully developed flow (red box, figure 1) corresponds to the value calculated previously. Finally, the relative pressure at the inlet of the single isolated finned module (red box, figure 1) is calculated from the sum of the pressure drop in this finned tube module and the proportional part (0.25) of the left finned tube module (green box, figure 1).

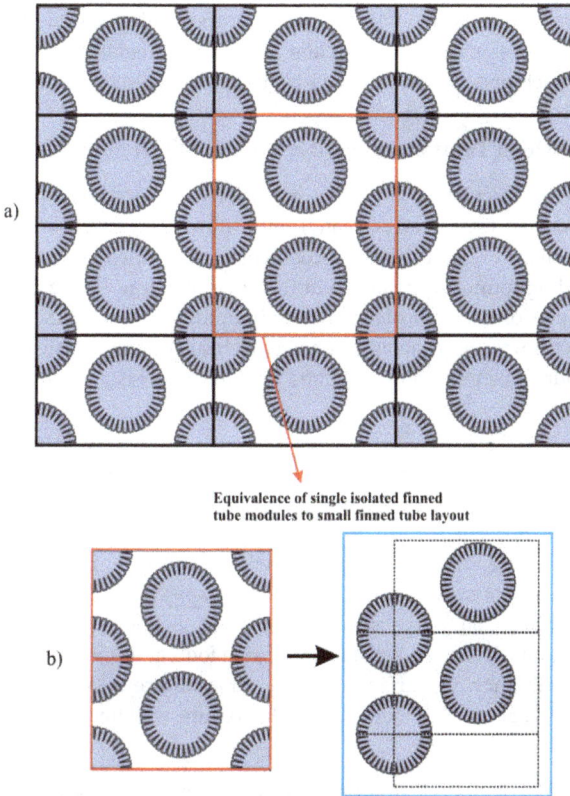

a)

Equivalence of single isolated finned
tube modules to small finned tube layout

b)

Figure 2. Single isolated finned tube modules in the finned tube bank.

The direct application of LMDT method for heat transfer in the equivalent small finned tube bank (blue box, figure 2b) is not recommended because the outside flow, in the single isolated finned tube module (red box, figure 2b), is cooled by tree different cooling sources (inside fluid temperature). While the equivalent small finned tube bank (blue box, figure 2) only has the influence of two cooling sources. One way to solve this problem is by means of temperature evaluation of outside flow for small finned tube layouts. The finned tube layouts considered have an initial arrangement of two finned tube rows, which are evaluated. Later, calculations over initial finned tube bank with additional finned tube rows are proposed (see section 2.3). This procedure is iterative because only initial conditions at the inlet of gas-phase flow and inside fluid in the complete finned tube bank (figure 1) are known. Once temperature evaluation for each finned tube layout is done, the inside fluid temperatures for each finned tube row are defined. The sensitivity analysis (section 4.1) of a finned tube bank shows that consideration applied for evaluating boundary conditions are satisfactory. These boundary conditions are employed in a numerical analysis of the finned tube bank showed in figure 1, which is described in section 2.3. Numerical predictions are compared with results from models for heat transfer and pressure drop in helically segmented finned tubes presented in section 2.2. The comparative analysis show close values between models and numerical results as discussed in results (section 4).

2.2. Logarithmic Mean Temperature Difference (LMTD) method

The methodology for calculating boundary conditions in the single isolated finned tube module is based on the Logarithmic Mean Temperature Difference method. This method allows the evaluation of heat transfer coefficients and friction factors for different geometries, according to models available in the open literature. These parameters permit the global evaluation of heat transfer and flow hydrodynamics of finned tube bank. The LMTD method considers the evaluation of overall heat transfer coefficient (U), which is based on the outside finned tube and is defined with the following equation:

$$U_o = \cfrac{1}{\cfrac{A_o + A_o R_{fo}(h_o + h_r)}{(h_o + h_r)(\eta_f A_f + A_t)} + \cfrac{e_w A_o}{k_w A_i} + \left(\cfrac{1}{h_i} + R_{fi}\right)\cfrac{A_o}{A_i}} \tag{1}$$

where h_o, h_i, and h_r are mean outside convective coefficient, mean inside convective coefficient, and radiation heat transfer coefficient, respectively. In the case of flue gas temperatures lower than 300ºC, the value of h_r could be negligible [11], and so this value is considered zero. R_{fo} and R_{fi}, are the outside and inside fouling factors, respectively. η_f, A_f, A_t, A_o, and A_i are fin efficiency, fin surface area, bare tube surface area, total surface area, and inside surface area, respectively. Finally, e_w and k_w are tube wall thickness and tube material thermal conductivity, respectively.

The mean convective coefficients are calculated for the inside and outside of finned tubes. The mean inside convective coefficient (h_i) considered in the evaluation of U is the Gnielinski's correlation [12]. This model has been validated with satisfactory results by Rane, et al. [13] and according to Bejan [14] is the best available in the open literature. The Gnielinski's correlation [12] is valid for $0.5 \leq Pr \leq 2000$ and $3000 \leq Re \leq 5\times10^6$, which is shown in the following equation:

$$Nu = \frac{h_i d_i}{k} = \frac{(f_i / 8)(Re - 1000)Pr}{1 + 12.7(f / 8)^{1/2}(Pr^{2/3} - 1)} \tag{2}$$

where d_i and k are inside diameter of tube and thermal conductivity of fluid. Re and Pr are the Reynolds Number and Prandtl Number. Finally, f_i is the friction factor, which is defined in the following equation:

$$f_i = \frac{1}{\left(1.82 \log_{10} Re - 1.64\right)^2} \tag{3}$$

There are many mean outside convective heat transfer coefficients (h_o) in the open literature such as models proposed by Weierman [15], ESCOA [16], Nir [17], and Kawaguchi et al. [18]. The models attributable to Weierman [15] and Kawaguchi et al. [18] are recommended by Martinez et al. [19], but Kawaguchi´s et al. [18] model is adopted. The model of Kawaguchi, et al. [18] is valid for $7000 \le Re_v \le 50000$ and $0.112 \le s_f/d_v \le 0.198$, which is shown in the next equation:

$$Nu = A_2 \, Re_v^{0.784} \, Pr^{1/3} \left(s_f / d_v\right)^{-0.062} = \frac{h_o d_v}{k_g} \tag{4}$$

where Re_v is the Reynolds number based on the volume-equivalent diameter. The terms A_2, s_f, and d_v are the experimental coefficients for tube rows, fin gap, and volume-equivalent diameter, respectively. The coefficient A_2 is obtained from Kawaguchi´s, et al. [18] model. Finally, k_g is the thermal conductivity of gases.

The volume-equivalent diameter, d_v, is defined by the following equation:

$$d_v = \left[t_f n_f \left\{ \left(d_o + 2l_f\right)^2 - d_o^2 \right\} + d_o^2 \right]^{1/2} \tag{5}$$

where n_f, t_f, l_f, and d_o are fin number per unit length, fin thickness, fin height, and outside diameter of bare tube, respectively.

Gas phase pressure drop can be calculated with models proposed by Weierman [15], ESCOA [16], Nir [17], and Kawaguchi et al. [18]. The model of Weierman [15] has been validated with satisfactory results by Martinez, et al. [19] and as a consequence, is adopted. In the analysis of compact heat recoveries a maximum pressure drop of 248.9 Pa [20] (1 inch (in) of water column (wc)) is considered in order to avoid technical problems such as backpressure. The pressure drop is calculated with the following empirical equation:

$$\Delta P_g = \frac{\left(f_o + \dfrac{\left(1 + B^2\right)\rho_{gp}}{4N_r} \right) G_o^2 N_r}{1.083 \times 10^9 \, \rho_{gp}} \tag{6}$$

where B, G_o, N_r, f_o, and ρ_{gp} are the contraction factor, the gas mass flux, the number of tube rows, the friction factor, and the gas-phase density at the average outside temperature, respectively.

The gas phase friction factor (f_o) is calculated with Weierman´s model [15], which is valid for tube diameters between 38.1-60.96 mm and mass velocity of the gas between 0.67-40.36 kg/m²s. The Weierman´s model [15] is shown in the following equation:

$$f_o = \left[0.07 + 8\,\mathrm{Re}_o^{-0.45}\right]0.11\left(\frac{0.05S_t}{d_o}\right)^{-0.7\left(\frac{l_f}{s_f}\right)^{0.23}}$$
$$\left[1.1 + \left(1.8 - 2.1e^{-0.15N_r^2}\right)e^{-\frac{2S_t}{S_t}} - \left(0.7 - 0.8e^{-0.15N_r^2}\right)e^{-\frac{0.6S_t}{S_t}}\right]\left(\frac{d_f}{d_o}\right)^{1/2} \tag{7}$$

where Re_o is the Reynolds number based on the outside bare and c_p is the specific heat capacity at constant pressure. Also, d_f and d_o are the outside diameter of the finned tube and the outside diameter of the bare tube, respectively. The terms S_l and S_t indicate the longitudinal pitch and the transverse pitch, respectively.

For more details about thermal evaluation of helically segmented finned tube banks see reference [21].

2.3. Sensitive analysis of finned tube bank

The methodology is based on the equivalence of a finned tube bank with single isolated finned tubes modules layouts. This methodology, which is based on the LMTD method, requires that predictions of friction factor and Nusselt number be independent of finned tube layout and number of finned tubes involved in compact heat recoveries. Then, a sensitive analysis of the finned tube bank is required in order to validate considerations in the methodology. The sensitive analysis is done for a small heat recovery (figure 1), in which results of friction factor and Nusselt number are compared for different finned tube layouts and finned tube lengths. The geometric characteristics of helically segmented finned tube and finned tube layout are shown in figure 3. In this figure, a front and side views of the finned tube used in this study and its staggered configuration is presented. The finned tube bank analyzed corresponds to the layout shown in figure 1, which is composed by 4 finned tubes per row and 6 finned tube rows. The small heat recovery (figure 1) is analyzed with LMTD method for different finned tube configurations.

The finned tube configurations used in the sensitive analysis considers an initial finned tube layout of 2 finned tubes per row and 2 finned tubes rows (red polygon in figure 4). Subsequently, one finned tube row is added to reach 6 finned tube rows (red dash boxes in figure 4). Then, another finned tube is added to the row (blue polygon, figure 4) in order to obtain 3 finned tubes per row and 2 finned tube rows. Previous procedures in initial finned tube layout are done until 6 finned tube rows (blue dash boxes, figure 4) are reached. Finally, one last finned tube row is added (green polygon in figure 4) in order to get 4 finned tubes per row and 2 finned tube rows. The same procedure for previous finned tube layouts are done until full finned tube bank (gray dashed box) is reached. Once finned tube bank layout analysis is done, a similar study is proposed for different finned tube length. The procedure is the same as previous analysis but at different finned tube length. The finned tube lengths proposed are 1m, 0.5 m, and 0.05194 m. The last finned tube length corresponds to the value used in the numerical simulation. The results (section 4.1) show variations lower than 1% for Nusselt number and 3.6% lower for friction factor. These results confirm that considerations employed in this methodology (boundary condition calculation) are appropriate. The analysis is obtained with the thermodynamic conditions presented in table 1 for different finned tubes per row and finned tubes length which, shows deviations on Reynolds number based in the outside diameter tube (Re_o) lower than 0.4%. The variations in Re_o are due to adjustment of mass flow at different finned tube configurations.

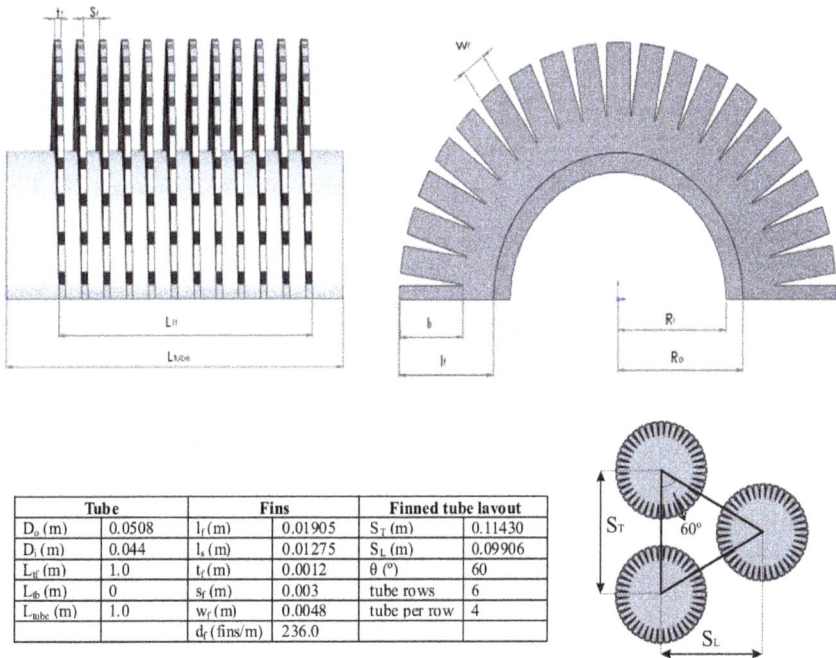

Tube		Fins		Finned tube layout	
D_o (m)	0.0508	l_f (m)	0.01905	S_T (m)	0.11430
D_i (m)	0.044	l_s (m)	0.01275	S_L (m)	0.09906
L_{tf} (m)	1.0	t_f (m)	0.0012	θ (°)	60
L_{fb} (m)	0	s_f (m)	0.003	tube rows	6
L_{tube} (m)	1.0	w_f (m)	0.0048	tube per row	4
		d_f (fins/m)	236.0		

Figure 3. Geometric characteristics of finned tubes and finned tube layout.

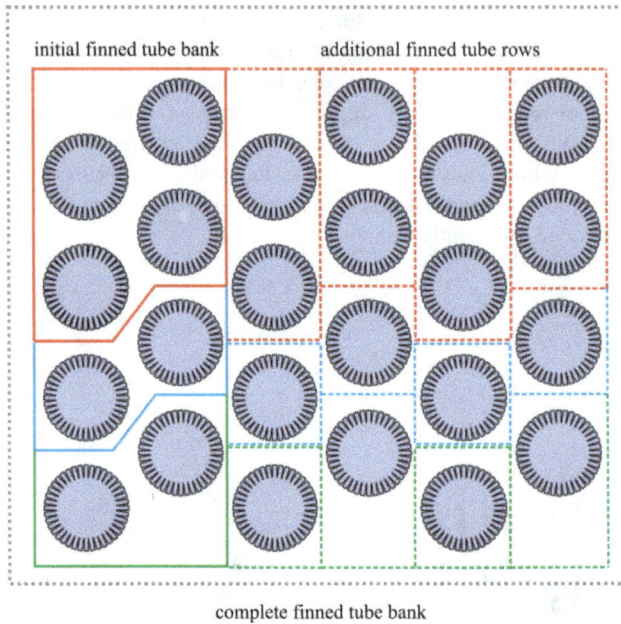

Figure 4. Finned tube layout configurations in sensitive analysis.

Finned tube length: 1 m			
	Finned tubes per row		
Parameter	2	3	4
mass flow (kg/s)	0.3624	0.05436	0.7248
Re_o	8562	8562	8562
Ma	< 0.03	< 0.03	< 0.03
Temperature (ºC)	60	60	60
Finned tube length: 0.5 m			
	Finned tubes per row		
Parameter	2	3	4
mass flow (kg/s)	0.1812	0.2718	0.3624
Re_o	8567	8567	8567
Ma	< 0.03	< 0.03	< 0.03
Temperature (ºC)	60	60	60
Finned tube length: 0.05194 m			
	Finned tubes per row		
Parameter	2	3	4
mass flow (kg/s)	9.411×10^{-3}	14.116×10^{-3}	18.822×10^{-3}
Re_o	8599	8599	8599
Ma	< 0.03	< 0.03	< 0.03
Temperature (ºC)	60	60	60

Table 1. Thermodynamic conditions.

3. Numerical simulation

In this section, the methodology for calculating boundary conditions in a single isolated finned tube module is employed in a numerical simulation of helically segmented finned tube bank in staggered layout. The objective is to apply the methodology in a numerical model that represents the interaction between the hydrodynamics and the heat transfer of a turbulent gas flow in complex systems. The simulation allows detailed analysis on compact heat recoveries that can be used for improving thermal behaviour. So, a correct implementation of boundary conditions in order to represent adequately physical phenomena is essential. The numerical model is focused to the outside finned tube, because the gas phase dominates the heat transfer [22] and also the pressure drop is critical [22]. However, the effect of the internal fluid is considered by means of an average inside temperature in each finned tube. Thus, the boundary conditions for outside flow and inside fluid are calculated with methodology described in section 2. In this numerical simulation, periodic boundary conditions are proposed for the single isolated finned tube module in the fully developed flow (Figure 1). Numerical predictions of Nusselt number and friction factor are compared with results obtained from correlations of Kawaguchi et al. [18] and Weierman [15], respectively in order to verify numerical results that depend on boundary conditions values.

3.1. Governing equations

The analysis of turbulent flows is complex, because fluid properties are irregular in space and time. Instantaneous variables (ϕ) are represented as a function of two terms: the mean ($\bar{\phi}$) and its fluctuation (φ'). These terms are used in the transport equations that govern the flow movement and the heat exchange. The transport equations are averaged by means of Favre [23] method, which are solved for mean values of the fluid properties. Thereby, the set of mass, momentum, and energy transport equations in Cartesian system is defined as follows:

$$\frac{\partial \bar{\rho}}{\partial t} + \nabla \cdot \left(\bar{\rho} \tilde{\vec{V}} \right) = 0 \tag{8}$$

$$\frac{\partial}{\partial t}\left(\bar{\rho} \tilde{\vec{V}} \right) + \nabla \cdot \left(\bar{\rho} \tilde{\vec{V}} \otimes \tilde{\vec{V}} \right) = -\nabla \cdot \bar{P} + \mu \nabla^2 \tilde{\vec{V}} - \nabla \cdot \left(\overline{\bar{\rho} \vec{V}'' \vec{V}''} \right) + \bar{\rho} \vec{g} \tag{9}$$

$$\frac{\partial}{\partial t}\left(\bar{\rho} \tilde{h} \right) + \nabla \cdot \left(\bar{\rho} \tilde{\vec{V}} \tilde{h} \right) = -\nabla \cdot \tilde{\vec{J}}_h - \nabla \cdot \left(\overline{\bar{\rho} \vec{V}'' h''} \right) \tag{10}$$

where $\bar{\rho}$, $\tilde{\vec{V}}$, $\vec{V}''$, and $\tilde{h}$ are the mean density, instantaneous velocity, fluctuating velocity, and mean enthalpy, respectively. μ and $\vec{g}$ are the viscosity and the gravitational acceleration, respectively. The terms $\overline{\bar{\rho} \vec{V}'' \vec{V}''}$ and $\overline{\bar{\rho} \vec{V}'' h''}$ are the apparent Reynolds stress tensor and the turbulent heat flux, respectively. $\bar{\bar{P}}$ is the pressure tensor. Finally, $\nabla \cdot \tilde{\vec{J}}_h$ is the diffusive heat flux, which is modeled with Fourier's law.

Equations (2) and (3) demand additional mathematical expressions to model $\overline{V''\varphi''}$ term. The closure of these equations requires modeling of Reynolds stress tensor and of the turbulent heat flux. Then, the Reynolds stress tensor is closed with turbulence models. In this work, the k-ε RNG (Renormalization Group) turbulence model developed by Yakhot and Orzag [24] is considered. The turbulent heat flux vector is obtained by means of an analogy between momentum transfer and thermal energy transfer. Under this concept, it is possible to establish a suitable articulation between the turbulent heat flux vector and the turbulent flow viscosity.

3.2. Boundary conditions

The implementation of periodic boundary conditions alleviates the computational resources, because the computational domain is reduced considerably. So, only a single isolated finned tube module in the fully developed flow region needs to be simulated (figure 1). Also, the tube length required (spatial direction y in figure 1) is minimum because only is necessary the length for the flow does not vary in this spatial direction because periodic conditions assume no influence of walls or position in any direction. Then, the computational domain is reduced 99%, as shown in table 2, because only is necessary the single isolated finned tube module (figure 1) with a tube length of 0.05194 m. However, the numerical simulation requires a correct inclusion of boundary conditions in order to adequately represent the physical phenomenon. So, the boundary conditions are applied to a turbulent air flow in a stationary and fully-developed flow regime. The methodology developed by Patankar et al. [25] and Kelkar and Patankar [26] is implemented in this work. This methodology was generated for laminar flows and as a consequence, only velocity, pressure and temperature are considered as a periodic behaviour. However, additional considerations for turbulent flows need to be implemented. These considerations should depend on the turbulence model selected.

The periodic velocity is based on the non slip condition on the boundary walls of the computational domain and negligible value of the velocity variation in every spatial direction. So, the analysis is presented for the flow direction (z), which can be generalized for the remaining spatial directions. Thereby, the periodic condition for the component velocity in the flow direction ($\tilde{w}$) is defined as follows:

$$\frac{\partial \tilde{w}}{\partial z} = 0 \qquad\qquad \tilde{u} = 0 \qquad\qquad \tilde{v} = 0 \qquad\qquad (11)$$

where $\tilde{u}, \tilde{v}, \tilde{w}$ are the mean velocity components in the respective spatial directions x, y, z.

In the case of fully developed flows, the velocity is the same at a characteristic length (L), and this is shown in the next equation:

$$\tilde{w}(x,y,z) = \tilde{w}(x,y,z+L) \qquad\qquad (12)$$

The periodic boundary velocities need an initial value w_b (bulk velocity), which is calculated with the following equation:

$$w_b = \frac{1}{A_{yz}} \int_0^x \int_0^y w_{med} dx dy \qquad (13)$$

where w_{med} is an initial velocity profile.

The pressure field is obtained from a periodic behavior taking into consideration a pressure drop in the flow direction. This implies that the pressure is defined by the sum of a periodic pressure term in the flow direction and an average pressure drop. This is represented in the following equation:

$$P(x,y,z) = \tilde{P}(x,y,z) - \beta z \qquad (14)$$

where $\tilde{P}$ and β are periodic pressure and average pressure-gradient in the flow direction.

	Dimensions:			
Layout	x (m)	y (m)	z (m)	cells
Finned tube bank	0.4572	1.0	0.59878	34200000
Single isolated finned tube module	0.1143	0.05194	0.20312	178512
	Computational domain reduction:			
Finned tube bank	reference			
Single isolated finned tube module	99.6			

Table 2. Finned tube layout dimensions.

These two terms are defined in the pair of equations (15) and (16):

$$\tilde{P}(x,y,z) = \tilde{P}(x,y,z+L) \qquad (15)$$

$$\beta = \frac{\tilde{P}(x,y,z) - \tilde{P}(x,y,z+L)}{L} = \frac{\Delta \tilde{P}}{L} \qquad (16)$$

where $\Delta \tilde{P}$ is the pressure drop over a finned tubes module, which can be calculated through the empirical equation (6).

The temperature field is obtained with constant wall heat flux boundary condition, which considers a constant variation of temperature in the flow direction; that is, the heat transfer magnitude is the same from one finned tube module to another finned tube module. This boundary condition can be appropriate for the present study because it can be applied to turbulent flows. However, it is appropriate if a uniform heat transfer is found in the small finned tube bank simulation. The numerical results in small finned tube bank show a quasi-constant mean temperature (plane xy) in the flow direction as shown in results section.

The temperature field is obtained from a periodic behavior and an adjustment term of the temperature in the flow direction. Thus, the temperature is defined by a periodic temperature term in the flow direction and an average temperature adjustment term expressed in the next expression:

$$T(x,y,z) = \tilde{T}(x,y,z+L) + \gamma z \tag{17}$$

where $\tilde{T}(x,y,z)$ and γ are the mean temperature field and the temperature-gradient term, respectively. The temperature-gradient term can be calculated with the following equation:

$$\gamma = \frac{\tilde{T}(x,y,z) - \tilde{T}(x,y,z+L)}{L} = \frac{\dot{Q}}{\dot{m}_g c_p L} \tag{18}$$

where $\dot{Q}$, $\dot{m}$, and c_p are heat addition, mass flow, and specific heat at constant pressure, respectively.

The heat addition in the flow direction for the single isolated module is determined with the next equation:

$$\dot{Q} = U_o A_T \Delta T_{ML} \tag{19}$$

where U_o is the overall heat transfer coefficient, A_T the overall finned surface and ΔT_{ML} the logarithmic mean temperature difference.

The value of U_o is obtained from equation (1) and A_T is obtained with geometry of finned tube. For more details about calculations of these parameters see reference [21].

The numerical simulation of turbulent flow on a single isolated module needs the implementation of periodic conditions for additional variables, according to the turbulence model. In the case of k-ε RNG turbulence model, the turbulent kinetic energy (k) and its dissipation rate (ε) show a periodic behavior, according to Martínez [27]. So, the periodic condition for the turbulent kinetic energy and its dissipation rate are defined as follows:

$$\tilde{k}(x,y,z) = \tilde{k}(x,y,z+L) \tag{20}$$

$$\tilde{\varepsilon}(x,y,z) = \tilde{\varepsilon}(x,y,z+L) \tag{21}$$

The inclusion of periodic boundary conditions in the simulation of single isolated module requires light changes in the governing equations. The mass conservation equation does not change and is evaluated by means of equation (1). In the case of pressure field, the momentum conservation equation needs to include the adjust term of average pressure-gradient in the flow direction. So, the equation (2) can be written as:

$$\frac{\partial}{\partial t}\left(\bar{\rho}\tilde{V}\right) + \nabla \cdot \left(\bar{\rho}\tilde{V} \otimes \tilde{V}\right) = \beta - \nabla \cdot \bar{P} + \mu \nabla^2 \tilde{V} - \nabla \cdot \left(\overline{\bar{\rho}V''V''}\right) + \bar{\rho}\vec{g} \tag{22}$$

The term β is included in the CFD code PHOENICS 3.5.1 [28] by means of an additional source term in the momentum equation. On the other hand, the evaluation of temperature field requires that energy conservation equation must be adjusted in order to include the cooling of gas phase in the flow direction. So, the equation (3) can be written as:

$$\frac{\partial}{\partial t}\left(\bar{\rho}\tilde{h}\right) + \nabla \cdot \left(\bar{\rho}\tilde{V}\tilde{h}\right) + \nabla \cdot (\tilde{V}\gamma) = -\nabla \cdot \bar{\vec{J}}_h - \nabla \cdot \left(\overline{\bar{\rho}V''h''}\right) \tag{23}$$

The term $\nabla \cdot (\tilde{V}\gamma)$ is included in the CFD code PHOENICS 3.5.1 [28] by means of an additional source term in the energy equation.

The turbulent kinetic energy and the dissipation turbulent rate are calculated directly from equations (20) and (21) because these equations do not change.

3.3. Numerical details

The numerical simulation is developed for a single isolated finned tube module in the fully developed flow, which is shown in Figure 5. In this figure, the single isolated finned tube module is presented with boundary conditions calculated with methodology proposed. The dimensions of computational domain and the mesh used for simulation are shown in table 2. This table exhibits a reduction in computational domain with single isolated finned tube module of 99% that represents finite calculation times. The used mesh in numerical simulation considers that numerical predictions are independent from it. The thermodynamic employed conditions in the simulation work are presented in table 1 for finned tube length of 0.05194 m and figure 5. On the other hand, the complex geometry is represented by cut-cell method [29], which allows the use of Cartesian grids. The numerical simulations consider a staggered grid under a hybrid discretization scheme of the convective term. It was also considered that the system is in a stationary state and is only exposed to one gravitational field (in heat recoveries, the fluid generally flows in vertical direction), and that the gas discharge (finned tube bank outlet) occurs in a sea level atmosphere. Finally, the geometric characteristics of the finned tube used in this study are shown concurrently in Figure 3.

The numerical results are used for evaluating average Nusselt Number (Nu) and average friction factor in small finned tube bank and single isolated module. The results are compared with values obtained from Correlations of Kawaguchi, et al [18] and Weierman [15] models, respectively.

The average Nusselt number is calculated with the next equation:

$$\overline{Nu} = \frac{\overline{h}d_v}{k_g} \tag{24}$$

where $\overline{h}$ is the average convective coefficient, which is defined as:

$$\overline{h} = \frac{\dot{Q}}{A_o \Delta T_{LMnum}} \tag{25}$$

where ΔT_{LMnum} is the numerical-logarithmic mean temperature difference, which is calculated as:

$$\Delta T_{LMnum} = \frac{\left[T_i(z+L) - T_b(z+L)\right] - \left[T_i(z) - T_b(z)\right]}{Ln \dfrac{\left[T_i(z+L) - T_b(z+L)\right]}{\left[T_i(z) - T_b(z)\right]}} \tag{26}$$

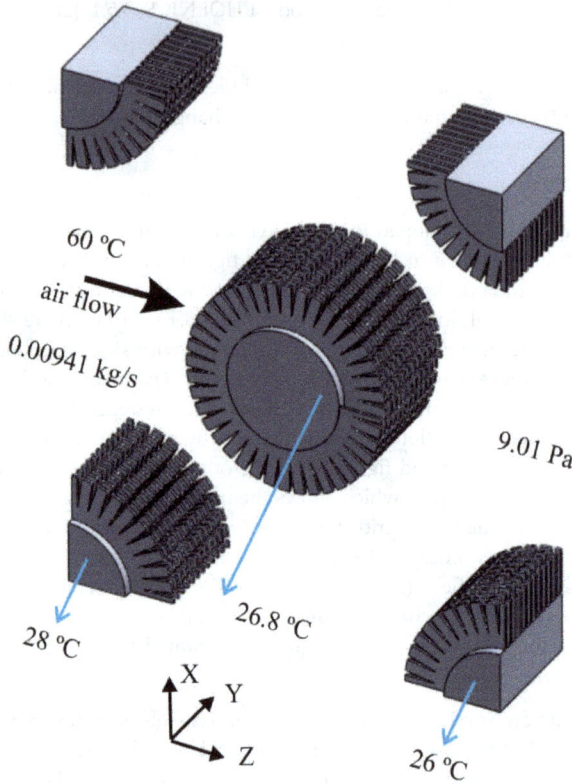

Figure 5. Computational domain in numerical simulation.

where T_i and T_b are the mean temperature inside finned tubes and mean boundary temperature, respectively. The conditions at z and $z+L$ correspond to the inlet and exit of the single isolated finned tube module.

The numerical-average friction factor is obtained directly from equation (6) in which numerical pressure drop is calculated from the next expression:

$$\Delta \tilde{P} = P_b(z) - P_b(z + L) \tag{27}$$

where P_b is the numerical-mean pressure.

4. Results

The methodology proposed is applied for evaluating boundary conditions in a single isolated finned tube module (figure 1). The predicted values are utilized in a numerical simulation in order to verify a correct representation of physical phenomena with these boundary conditions. These values require independence of results in heat transfer (Nusselt number) and pressure drop (friction factor) for different finned tube configurations, number of finned tubes, and finned tube lengths. So, a sensitive analysis is necessary in order to support previous considerations. The results of sensitive analysis shows that assumptions of methodology are correct, which are discussed in section 4.1. On the other hand, numerical predictions (focused in the outside flow) show a correct representation of interaction of heat transfer and flow hydrodynamics, which is presented in section 4.2. The comparative analysis of results between numerical predictions and results obtained from Kawaguchi et al [18] and Weierman [15] models for Nusselt number and friction factor, respectively, show close values as discussed in section 4.2.

4.1. Sensitive analysis

The LMTD method is applied to different finned tube layouts and finned tube lengths, which are described in section 2.3. The main goal of the analysis is to verify that friction factor and Nusselt number are independent of finned tube rows, number of finned tube per row, and finned tube length for the finned tube bank showed in figure 1. The comparative analysis at different finned tube configurations and finned tube lengths with thermodynamic conditions presented in table 1 are shown in table 4. In this table, Nusselt number and friction factor are presented at different finned tube configurations from 2 finned tubes per row and 2 finned tubes rows (red polygon in figure 4) until reach 4 finned tubes per row and 6 finned tubes rows (gray dashed box, figure 4). Evaluations are done for tube lengths of 1 m, 0.5 m y 0.05194m as shown in table 3. Predictions show constant values of Nusselt number and friction factor for the same finned tube rows. So, the Nusselt number varies from 100.3 to 99.3 and the friction factor changes from 0.31 to 0.321 for different finned tube rows (2 to 6) of the small heat recovery. These results exhibit independence of predictions in dimensionless parameters with finned tube length and finned tubes per row in a heat recovery. The only variation of results is for configurations at different finned tube rows as shown in table 3. The main reason of dimensionless parameters variation is temperature reached by the gas phase (outside flow). These temperature variations slightly affect results because the gas phase is cooled in 3 °C temperature difference for configurations analyzed at different finned tube rows. The results exhibit variations lower than 1% for Nusselt number and 3.6% for friction factor. These deviations are calculated for different finned tube rows, which is not representative of the same physical phenomenon because different finned tube layouts represent different gas cooling. In spite of this situation, the results show close values in Nusselt number and friction factor. Therefore, assumptions considered in the methodology are correct because finned tube bank performance is independent of finned tube rows, number of finned tubes per row, and finned tube length in a finned tube bank. This methodology allows numerical studies in heat recoveries at industrial scale because the computational domain can be reduced in 99% as shown in table 2. So, the calculation times are finites due to the analysis is focused in a single isolated finned tube module in the fully developed flow.

Finned tube length: 1 m						
Finned tubes per row						
	2		3		4	
Finned tube rows	Nu	f	Nu	f	Nu	f
2	99.3	0.321	99.3	0.321	99.3	0.321
3	99.6	0.319	99.6	0.319	99.6	0.319
4	99.8	0.313	99.8	0.313	99.8	0.313
5	100.01	0.311	100.01	0.311	100.01	0.311
6	100.3	0.31	100.3	0.31	100.3	0.31
Finned tube length: 0.5 m						
Finned tubes per row						
	2		3		4	
Finned tube rows	Nu	f	Nu	f	Nu	f
2	99.3	0.321	99.3	0.321	99.3	0.321
3	99.6	0.319	99.6	0.319	99.6	0.319
4	99.8	0.313	99.8	0.313	99.8	0.313
5	100.01	0.311	100.01	0.311	100.01	0.311
6	100.3	0.31	100.3	0.31	100.3	0.31
Finned tube length: 0.05194 m						
Finned tubes per row						
	2		3		4	
Finned tube rows	Nu	f	Nu	f	Nu	f
2	99.3	0.321	99.3	0.321	99.3	0.321
3	99.6	0.319	99.6	0.319	99.6	0.319
4	99.8	0.313	99.8	0.313	99.8	0.313
5	100.01	0.311	100.01	0.311	100.01	0.311
6	100.3	0.31	100.3	0.31	100.3	0.31

Table 3. Sensitivity analysis results.

4.2. Numerical results

The predictions for velocity field, pressure field, and temperature field are shown in Figure 6. In all figures, profiles of the variables are presented on the x-z plane, because this is the plane that exhibits most changes in properties. The velocity contours (figure 6a) reflect an apparently symmetric behaviour but the field is slight asymmetry due to the fin helical layout. The recirculation zones are observed at the rear portion of the tubes, with reference to air flow direction. These recirculation zones are narrow because the flow tends to stick to the contour of finned tubes by turbulence generated. The recirculation region has amplitude and length of approximately 49.85 mm and 52.61 mm, respectively. The flow tends to accelerate in the free zones where the cross-section area is smaller (central region of computational domain and at the side portions of the central finned tube) with a maximum velocity of 5.017 m/s. Finally, the backwater area where the flow is stopped abruptly at the central portion of the finned tube exhibit has a width and a length of about 9.75 mm and 6.57 mm, respectively.

The pressure contours (figure 6b) indicate an apparently symmetric pressure profile, similar to the velocity profile, which is not symmetric due to the helical fin. The results show a high-pressure zone at the front portion of the tube, taking the flow direction as the point of reference. This high-pressure zone is created due to flow stopping abruptly at the central portion of the finned tube, producing a backwater zone. At the rear of the finned tube, there is a low-pressure area that is stratified at the outlet of single isolated module. The mean pressure at the inlet and outlet of the module is 23.184 Pa and 9.019 Pa, respectively, as shown in table 4. The pressure drop is 14.165 Pa, which represents a friction factor of 0.311 (table 4). The deviation of numerical results is 3.13% and 0.32% for pressure drop and friction factor, respectively. So, numerical predictions with boundary conditions calculated from methodology correspond to results obtained with Weierman's [15] model.

Numerical predictions of temperature contours (figure 6c) show an apparently symmetric profile, which are not symmetric due to the helical fin. The results exhibit that the high-temperature region is located at the central region of the computational domain, whereas the low-temperature areas are located at the rear of the finned tubes, viewed with respect to the flow direction. So, the most important heat transfer effects occur at the front of the extended surfaces and at the sides of the finned tube between the backwater zone and the recirculation zone. This pattern is attributed to the air flow coming into abrupt contact with the finned tubes, and to turbulence being created at the sides of the finned tubes. On the other hand, the temperature contours exhibit that flow temperature is dominated for inside fluid temperature in the recirculation zone. The mean temperature at inlet and outlet of module is 43.76°C and 34.73°C (table 4). The temperature difference is 9.03°C, which represents a numerical-mean Nusselt number of 95.1, as shown in table 4. Numerical results show deviations of 4.25% and 5.18% for temperature difference and Nusselt number, respectively. These results show that numerical predictions are close to results obtained with Kawaguchi's et al. [18] model. So, the numerical simulation represents adequately heat transfer in a single isolated finned tube module with boundary conditions calculated from methodology.

	Numerical Simulation	model
		Weierman
P_{inlet} (Pa)	24.44	na
P_{outlet} (Pa)	10.27	na
ΔP (Pa)	14.17	13.74
deviation	3.13%	reference
f	0.311	0.31
deviation	0.32%	reference
		Kawaguchi
T_{inlet} (°C)	40.78	na
T_{outlet} (°C)	31.78	na
ΔT (°C)	9.0	9.4
deviation	4.25%	reference
Nu	95.1	100.3
deviation	5.18%	reference

Table 4. Comparative results.

a) velocity contours

b) pressure contours

c) temperature contours

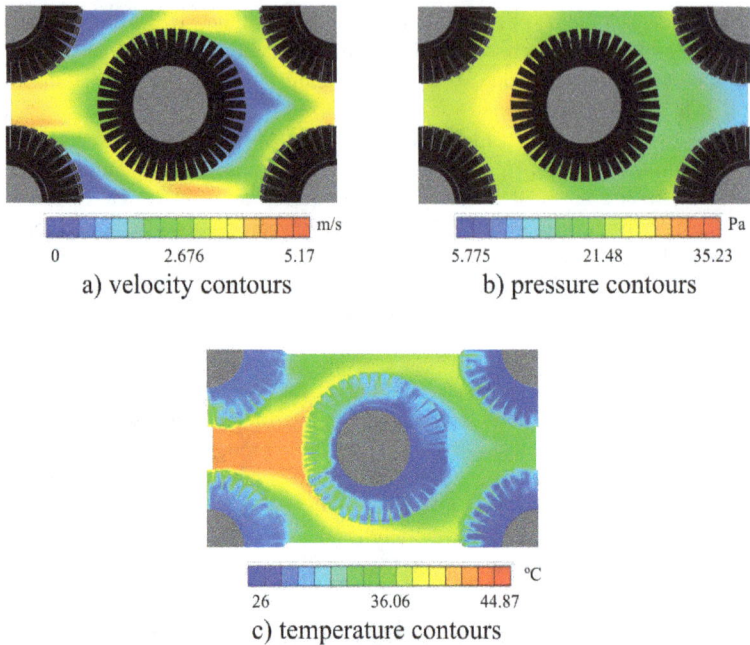

Figure 6. Numerical predictions.

5. Conclusion

The methodology, based on the LMTD method, for calculating boundary conditions in a single isolated finned tube module is adequate because the Nusselt number and friction factor show a quasi-constant behaviour for different finned tube length and finned tubes per row. The maximum deviations for Nusselt number and friction factor are lower than 1% and 3.6%, respectively. Therefore, the methodology can be used in numerical analysis of heat recoveries at industrial scale because the computational domain can be reduced in 99% which allows finite computational times. Numerical predictions in the single isolated finned tube module show that pressure contours are adequate because mean pressure drop and mean friction factor exhibit a deviation of 3.13% and 0.32% with respect to model developed by Weierman [15]. On the other hand, the pressure contours exhibit that the high pressure values are located in backwater zone and the main dissipative effect of flow energy is located at this region. Therefore, the pressure drop is dominated for backwater zone. The temperature field show a deviation of 4.25% and 5.18% for temperature difference and Nusselt number with respect to Kawaguchi´s, et al [18] model. The temperature contours exhibit that bare tube temperature is dominated by inside fluid temperature. Therefore, the inside fluid temperature must be considered as a lower limit temperature for cooling of flue gases in order to avoid dew point of acid gases.

Nomenclature

Symbols

A	surface area
A_2	tube row coefficient
B	contraction factor
c_p	specific heat at constant pressure
d	diameter
e	thickness
f	friction factor
G	gas mass flux
$\bar{\bar{g}}$	gravitational acceleration
$\bar{h}$	average convective coefficient
$\tilde{h}$	mean enthalpy
h	mean convective coefficient
k	thermal conductivity
$\tilde{k}$	turbulent kinetic energy
L	characteristic length
l	height
$\dot{m}$	mass flow
N	Number of tubes
n	number
Nu	Nusselt Number
$\overline{Nu}$	average Nusselt number
$\tilde{P}$	periodic pressure
P_b	mean pressure
Pr	Prandtl number
$\dot{Q}$	heat addition
Re	Reynolds number
R_f	fouling factor
S	pitch
s	gap
$\tilde{T}$	periodic temperature
T	temperature
t	thickness
U	overall heat transfer coefficient
$\tilde{u},\tilde{v},\tilde{w}$	mean velocity components
$\tilde{\bar{V}}$	instantaneous velocity
$\bar{V}''$	fluctuating velocity
w	velocity
$\Delta\tilde{P}$	pressure drop

ΔT_{ML} logarithmic mean temperature difference.

Subscripts

b   bulk, boundary
f   fin
g   gases
gp gas-phase
i   inside
l longitudinal
med   Initial profile
num   numeric
o   outside (tube diameter), overall
r   Radiation, rows
t   bare tube, transverse
v   volume-equivalent diameter
w   wall, tube material

Greek letters

β average-pressure gradient
$\bar{\rho}$ mean density
ϱ density
$\bar{\varepsilon}$ dissipation turbulent rate
η efficiency
μ viscosity
γ temperature-gradient term

Author details

E. Martínez and G. Soto
Universidad Autónoma Metropolitana – Azcapotzalco, Mexico City, Mexico

W. Vicente and M. Salinas
Instituto de Ingeniería, Universidad Nacional Autónoma de México,
Ciudad Universitaria, Mexico City, Mexico

A. Campo
Department of Mechanical Engineering, University of Texas at San Antonio, San Antonio,
TX, USA

Acknowledgement

We appreciate the support given to the research presented here by Universidad Nacional Autonoma de Mexico (Direccion General de Asuntos del Personal Academico, PAPIIT-IN106112-3), and Universidad Autonoma Metropolitana Azcapotzalco.

6. References

[1] Beale, S. B, and Spalding, D. B. (1999). A Numerical Study of Unsteady Fluid Flow in In-line and Staggered Tube Banks. *Journal of Fluids and Structures*, Vol.13, No.6, pp. 723-754, ISSN 0889-9746.

[2] Comini, G, and Croce, G. (2003). Numerical Simulation of Convective Heat and Mass Transfer in Banks of Tubes. *International Journal for Numerical Methods in Engineering*, Vol.57, No.12, pp. 1755-1773, ISSN 1755-1773.

[3] Beale, S. B. (2007). Use of Streamwise Periodic Boundary Condition for Problems in Heat and Mass Transfer. *ASME Journal of Heat Transfer*, Vol.129, No.4, pp. 601-605, ISSN 00221481.

[4] Beale, S. B. (2008). Benchmark Studies for the Generalized Streamwise Periodic Heat Transfer Problem. *ASME Journal of Heat Transfer*, Vol.130, No.11, pp. 114502_1-114502_4, ISSN 00221481.

[5] Benhamadouche, S, and Laurence, D. (2003). LES, Coarse LES, and Transient RANS Comparisons on the Flow Across a Tube Bundle. *International Journal of Heat and Fluid Flow*, Vol.24, No.4, pp. 470-479, ISSN 0142-727X.

[6] M. Salinas-Vázquez, M.A. de la Lama, W. Vicente, E. Martínez, (2011). Large Eddy Simulation of a Flow through Circular Tube Bundle. *Applied Mathematical Modelling*, Vol.35, No.9, pp. 4393–4406. ISSN 0307-904X.

[7] Hofmann, R, and Ponweiser, K. (2008). Experimental and Numerical Investigations of Serrated-Finned Tubes in Cross-Flow. Available from:
www.zid.tuwien.ac.at/fileadmin/files_zid/projekte/2008/08-302-2.pdf.

[8] Mcilwain S. R. (2010). A Comparison of Heat Transfer Around a Single Serrated Finned Tube and a Plain Finned Tube. *IJRAAS*, Vol.2, No.2, pp. 88-94, ISSN

[9] Mcilwain S. R. (2010). A CFD Comparison of Heat Transfer and Pressure Drop Across Inline Arragement Serrated Finned Tube Heat Exchangers with an Increasing Number of Rows. *IJRAAS*, Vol.4, No.2, pp. 162-169, ISSN

[10] Lemouedda, A, Schmid, A, Franz, E, Breuer, M, Delgado, A. (2011). Numerical Investigations for the Optimization of Serrated Finned-Tube Heat Exchangers. *Applied Thermal Engineering*, Vol.31, No. 8-9, pp. 1393-1401, ISSN 1359-4311.

[11] Ganapathy, V. (2002). Industrial Boilers, Heat Recovery and Steam Generators: Design, Applications and Calculations. In Marcel Dekker, (Ed), 333-507, ISBN 0-8247-0814-8, New York.

[12] Gnielinski, V. (1976). New equations for heat and mass transfer in turbulent pipe and channel flow. *Int. Chem. Eng.* Vol.16, No.1, pp. 359-366, ISSN 0020-6318.

[13] Rane, M. V., Tandale, S. (2005). Water-to-water heat transfer in tube-tube heat exchanger: Experimental and analytical study. *Applied Thermal Engineering*, Vol.25, No.17 , pp. 2715-2729, ISSN 1359-4311.

[14] Bejan, A. (1995). Convection Heat Transfer. In Wiley, (Ed), pp. 391-395, ISBN 0-471-27150-0.

[15] Weierman, C. (1976). Correlations Ease The Selection of Finned Tubes. *Oil and Gas Journal*, Vol.74, No.36, pp. 94-100, ISSN 0030-1388.

[16] ESCOA Turb-X HF Rating Instructions. (1979). Extended Surface Corporation of America (ESCOA), Pryor, OK.

[17] A. Nir. (1991). Heat Transfer and Friction Factor Correlations for Crossflow over Staggered Finned Tube Banks. *Heat Transfer Engineering*, Vol.12, No.1, pp. 43-58, ISSN 0145-7632.

[18] Kiyoshi Kawaguchi, Kenichi Okui, Takaharu Kashi. (2005). Heat transfer and pressure drop characteristics of finned tube banks in forced convection. *Journal of Enhanced Heat Transfer*, Vol.12, No.1, pp. 1-20, ISSN 1065-5131.

[19] E. Martinez, G. Soto, W. Vicente, M. Salinas. (2005). Comparative Analysis of Heat Transfer and Pressure Drop in Helically Segmented Finned Tube Heat Exchangers. *Applied Thermal Engineering*, Vol.30, No.11-12, pp. 1470-1476, ISSN 1359-4311.

[20] Jonh Weale, P.E., Peter H. Rumsey, P.E., Dale Sartor, P.E., and Lee Eng Lock. (2002). Laboratory Low-Pressure Drop Design. *ASHRAE Journal*, Vol. August pp. 38-42. ISSN 0001-2491.

[21] E. Martinez, W. Vicente, M. Salinas, G. Soto. (2010). Thermal Design Methodology of Industrial Compact Heat Recovery with Helically Segmented Finned Tubes. Heat Exchangers; design, types and applications. In Nova Publishers, (Ed), Series: Energy Science, Engineering and Technology, pp. 215-228. ISBN: 978-1-61761-308-1.

[22] Martínez, E, Vicente, W, Salinas, M, Soto, G. (2009). Single-phase experimental analysis of heat transfer in helically finned heat exchangers, *Applied Thermal Engineering*, Vol.29, No. 11-12, pp. 2205-2210, ISSN 1359-4311.

[23] Favre, A. (969). Problems of Hydrodynamics and Continuum Mechanics, SIAM.

[24] Yakhot, V, and Orszag, S. (1986). Renormalization Group Analysis of Turbulence. Basic Theory. *Journal of Scientific Computing*, Vol.1, No.1, pp. 3-51, ISSN 08857474.

[25] Patankar, S. V, Liu, C. H, Sparrow, E. M, 1977, Fully Developed Flow and Heat Transfer in Ducts Having Streamwise-Periodic Variations of Cross-Sectional Area, ASME Journal of Heat Transfer, 99, pp. 180-186.

[26] Kelkar, K. M, and Patankar, S. V, 1987, Numerical Prediction of Flow and Heat Transfer in a Parallel Plate Channel with Staggered Fins, ASME Journal of Heat Transfer, 109, pp. 25-30.

[27] Martínez, E. E. (2011). Simulación Numérica de un Flujo de Gases Turbulentos en un Banco de Tubos aletdos en Geometría Compleja. Ph.D Thesis. Universidad Autónoma Metropolitana, México City, México.

[28] J. C. Ludwing, H. Q. Qin, D. B. Spalding. (1989). The PHOENICS Reference Manual. *Technical Report CHAM TR/200*, CHAM Ltd, London.

[29] Y. Choi, J. Hong, H. Hwang, J. Choi. (2008). Cartesian Grid Method with Cut Cell in the Mold Filling Simulation. *Journal Materials Science British Technology*, Vol.24, No.3, pp. 379-382, ISSN 1005-0302.

Chapter 3

Homotopy Perturbation Method to Solve Heat Conduction Equation

Anwar Ja'afar Mohamed Jawad

Additional information is available at the end of the chapter

1. Introduction

Fins are extensively used to enhance the heat transfer between a solid surface and its convective, radiative, or convective radiative surface. Finned surfaces are widely used, for instance, for cooling electric transformers, the cylinders of air-craft engines, and other heat transfer equipment. In many applications various heat transfer modes, such as convection, nucleate boiling, transition boiling, and film boiling, the heat transfer coefficient is no longer uniform. A fin with an insulated end has been studied by many investigators [Sen, S. Trinh(1986)]; and [Unal (1987)]. Most of them are immersed in the investigation of single boiling mode on an extended surface. Under these circumstances very recently, [Chang (2005)] applied standard Adomian decomposition method for all possible types of heat transfer modes to investigate a straight fin governed by a power-law-type temperature dependent heat transfer coefficient using 13 terms. [Liu (1995)] found that Adomian method could not always satisfy all its boundaries conditions leading to boundaries errors.

The governing equations for the temperature distribution along the surfaces are nonlinear. In consequence, exact analytic solutions of such nonlinear problems are not available in general and scientists use some approximation techniques to approximate the solutions of nonlinear equations as a series solution such as perturbation method; see [Van Dyke M. (1975)], and Nayfeh A.H. (1973)], and homotopy perturbation method; see [He J. H. (1999, (2000), and(2003)].

In this chapter, we applied HPM to solve the linear and nonlinear equations of heat transfer by conduction in one-dimensional in two slabs of different material and thickness L.

2. The perturbation method

Many physics and engineering problems can be modelled by differential equations. However, it is difficult to obtain closed-form solutions for them, especially for nonlinear ones. In most cases, only approximate solutions (either analytical ones or numerical ones) can be expected. Perturbation method is one of the well-known methods for solving nonlinear problems analytically.

In general, the perturbation method is valid only for weakly nonlinear problems[Nayfeh (2000)]. For example, consider the following heat transfer problem governed by the nonlinear ordinary differential equation, see [Abbasbandy (2006)]:

$$(1+\varepsilon u)u' + u = 0, \qquad u(0) = 1 \tag{1}$$

where $\varepsilon > 0$ is a physical parameter, the prime denotes differentiation with respect to the time t. Although the closed-form solution of u(t) is unknown, it is easy to get the exact result $u'(0) = -1/(1+\varepsilon)$, as mentioned by [Abbasbandy (2006)]. Regard that ε as a perturbation quantity, one can write $u(t)$ into such a perturbation series

$$u(t) = u_0(t) + \varepsilon u_1(t) + \varepsilon^2 u_2(t) + \varepsilon^3 u_3(t) + \ldots\ldots\ldots\ldots \tag{2}$$

Substituting the above expression into (1) and equating the coefficients of the like powers of ε, to get the following linear differential equations

$$u_0' + u_0 = 0, \qquad u_0(0) = 1, \tag{3}$$

$$u_1' + u_1 = -u_0 u_0', \qquad u_1(0) = 0, \tag{4}$$

$$u_2' + u_2 = -(u_0 u_1' + u_1 u_0'), \qquad u_2(0) = 0, \tag{5}$$

$$u_3' + u_3 = -(u_0 u_2' + u_1 u_1' + u_2 u_0'), \qquad u_3(0) = 0, \tag{6}$$

Solving the above equations one by one, one has

$$u_0(t) = e^{-t}$$

$$u_1(t) = e^{-t} - e^{-2t}$$

$$u_2(t) = \frac{1}{2}e^{-t} - 2e^{-2t} + \frac{3}{2}e^{-3t} \tag{7}$$

Thus, we obtain $u(t)$ as a perturbation series

$$u(t) = e^{-t} + \varepsilon(e^{-t} - e^{-2t}) + \varepsilon^2(\frac{1}{2}e^{-t} - 2e^{-2t} + \frac{3}{2}e^{-3t}) + \ldots\ldots\ldots \tag{8}$$

which gives at t = 0 the derivative

$$u'(0) = -1 + \varepsilon - \varepsilon^2 + \varepsilon^3 - \varepsilon^4 + \varepsilon^5 - \varepsilon^6 + \varepsilon^7 - \varepsilon^8 + \varepsilon^9 - \varepsilon^{10} + \ldots\ldots\ldots \tag{9}$$

Obviously, the above series is divergent for $\varepsilon \geq 1$, as shown in Fig. 1. This typical example illustrates that perturbation approximations are valid only for weakly nonlinear problems in general. In view of the work by [Abbasbandy (2006)], the HAM extends a series approximation beyond its initial radius of convergence.

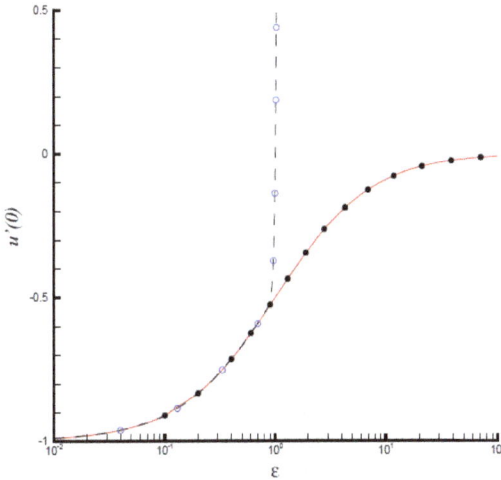

Figure 1. Comparison of the exact and approximate solutions of (1). Solid line: exact solution $u'(0) = -1/(1+\varepsilon)$; Dashed-line: 31th-order perturbation approximation; Hollow symbols: 15th-order approximation given by the HPM; Filled symbols: 15th-order approximation given by the HAM when $h = -1/(1+2\varepsilon)$.

To overcome the restrictions of perturbation techniques, some non-perturbation techniques are proposed, such as the Lyapunov's artificial small parameter method [Lyapunov A.M. (1992)], the δ-expansion method [Karmishin et al(1990)] , the homotopy perturbation method [He H., J.,(1998)], and the variational iteration method (VIM), [He H., J.,(1999)],. Using these non-perturbation methods, one can indeed obtain approximations even if there are no small/large physical parameters. However, the convergence of solution series is not guaranteed. For example, by means of the HPM, we obtain the same and exact approximation of Eq.(1), as the perturbation result in Eq.(9), that is divergent for ε > 1, as shown in Fig.1. ; For details, see [Abbasbandy (2006)]. This example shows the importance of the convergence of solution series for all possible physical parameters. From physical points of view, the convergence of solution series is much more important than whether or not the used analytic method itself is independent of small/large physical parameters. If one does not keep this in mind, some useless results might be obtained. For example, let us consider the following linear differential equation [Ganji et al (2007)]:

$$u_t + u_x = 2u_{xxt}, \qquad x \in R, t > 0 \tag{10}$$

$$u(x,0) = e^{-x} \tag{11}$$

Its exact solution reads

$$u_{exact}(x,t) = e^{-x-t} \tag{12}$$

By means of the homotopy perturbation method, [Ganji et al (2007)] wrote the original equation in the following form:

$$(1-p)\frac{\partial \varphi(x,t:p)}{\partial t} + p[\frac{\partial \varphi(x,t:p)}{\partial t} + \frac{\partial \varphi(x,t:p)}{\partial x} - \frac{\partial^3 \varphi(x,t:p)}{\partial x^2 \partial t}] = 0 \tag{13}$$

subject to the initial condition

$$\varphi(x,0:p) = e^{-x} \tag{14}$$

where $p \in [0;1]$ is an embedding parameter. Then, regarding p as a small parameter, [Ganji et al (2007)] expanded $\varphi(x,t:p)$ in a power series

$$\varphi(x,t:p) = u_0(x,t) + \sum_{m=1}^{+\infty} u_m(x,t).p^m \tag{15}$$

which gives the solution. For $p = 1$, and substitute (15) into the original equation (13) and initial condition in (14), then equating the coefficients of the like powers of p, one can get governing equations and the initial conditions for $u_m(x,t)$. In this way, [Ganji et al (2007)] obtained the mth-order approximation

$$u(x,t) \approx u_0(x,t) + \sum_{k=1}^{m} u_k(x,t) \tag{16}$$

and the 5th-order approximation reads

$$u_{HPM}(x,t) \approx \frac{e^{-x}}{720}[t^6 + 66t^5 + 1470t^4 + 13320t^3 + 47440t^2 + 45360t + 720] \tag{17}$$

However, for any given $x \geq 0$, the above approximation enlarges monotonously to the positive infinity as the time t increases, as shown in Fig.2. Unfortunately, the exact solution monotonously decreases to zero! Let

$$\delta(t) = \left| \frac{u_{exact} - u_{HAM}}{u_{exact}} \right| \tag{18}$$

where $\delta(t)$ denotes the relative error of the HPM approximation (17). As shown in Fig. 2, the relative error $\delta(t)$ monotonously increases very quickly:

In fact, it is easy to find that the HPM series solution (16) is divergent for all x and t except t = 0 which however corresponds to the given initial condition in (11). In other words, the convergence radius of the HPM solution series (17) is zero. It should be emphasized that, the variational iteration method (VIM) obtained exactly the same result as (17) by the 6th iteration see; [He H. J., (1999)], and [Ganji et al (2007)]. This example illustrates that both of the HPM and the VIM might give divergent approximations. Thus, it is very important to ensure the convergence of solution series obtained.

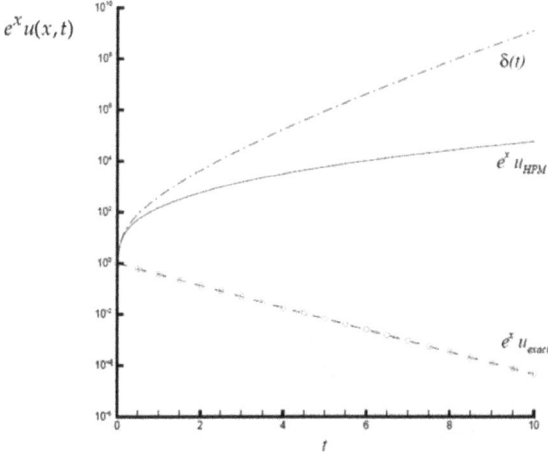

Figure 2. Approximations of (10) given by the homotopy perturbation method. Dashed-line: exact solution(12); Solid line: the 5th-order HPM approximation(17); Dash-dotted line: the relative error $\delta(t)$ defined by(18).

3. Outline of Homotopy Perturbation Method (HPM)

The homotopy analysis method (HAM) has been proposed by Liao in his PhD dissertation in [Liao (1992)]. Liao introduced the so-called auxiliary parameter in [Liao (1997a)] to construct the following two-parameter family of equation:

$$(1-p)L(u-u_0) = hpN(u) \tag{19}$$

where u_0 is an initial guess. [Liao (1997a)] pointed out that the convergence of the solution series given by the HAM is determined by h, and thus one can always get a convergent series solution by means of choosing a proper value of h. Using the definition of Taylor series with respect to the embedding parameter p (which is a power series of p), [Liao (1997b)] gave general equations for high-order approximations.

[He J. H. (1999)] followed Dr. Liao's early idea of Homotopy Perturbation Method (HPM) when he constructed the one-parameter family of equation:

$$(1-p)L(u) + pN(u) = 0 \tag{20}$$

where Eq.(20) represented special case of Eq.(19) for convergent solution of (HAM) at $h = -1$. To illustrate the basic ideas of this method, consider the following general nonlinear differential equation [see Ghasemi et al (2010)].

$$A(u) - f(r) = 0, \qquad r \in \Omega \tag{21}$$

With boundary conditions

$$B(u, \frac{\partial u}{\partial n}) = 0, \qquad r \subset \Gamma \tag{22}$$

where A is a general differential operator, B is a boundary operator, $f(r)$ is a known analytic function, and is the boundary of the domain .

The operator A can be generally divided into linear and nonlinear parts, say L and N. Therefore (21) can be

written as

$$L(u) + N(u) - f(r) = 0 \tag{23}$$

[He (1999)] constructed a homotopy $v(r,p) : \Omega \times [0,1] \rightarrow R$ which satisfies:

$$H(v, p) = (1 - p)[L(v) - L(v_0)] + p[A(v) - f(r)] = 0 \tag{24}$$

where $r \in \Omega$, $p \in [0,1]$ that is called homotopy parameter, and v_0 is an initial approximation of (19). Hence, it is obvious that:

$$H(v, 0) = L(v) - L(v_0) = 0 \tag{25}$$

and

$$H(v, 1) = [A(v) - f(r)] = 0 \tag{26}$$

In topology, $L(v) - L(v_0)$ is called deformation, and $[A(v) - f(r)]$ is called homotopic. The embedding parameter p monotonically increases from zero to unit as the trivial problem $H(v, 0) = 0$ in (25) is continuously deforms the original problem in (26), $H(v, 1) = 0$. The embedding parameter $p \in [0,1]$ can be considered as an expanding parameter. [Nayfeh A.H. (1985)] Apply the perturbation technique due to the fact that $0 \leq p \leq 1$, can be considered as a small parameter, the solution of (21) or (23) can be assumed as a series in p, as follows:

$$v = v_0 + pv_1 + p^2v_2 + p^3v_3 + ... \tag{27}$$

when $p \rightarrow 1$, the approximate solution, i.e.,

$$u = \lim_{p \to 1} v = v_0 + v_1 + v_2 + v_3 + ... \tag{28}$$

The series (28) is convergent for most cases, and the rate of convergence depends on $A(v)$, [He, L. (1999)].

4. Application of Homotopy Perturbation Method HPM

An analytic method for strongly nonlinear problems, namely the homotopy analysis method (HAM) was proposed by Liao in 1992, six years earlier than the homotopy perturbation method by [He H., J.,(1998)], and the variational iteration method by [He H., J.,(1999)]. Different from perturbation techniques, the HAM is valid if a nonlinear problem contains small/large physical parameters.

More importantly, unlike all other analytic techniques, the HAM provides us with a simple way to adjust and control the convergence radius of solution series. Thus, one can always get accurate approximations by means of the HAM. In the next section, HPM is applied to solve the linear and nonlinear equations of heat transfer by conduction in one-dimensional in a slab of thickness (L). [Anwar (2010)] solved the linear and non-linear heat transfer equations by means of HPM.

4.1. Non-Linear Heat transfer equation

Consider the heat transfer equation by conduction in one-dimensional in a slab of thickness L. The governing equation describing the temperature distribution is:

$$\frac{d}{dx}(k\frac{dT}{dx}) = 0, \qquad x \in [0,L] \tag{29}$$

Where the two faces are maintained at uniform temperatures T_1 and T_2 with $T_1 > T_2$ the slab make of a material with temperature dependent thermal conductivity $k = k(T)$; see [Rajabi A.(2007)]. The thermal conductivity k is assumed to vary linearly with temperature, that is:

$$k = k_2[1 + \varepsilon\frac{T - T_2}{T_1 - T_2}] \qquad T(0) = T_1, \qquad T(L) = T_2 \tag{30}$$

where ε is a constant and k_2 is the thermal conductivity at temperature T_2 . Introducing the dimensionless quantities

$$\theta = \frac{T - T_2}{T_1 - T_2}, \quad \varepsilon = \frac{k_1 - k_2}{k_2} \quad X = \frac{x}{L} \quad X \in [0,1]$$

where k_1 is the thermal conductivity at temperature T_1 , then (29) reduces to

$$\frac{d^2\theta}{dX^2} + \varepsilon\frac{(\frac{d\theta}{dX})^2}{(1 + \varepsilon\theta)} = 0, \qquad X \in [0,1] \tag{31}$$
$$\theta(0) = 1, \qquad \theta(1) = 0$$

The problem is formulated by using (19) as:

$$(1 - p)L[\hat{\theta}(X,p) - \theta_0(X)] + pN[\hat{\theta}(X,p)] = 0 \tag{32}$$

Where the Linear operator:

$$L(\theta) = \theta'' \tag{33}$$

and,

$$\theta_{0XX} = 0 \tag{34}$$

from Eq.(31), the initial guess is:

$$\theta_0(X) = 1 - X \tag{35}$$

and the linear operator:

$$L[C_1 X + C_2] = 0 \tag{36}$$

and the nonlinear operator of $\hat{\theta}(X,p)$ is:

$$N[\hat{\theta}(X,p)] = \hat{\theta}(X,p)_{XX} + \varepsilon[\hat{\theta}(X,p).\hat{\theta}(X,p)_{XX} + (\hat{\theta}(X,p)_X)^2] = 0 \tag{37}$$

and:

$$\hat{\theta}(0,p) = 1, \qquad \hat{\theta}(1,p) = 0 \tag{38}$$

where $p \in [0,1]$ is an embedding parameter. For p = 0 and 1, we have

$$\hat{\theta}(X,0) = \theta_0(X) \qquad \hat{\theta}(X,1) = \theta(X) \tag{39}$$

$\theta_0(X)$ tends to $\theta(X)$ as p varies from 0 to 1. Due to Taylor's series expansion:

$$\hat{\theta}(X,p) = \theta_o(X) + \sum_{s=1}^{\infty} \frac{1}{s!} \frac{\partial^s \hat{\theta}(X,p)}{\partial p^s} \bigg|_{p=0} \tag{40}$$

and the convergence of series (40) is convergent at $p = 1$. Then by using (35) and (36) one obtains

$$\theta(X) = \theta_0(X) + \sum_{s=0}^{\infty} \theta_s(X) \tag{41}$$

For the s-th- order problems, if we first differentiate Eq.(32) s times with respect to p then divide by s! and setting p = 0 we obtain:

$$L[\theta_s(X) - u_s \theta_{s-1}(X)] + [\theta_{s-1}''(X) + \varepsilon \sum_{n=0}^{s-1} (\theta_{s-1-n}'.\theta_n' + \theta_{s-1-n}\theta_n'')] = 0 \tag{42}$$

Where:

$$\theta_s(0) = \theta_s(1) = 0 \tag{43}$$

$$u_s = \begin{cases} 0, & s \leq 1 \\ 1, & s > 1 \end{cases} \tag{44}$$

The general solutions of (42) can be written as:

$$\theta_s(X) = \theta_0(X) + \theta_s^*(X) \tag{45}$$

where $\theta_s^*(X)$ is the particular solution.

The linear non-homogeneous (42) is solved for the order s = 1, 2, 3,..., for s=1, (42) becomes:

$$\frac{d^2\theta_1}{dX^2} + \varepsilon(\frac{d\theta_0}{dX}\frac{d\theta_0}{dX} + \theta_0\frac{d^2\theta_0}{dX^2}) = 0 \qquad \theta_1(0) = 0, \qquad \theta_1(1) = 0 \tag{46}$$

Then

$$\frac{d^2\theta_1}{dX^2} + \varepsilon = 0 \tag{47}$$

the solution of (47) gives :

$$\theta_1 = \frac{\varepsilon}{2}(X - X^2) \tag{48}$$

For s = 2, Eq.(42) becomes:

$$\frac{d^2\theta_2}{dX^2} + \varepsilon(2\frac{d\theta_1}{dX}\frac{d\theta_0}{dX} + \theta_1\frac{d^2\theta_0}{dX^2} + \theta_0\frac{d^2\theta_1}{dX^2}) = 0, \qquad \theta_2(0) = 0, \qquad \theta_2(1) = 0 \tag{49}$$

Solution of (49) gives :

$$\theta_2 = \frac{\varepsilon^2}{2}[2X^2 - X^3 - X] \tag{50}$$

Then, solution of (31) is:

$$\theta(X) = (1 - X) + \frac{\varepsilon}{2}(X - X^2) + \frac{\varepsilon^2}{2}[2X^2 - X^3 - X] \tag{51}$$

Results of θ obtained for different values of ε are presented in Table 1 and Fig. 3. Clearly, for small value for $0 \leq \varepsilon \leq 1$ then (51) is a good approximation to the solution. That means for $\varepsilon = 0$, then $k_1 = k_2$, for $\varepsilon = 1$, then $k_1 = 2k_2$. However, as ε increases, (51) produces inaccurate divergent results. The results obtained via HPM are compared to those via General Approximation Method GAM obtained by Khan R. A.(2009). For this problem, it is found that HPM produces agreed results compared to GAM.

X	ε =.5	ε =.8	ε =1.0	ε =1.5	ε =2
0.1	0.912375	0.91008	0.9045	0.876375	0.828
0.2	0.824	0.82304	0.816	0.776	0.704
0.3	0.734125	0.73696	0.7315	0.692125	0.616
0.4	0.642	0.64992	0.648	0.618	0.552
0.5	0.546875	0.56	0.5625	0.546875	0.5
0.6	0.448	0.46528	0.472	0.472	0.448
0.7	0.3446249	0.36384	0.3735	0.386625	0.384
0.8	0.2359999	0.2537599	0.2639999	0.2839999	0.2959999
0.9	0.1213749	0.1331199	0.1404999	0.1573749	0.1719999

Table 1. Solutions θ via X for different values of ε.

Special computer program was used as special case, the temperature distribution along a road of length (L = 1 m) when T_1 = 100 C^0 and T_2 = 50 C^0, are presented in Table 2 and Fig.4.

X	ε =.5	ε =.8	ε =1.0	ε =1.5	ε =2
0.0	100	100	100	100	100
0.1	95.61875	95.504	95.225	93.81875	91.4
0.2	91.2	91.152	90.8	88.8	85.2
0.3	86.70625	86.848	86.575	84.60625	80.8
0.4	82.1	82.496	82.4	80.9	77.6
0.5	77.34375	78	78.125	77.34375	75
0.6	72.4	73.264	73.6	73.6	72.4
0.7	67.23125	68.192	68.675	69.33125	69.2
0.8	61.8	62.688	63.2	64.2	64.8
0.9	56.06874	56.656	57.025	57.86874	58.6
1.0	50	50	50	50	50

Table 2. Solutions T via X for different values of ε

4.2. Linear Heat transfer equation

In this section we consider the linear one-dimensional equation of heat transfer by conduction (diffusion equation) [Anderson (1984)]:

$$\frac{\partial T}{\partial t} - \alpha \frac{\partial^2 T}{\partial x^2} = 0 \qquad 0 \le x \le 1, \; t > 0 \tag{52}$$

for initial condition

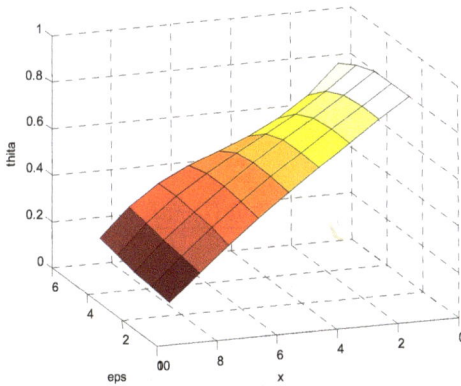

Figure 3. Graphical results θ via X obtained by HPM for different values of ε.

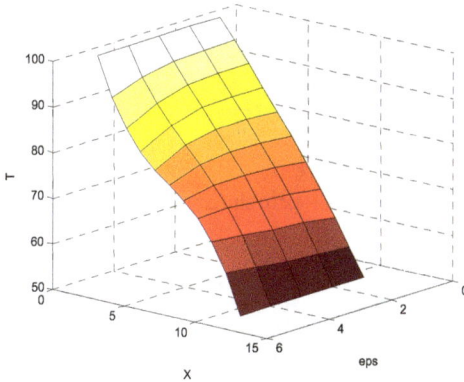

Figure 4. Graphical results of Temperature via X obtained by HPM for different values of ε.

$$T(x,0) = g(x) = \sin(2\pi x) \tag{53}$$

and boundary condition

$$T(0,t) = T(1,t) = 0 \tag{54}$$

α is thermal conductivity that is assumed constant with temperature. To solve the parabolic partial differential equation (52) using HPM, we consider a correction functional equation as:

$$(1-p)[\frac{\partial T}{\partial t} - \frac{\partial u_0}{\partial t}] + p[\frac{\partial T}{\partial t} - \alpha\frac{\partial^2 T}{\partial x^2}] = 0 \tag{55}$$

Then:

$$\frac{\partial T}{\partial t} - \frac{\partial u_0}{\partial t} - p\frac{\partial u_0}{\partial t} - \alpha p\frac{\partial^2 T}{\partial x^2} = 0 \tag{56}$$

$$\frac{\partial T}{\partial t} - \alpha p\frac{\partial^2 T}{\partial x^2} = 0 \tag{57}$$

$$\frac{\partial(T_0 + pT_1 + p^2T_2 + p^3T_3 + ...)}{\partial t} - \alpha p\frac{\partial^2(T_0 + pT_1 + p^2T_2 + p^3T_3 + ..)}{\partial x^2} = 0 \tag{58}$$

For zeroth order of p:

$$\frac{\partial T_0}{\partial t} = 0 \tag{59}$$

Then $T_0(x,t) = \sin(2\pi.x)$

For first order of p:

$$\frac{\partial T_1}{\partial t} - \alpha\frac{\partial^2 T_0}{\partial x^2} = 0 \tag{60}$$

$$\frac{\partial T_1}{\partial t} + 4\pi^2\alpha\sin(2\pi.x) = 0 \tag{61}$$

$$T_1(x,t) = \sin(2\pi.x) - 4\pi^2\alpha\sin(2\pi.x)t \tag{62}$$

For second order of p:

$$\frac{\partial T_2}{\partial t} - \alpha\frac{\partial^2 T_1}{\partial x^2} = 0 \tag{63}$$

$$T_2(x,t) = \sin(2\pi.x) - 4\pi^2\alpha\sin(2\pi.x)t + 8\pi^4\alpha^2\sin(2\pi.x)t^2 \tag{64}$$

Using equation (56) for other orders of p, we can obtain the following results:

$$T(x,t) = \sin(2\pi.x)[1 - (4\pi^2\alpha.t) + \frac{1}{2}(4\pi^2\alpha.t)^2 - .....] \tag{65}$$

It is obvious that $T(x,t)$ converge to the exact solution as increasing orders of p:

$$T(x,t) = \sin(2\pi.x).\exp(-4\pi^2\alpha.t) \tag{66}$$

Fig.5 and Fig.6 represent the HPM solution T(x, t) for $\alpha = 0.05$, and $\alpha = 0.1$ respectively for $0 \le x \le 1,\ 0 \le t \le 0.4$.

X	t=0	t = 0.1	t = 0.2	t = 0.3	t = 0.4
0	0	0	0	0	0
0.05	0.3091373	0.2082383	0.1402717	9.448861E-2	6.364859E-2
0.1	0.5879898	0.3960766	0.2668017	0.1797206	0.1210618
0.15	0.8092399	0.5451131	0.3671944	0.2473463	0.1666153
0.2	0.9512127	0.6407476	0.4316148	0.2907406	0.1958462
0.25	0.9999998	0.6736112	0.4537521	0.3056525	0.205891
0.3	0.9508218	0.6404843	0.4314375	0.2906211	0.1957657
0.35	0.8084964	0.5446123	0.366857	0.247119	0.1664622
0.4	0.5869664	0.3953872	0.2663373	0.1794078	0.1208511
0.45	0.3079342	0.207428	0.1397258	0.0941209	0.0634009
0.5	0	0	0	0	0
0.55	-0.3103399	-0.2090485	-0.1408174	-0.0948562	-6.389621E-2
0.60	-0.5890125	-0.3967655	-0.2672657	-0.1800332	-0.1212724
0.65	-0.8099824	-0.5456133	-0.3675313	-0.2475732	-0.1667681
0.70	-0.9516023	-0.64101	-0.4317916	-0.2908597	-0.1959264
0.75	-0.9999982	-0.6736101	-0.4537514	-0.3056521	-0.2058907
0.80	-0.9504291	-0.6402198	-0.4312593	-0.2905011	-0.1956849
0.85	-0.8077511	-0.5441103	-0.3665189	-0.2468912	-0.1663087
0.90	-0.5859417	-0.3946969	-0.2658723	-0.1790946	-0.1206401
0.95	-0.3067303	-0.206617	-0.1391795	-9.37529E-2	-6.315302E-2
1	0	0	0	0	0

Table 3. Solution $T(x, t)$ for $0 \leq x \leq 1$, $0 \leq t \leq 0.4$ at $\alpha = 0.05$.

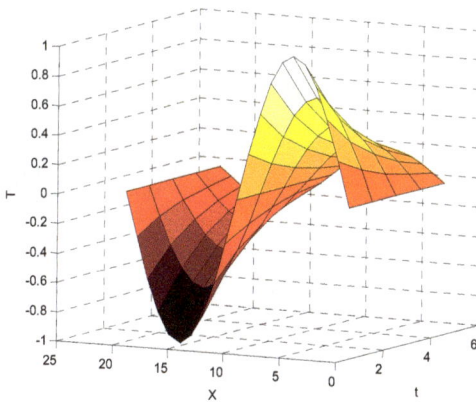

Figure 5. Solution $T(x, t)$ for $0 \leq x \leq 1$, $0 \leq t \leq 0.4$ and $\alpha = 0.05$.

X	t=0	t = 0.1	t = 0.2	t = 0.3	t = 0.4
0	0	0	0	0	0
0.05	0.3091373	0.2537208	0.2082383	0.1709092	0.1402717
0.1	0.5879898	0.4825858	0.3960766	0.3250752	0.2668017
0.15	0.8092399	0.6641742	0.5451131	0.4473952	0.3671944
0.2	0.9512127	0.7806966	0.6407476	0.5258861	0.4316148
0.25	0.9999998	0.8207381	0.6736112	0.5528585	0.4537521
0.3	0.9508218	0.7803758	0.6404843	0.52567	0.4314375
0.35	0.8084964	0.6635639	0.5446123	0.4469841	0.366857
0.4	0.5869664	0.4817458	0.3953872	0.3245094	0.2663373
0.45	0.3079342	0.2527334	0.207428	0.1702441	0.1397258
0.5	0	0	0	0	0
0.55	-0.3103399	-0.2547078	-0.2090485	-0.1715741	-0.1408174
0.60	-0.5890125	-0.4834251	-0.3967655	-0.3256406	-0.2672657
0.65	-0.8099824	-0.6647836	-0.5456133	-0.4478057	-0.3675313
0.70	-0.9516023	-0.7810164	-0.64101	-0.5261015	-0.4317916
0.75	-0.9999982	-0.8207368	-0.6736101	-0.5528576	-0.4537514
0.80	-0.9504291	-0.7800536	-0.6402198	-0.5254529	-0.4312593
0.85	-0.8077511	-0.6629522	-0.5441103	-0.4465721	-0.3665189
0.90	-0.5859417	-0.4809047	-0.3946969	-0.3239429	-0.2658723
0.95	-0.3067303	-0.2517453	-0.206617	-0.1695785	-0.1391795
1	0	0	0	0	0

Table 4. Solution $T(x, t)$ for $0 \leq x \leq 1$, $0 \leq t \leq 0.4$ at $\alpha = 0.1$

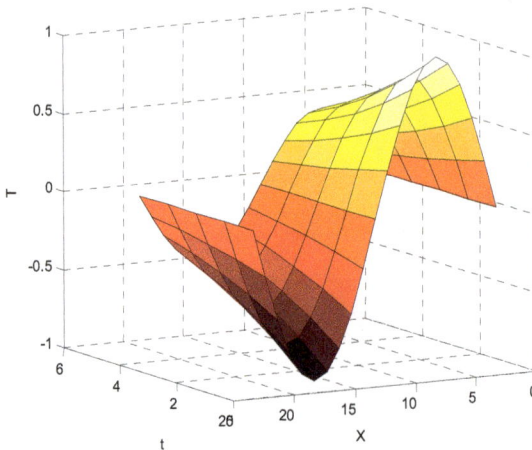

Figure 6. Solution $T(x, t)$ for $0 \leq x \leq 1$, $0 \leq t \leq 0.4$ and $\alpha = 0.1$

5. Discussion

For example.1, clearly, for $0 \leq \varepsilon \leq 1$, (51) is a good approximation to the solution. That means for $\varepsilon = 0$, then $k_1 = k_2$, and for $\varepsilon = 1$, then $k_1 = 2k_2$. However, as ε increases, (51) produces inaccurate divergent results. For example 2, (66) is a good approximation to the solution as α values decreased. That means as α increase, (66) produces inaccurate divergent results.

6. Conclusion

Homotopy Perturbation Method HPM is applied to solve the linear and nonlinear partial differential equation. Two numerical simulations are presented to illustrate and confirm the theoretical results. The two problems are about heat transfer by conduction in two slabs. Results obtained by the homotopy perturbation method are presented in tables and figures. Results are compared with those studied by the generalized approximation method by [Sajida et al (2008)]. Homotopy Perturbation Method is considered as effective method in solving partial differential equation.

Author details

Anwar Ja'afar Mohamed Jawad
Al-Rafidain University College, Baghdad, Iraq

7. References

Abbasbandy S. (2006), The application of homotopy analysis method to nonlinear equations arising in heat transfer. Phys. Lett. A 360, 109-113.

Anderson D. A. (1984), J. C. Tannehill, R. H. Pletcher, Computational Fluid Mechanics and Heat Transfer, McGraw-Hill company.

Anwar J. M. Jawad,(2010)"Application of Homotopy Perturbation Method in Heat Conduction Equation", Al-ba'ath University journal , Syria vol. 32.

Chang, M.H., (2005). A decomposition solution for fins with temperature dependent surface heat flux, Int.J. Heat Mass Transfer 48: 1819-1824.

Ganji D. D., H. Tari, M.B. Jooybari (2007), Variational iteration method and homotopy perturbation method for nonlinear evolution equations. Comput. Math. Appl. 54, 1018-1027.

Ghasemi, M., Tavassoli Kajani M.(2010), Application of He's homotopy perturbation method to solve a diffusion-convection problem, Mathematical Sciences Vol. 4, No. 2 171-186.

He H., J.,(1998), Approximate analytical solution for seepage flow with fractional derivatives in porous media. Comput. Meth. Appl. Mech. Eng. 167, 57-68.

He H. J., (1999),Variational iteration method: a kind of nonlinear analytical technique: some examples. Int. J. Non-Linear Mech. 34, 699-708.

He J. H. (2003), Homotopy perturbation method a new nonlinear analytical technique, Appl. Math. Comput. 135 , 73–79.

He J. H. (2000), A coupling method of a homotopy technique and a perturbation technique for non-linear problems, Int. J. Nonlinear Mech., 35, (37).

He J. H.(1999), Homotopy perturbation technique, J. Compt. Methods Appl. Mech. Eng., 178 , 257–262.

Karmishin A. V., A.I. Zhukov, V.G. Kolosov(1990), Methods of Dynamics Calculation and Testing for Thin-Walled Structures. Mashinostroyenie, Moscow.

Khan R. A.(2009), The generalized approximation method and nonlinear Heat transfer equations, Electronic Journal of Qualitative Theory of Differential Equations, No. 2, 1-15;

Lyapunov A. M. (1992), General Problem on Stability of Motion. Taylor & Francis,London, (English translation).

Liao S. J. (1992), On the proposed homotopy analysis technique for nonlinear problems and its applications, Ph.D. dissertation, Shanghai Jio Tong University, (1992).

Liao S. J. (1997a), Numerically solving nonlinear problems by the homotopy analysis method, Compt. Mech. 20 , 530–540.

Liao S. J. (1997b), A kind of approximate solution technique which does not depend upon small parameters (II): an application in fluid mechanics, Int. J. Non-linear Mech. 32 , 815–822.

Liu, G.L., (1995). Weighted residual decomposition method in nonlinear applied mathematics, in: Proceedings of the 6th congress of modern mathematical and mechanics, Suzhou, China: 643-649.

Nayfeh A.H. (1973), Perturbation Methods, Wiley, New York.

Nayfeh A.H. (1985), Problems in Perturbation, John Wiley, New York.

Nayfeh A.H. (2000), Perturbation Methods. Wiley, New York.

Rajabi A.(2007), D. D. Gunji and H. Taherian, Application of homotopy perturbation method in nonlinear heat conduction and convection equations, Physics Letter A, 360 , 570–573.

Sajida M. , T. Hayatb (2008), Comparison of HAM and HPM methods in nonlinear heat conduction and convection equations, Nonlinear Analysis, Real World Applications, 9 2296–2301.

Sen, A.K. S. Trinh, (1986). An exact solution for the rate of heat transfer from a rectangular fin governed by a power law type temperature dependence, Int. J. Heat Mass Transfer, 108: 457-459.

Songxin Liang , David J. Jeferey, (2009),Comparison of homotopy analysis method and homotopy perturbation method through an evolution equation , Communications in Nonlinear Science and Numerical Simulation, Volume 14, Issue 12, p. 4057-4064.

Unal, H.C., (1987). A simple method of dimensioning straight fins for nucleate pool boiling, Int. J. Heat Mass Transfer, 29: 640-644.

Van Dyke M. (1975), Perturbation Methods in Fluid Mechanics, Annotated Edition, Parabolic Press, Stand ford, CA.

Numerical Analysis of Heat Transfer

Chapter 4

3-D Numerical Simulation of Heat Transfer in Biomedical Applications

Aleksandra Rashkovska, Roman Trobec, Matjaž Depolli and Gregor Kosec

Additional information is available at the end of the chapter

1. Introduction

Computational modelling has received significant attention in the research community over the past few decades due to its pronounced contribution in better understanding of the nature, as well as in the development of advanced technologies. The modelling of more and more complex transport physical systems helps the community to address important issues, like identification of environmental problems, improvement of technological processes, development of biomedical applications, etc. However, the physical modelling is only one part of the entire problem. In most realistic scientific computing applications, a physical model cannot be solved in a closed form, thus adequate numerical approach is required. The continuous physical domain is replaced by its discrete form. In the majority of the simulations, the classical numerical methods, such as the Finite Volume Method (FVM) (Ferziger and Perić, 2002), Finite Difference Method (FDM) (Őzisik, 2000, Shashkov, 1996), Boundary Element Method (BEM) (Białecki, et al., 2006, Wrobel, 2002) or the Finite Element Method (FEM) (Rappaz, et al., 2003), are used to solve these problems.

An important part of the numerical approach is the effective implementation of the solution procedure on modern computer architectures. Realistic computations might take vast amount of time to complete, especially in 3-D domains with high number of computational nodes. The computing performance can be improved by numerous approaches, such as network computing, grid computing, cloud computing and other variants of "distributed computing" (Trobec, et al., 2009).

In recent decades, computer simulations have proved a great help in understanding and solving a variety of problems in science. Advances in computer technology enable simulation of natural phenomena that cannot be subject to experiment in reality since experiments would be ecologically problematic, dangerous to humans, or would cost vast amounts of money (Martino, et al., 1994). Especially in medicine, experiments are often difficult to perform because human subjects are involved. Measurements during clinical procedures are time consuming and often not as accurate as desired, because many parameters are difficult to control (McMaster, *et al.*, 1978). Many examples can be found in medicine, where in vivo experiments and measurements are often difficult, dangerous or even impossible (Trunk, *et al.*, 2003), especially if deep tissues or vital organs are in question (Tikuisis and Ducharme, 1991). However, simulation can provide the only safe and inexpensive insight into physiological processes. With the use of computer simulation, it is possible to calculate, analyse, and visualize both stationary temperature fields and the changes of temperature in time in living biological tissues (Grana, 1994). The temperature of the human tissue is an important factor in many fields of physiology, sports (Olin and Huljebrant, 1993), cryotherapy (Trobec, *et al.*, 1998), etc. It has been recognized that the temperature field is influenced by the environmental conditions, the temperatures of neighbouring tissues, the muscle metabolism and the blood circulation. Different tissues have different physical and thermodynamic properties and respond diversely to temperature changes (Bernardi, *et al.*, 2003, Pennes, 1948, Trunk, *et al.*, 2003). The temperature field varies in space and time in different parts of the investigated domain.

In this chapter we present numerical solution procedure for the treatment of heat transfer in biological tissues. Different body parts are modelled and given as numerical examples. All cases are focused on realistic three dimensional (3-D) models and realistic process parameters. In all numerical examples we use 3-D closed cavity as a computational domain. Within the domain, high resolution models of body parts are used. The models are obtained through expert assisted segmentation of human body parts from the Visible Human Dataset (VHD) (Ackerman, 1998). The body parts are composed of different tissues (muscles, bones, cartilage, fat, vessels and nerves). To respect the complexity of the modelled body parts, a one-domain approach is employed through spatial dependent thermo-physical properties. The final geometrical domain is composed of small cubic voxels, resulting in millions of discretization points.

The Pennes' Bio-heat equation (Pennes, 1948) represents the basis for the physical model. It describes the energy continuity within biological tissues by incorporating heat convection, heat transfer between blood and tissues, and heat production by metabolism. Some important extensions have been introduced, in particular, an inhomogeneous spatial model composed of tissues with different characteristics and modelling of the heat contributions from arteries, blood perfusion, and metabolism as functions of the surrounding tissue temperature. This improved Bio-heat equation was evaluated in terms of sensitivity and accuracy and solved numerically on single and parallel computers (Akl, 1997, Martino, *et al.*, 1994). Outside biological tissues, the Bio-heat equation is reduced to diffusion equation as both additional source terms (heat transfer between blood and tissues, and metabolism) vanish. The flow outside tissues is modelled as the Navier Stokes fluid (Ferziger and Perić, 2002), while in the tissue no flow is allowed. We consider only incompressible fluids.

A substantial amount of work on closed form and numerical solutions of the Bio-heat equation has been published (Bernardi, *et al.*, 2003, Bowman, *et al.*, 1975, Charny, 1992, Pennes, 1948). In our examples, the governing system of partial differential equations

(PDEs) is numerically solved through spatial discretization with the FDM and the Euler time stepping (Özisik, 2000). We present three different numerical studies regarding different topics in medicine, where results are computed using the proposed physical model and numerical solution procedure.

The first numerical example is dedicated to a study of the temperature changes in a resting proximal human forearm. By computer simulation, we explained the variations in the Pennes' well-known in vivo measurements of the steady-state temperatures along the transverse axis of the proximal forearm (Pennes, 1948). Such in vivo measurements are no longer practiced and are very valuable, as are their additional analyses. The domain of the simulation is a region of the forearm 5 cm – 15 cm below the elbow. We use Robin boundary conditions on the contact between the skin and the ambient air, while on the domain boundaries the heat flow is set to zero. The heat production by tissue metabolism is modelled using the Q10 rule (Tikuisis and Ducharme, 1991), while the heat exchange between the blood and the tissue is modelled as a function of the local temperature and the regional blood flow. We confirm the stability and accuracy of the method by varying the simulation parameters, the initial and boundary values, and the model dimensions, with subsequent analysis of the results. Our simulations indicate that the fluctuations of the measured steady-state temperatures are a natural consequence of a complex interplay between the position of the measuring probes, the anatomical position of the main arteries, the dimensions of the forearm, the blood flow, the inhomogeneity of tissues, and the environmental temperature.

The second illustrative example is simulation of a human knee exposed to topical cooling after anterior cruciate ligament (ACL) surgery (Martin, et al., 2001, McMaster, et al., 1978, Tikuisis and Ducharme, 1991, Trobec, et al., 2008). The computational domain is composed of the knee model and surrounding layers: protective bandage, cooling layer, isolating blanket and ambient air. The ambient air is held at constant initial temperature as it is assumed to be well mixed. The heat flux from the first and last slices is kept constant in order to imitate the influence of the leg not exposed to the cooling. Simulations are performed on the described model using realistic process parameters gathered from measurements. Two different methods of topical cooling are compared; first, the use of a gel-pack filled with refrigerated gel, which is exposed to ambient temperature and therefore becomes less and less effective as the gel is warming; and second, the use of a cryo-cuff cooled by a liquid at constant temperature maintained by an external cooling device. The results are presented in terms of temporal temperature development in critical points of the knee. The temperature dynamics is also validated by in vivo temperature measurements during cryotherapy for different regimes of cooling. We show that the second cooling approach gives much better results regarding the desired temperature distribution within the knee.

The cooling of a human heart during surgery is taken as a final, most complex example. In the last example full model is incorporated. Besides the Bio-heat equation Navier Stokes fluid flow is solved as well. The topical cooling (Olin and Huljebrant, 1993) is simulated through consideration of partially submerged heart in the cooling liquid. The effect of natural convection in the cooling liquid is tackled. We show that considering solely diffusive heat transport, as done in present (Šterk and Trobec, 2005, Trobec, et al., 1998, Trunk, et al., 2003), simulation does not supply adequate results.

The presentation of the numerical examples is organized as follows. First, a short historical overview of the tackled subject is introduced. Next, the spatial 3-D geometrical model of the designated body part is described. Then, different simulation modalities and parameters are stated. Afterwards, analysis of the simulated results and their comparison with the measured values, if available, is presented. Finally, conclusions are presented.

Before presenting the numerical examples, the considered physical model and solution procedure are introduced.

2. Physical model

The present chapter is focused on the formulation of the physical model suitable for describing biological tissues. The main part of the modelling is heat continuity, where biological response is treated as volumetric heat source. The power of such sources depends on the temperature of the tissue. Besides bio-heat transfer within tissues, the surrounding is modelled as a fluid in order to simulate the heat exchange with the ambient. The surrounding is treated as a Newtonian fluid described by Navier-Stokes equation. The fluid is assumed to be incompressible.

The model in the present chapter is developed from the classical continuum conservation equations for energy, momentum and mass. The differential forms of the conservation equations are:

$$\frac{\partial}{\partial t}\left(\rho c_p T\right) + \nabla \cdot \left(\rho c_p \mathbf{v} T\right) = -\nabla \cdot \mathbf{j} + S, \tag{1}$$

$$\frac{\partial}{\partial t}\left(\rho \mathbf{v}\right) + \nabla \cdot \left(\rho \mathbf{v} \mathbf{v}\right) = -\nabla P + \nabla \cdot \boldsymbol{\tau} + \mathbf{b}, \tag{2}$$

$$\frac{\partial \rho}{\partial t} + \nabla \cdot (\rho \mathbf{v}) = 0, \tag{3}$$

for energy (1), momentum (2) and mass (3), where $\rho, t, \mathbf{v}, P, \boldsymbol{\tau}, \mathbf{b}, S, c_p, T$ and $\mathbf{j}$ stand for density, time, velocity, pressure, deviatory stress tensor, body force, volumetric heat sources, specific heat, temperature and heat flux, respectively.

2.1. Model assumptions

To construct an accurate and stable model, on the one hand, and still solvable in a reasonable way, on the other hand, some basic assumptions are made:

- all material properties are assumed to be temperature independent,

- no flow is allowed in the tissues,
- Newtonian fluid is considered,
- the biological response of the tissues on temperature change is modelled with the Pennes' Bio-heat equation,
- the metabolic heat production per unit mass h_m is assumed to obey the Q10 rule,
- all materials are isotropic.

2.2. Bio-heat transport

The local mass heat flux is modelled by the Fourier law

$$\mathbf{j} = -\lambda \nabla T \,, \tag{4}$$

with λ standing for the thermal conductivity tensor. Considering the model assumption for material isotropy, the thermal conductivity diffusion tensor is rewritten in the following form:

$$\lambda = \lambda \mathbf{I} \,, \tag{5}$$

where $\mathbf{I}$ and λ stand for identity matrix and thermal conductivity.

The biological response to the conditions within the tissue is modelled through the volumetric sources introduced by Pennes (Pennes, 1948, Wissler, 1998):

$$S = v_r \underbrace{\frac{\rho^b c_p^b}{\rho c_p} V_r(T)(T_a - T)}_{\text{blood perfusion}} + \underbrace{h_r \frac{1}{c_p} h_m(T)}_{\text{metabolic heat production}} \,, \tag{6}$$

where ρ^b and c_p^b are the density and the specific heat of the blood, and S stands for the heat generated from the biological response. The dimensionless parameters h_r and v_r are used to control the spatial/material dependency of the biological response and were not included in the original paper. The functions V_r and h_m characterize the behaviour of heat generation with respect to the temperature of the tissue and are defined latter in this chapter. The temperature $T_a = 36.8\ ^\circ\text{C}$ stands for the reference tissue temperature.

Combining the equations (1),(4),(5) and (6) yields the well-known Pennes Bio-heat equation (Pennes, 1948, Wissler, 1998):

$$\frac{\partial T}{\partial t} = \underbrace{\frac{1}{\rho c_p} \nabla \cdot \left(\lambda \nabla T \right) - \nabla \cdot \left(\mathbf{v} T \right)}_{\text{heat transport}} + \underbrace{v_r \frac{\rho^b c_p^b}{\rho c_p} V_r(T)(T_a - T) + h_r \frac{1}{c_p} h_m(T)}_{\text{biological response}} \,. \tag{7}$$

Equation (7) models the heat transfer between the arterial blood and the tissues, and the heat production arising from tissue metabolism, in addition to the classical heat transport.

To close the energy transport model, both biological response functions have to be defined.

Published measurements of the blood flow rate (V_r) show that it increases with the tissue temperature (Ducharme and Frim, 1988). Earlier measurements of the forearm blood flow were limited to plethysmography that measures the mean total blood flow through a given region of the forearm (Barcroft and Edholm, 1942). Later, more data became available also for regional blood flow of specific tissues (Cooper, et al., 1955, McElfresh and Kelly, 1974, Sugimoto and Monafo, 1987). Several studies have established that the muscle blood flow does not increase with temperature; most of the increase takes place in the skin (Detry, et al., 1972). The value of the mean blood flow rate V as a function of the skin temperature T_S has been reported in several publications (Barcroft and Edholm, 1946, Cooper, et al., 1955, Ducharme and Tikuisis, 1994, Wenger, et al., 1986, Yue, et al., 2007). We use approximate exponential function:

$$V(T) = \left(V_1 e^{V_2 T} + V_3\right) V_0,$$ (8)

with newly introduced case dependant dimensionless parameters V_1, V_2 and V_3. The reference value is set to $V_0 = 1.667 \ 10^{-4} \ s^{-1}$. Although a more detailed blood flow model could be used in our simulation, for example, incorporating the autonomic thermoregulation of the body, its development is beyond the scope of this chapter.

The metabolic heat production in the human body can be divided into unregulated heat production from voluntary muscle contraction and normal metabolic pathways, and into regulated heat production for maintaining temperature homeostasis at lower ambient temperatures (Bullock and Rosendahl, 1992, Proulx, et al., 2003). The rate of metabolic heat production per unit mass h_m is assumed to obey the Q10 rule (Bullock and Rosendahl, 1992), which is expressed as a function of the tissue temperature:

$$h_m(T) = h_0 2^{\frac{T-T_r}{10}}$$ (9)

where $h_0 = 1 \ \text{W/kg}$ stands for the reference metabolic heat production of a tissue at reference temperature $T_r = 35 \ °C$. An example of the biological response against temperature change is presented in Figure 1, where the parameters are taken as in the following first numerical example.

Outside the tissue, the Bio-heat equation is reduced to the classical heat transfer equation (Amimul, 2011), as both biological sources vanish:

$$\frac{\partial T}{\partial t} = \frac{1}{\rho c_p} \nabla \cdot (\lambda \nabla T) - \nabla \cdot (\mathbf{v} T).$$ (10)

The model assumes no flow in the tissues and, as a consequence, the advection term vanishes. In future discussions, a tissue is referred as a solid material.

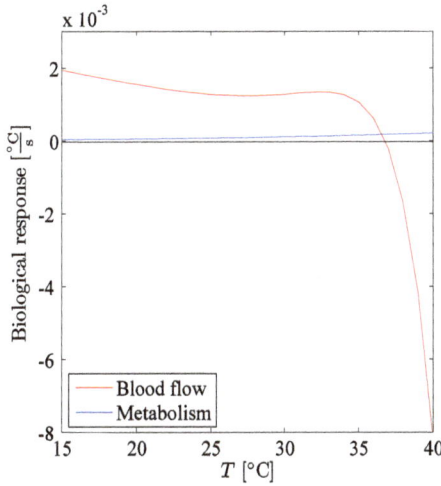

Figure 1. The modelled biological response composed of the blood perfusion and the metabolism as a function of the muscle temperature.

2.3. Fluid flow

In order to specify the momentum equation (2), the stress tensor has to be specified. It is constituted as a Newtonian fluid:

$$\tau_{ij} = \mu\left(\frac{\partial v_i}{\partial p_j} + \frac{\partial v_j}{\partial p_i}\right), \tag{11}$$

where μ stands for the dynamic viscosity. Further, the body force in the momentum conservation equation is modelled by the Boussinesq approximation for temperature dependent density:

$$\mathbf{b} = \mathbf{g}\rho_{ref}\left[1 + \beta_T\left(T - T_{ref}\right)\right], \tag{12}$$

where ρ_{ref} and T_{ref} stand for the reference values and $\mathbf{g}$ for the gravitational acceleration. Combining the equations (2), (11) and (12) yields:

$$\frac{\partial}{\partial t}\left(\rho\mathbf{v}\right) + \nabla\cdot\left(\rho\mathbf{v}\mathbf{v}\right) = -\nabla P + \nabla\cdot\left(\mu\nabla\mathbf{v}\right) + \mathbf{g}\rho_{ref}\left[1 + \beta_T\left(T - T_{ref}\right)\right]. \tag{13}$$

To close the system, the mass continuity (3) is considered. The model assumes incompressible fluids, thus the velocity is divergent free:

$$\nabla\cdot\mathbf{v} = 0. \tag{14}$$

2.4. Thermo-physical properties

The thermo-physical properties of living tissues are still not uniformly defined. For present simulation purposes, the average of published data is used (Bowman, *et al.*, 1975, Moses and Geddes, 1990, Ponder, 1962, Poppendiek, *et al.*, 1967)]. For muscles at rest, h_r was taken to be about half the human basal metabolic rate, which equals 0.6 W/kg (Harris and Benedict, 1918, Ho, *et al.*, 1994, Silverthorn, 2001). The impact of the metabolic rate production is almost negligible for the temperature field in the forearm; even so, it can be adjusted in the model for each tissue independently.

	$\lambda\left[\dfrac{W}{mK}\right]$	$c_p\left[\dfrac{Jkg}{K}\right]$	$\rho\left[\dfrac{kg}{m^3}\right]$	v_r	h_r
Skin	0.510	3431	1200	0.4	0.3
Subcutaneous/adipose tissue	0.550	2241	812	0.4	0.3
Muscle	1.030	4668	1179	0.8	0.6
Cortical bone	2.280	1260	1700	0.2	0.1
Cancellous bone	0.500	2260	900	0.6	0.3
Nerve	0.500	3277	1190	1.1	0.3
Blood	0.670	3890	1057	/	/
Ambient air	0.025	1012	1.29	/	/
Water	0.580	4204	1000	/	/

Table 1. Thermo-physical properties

3. Solution procedure

3.1. Geometric model preparation

The computer model domain of the forearm, the knee, and the heart is derived from the axial anatomic images of a male human body that are available from the Visible Human Dataset (VHD) (Ackerman, 1998). A similar model of the whole body at a resolution of 1 mm³ has been developed within the VHD project, but was not available for public use at full extent when this chapter was written. The axial anatomic images are 24-bit colour cross-sections photographs of a frozen male human cadaver, with resolution of 0.33 mm ×0.33 mm, and taken at spatial interval of 1 mm.

To construct the three-dimensional model, several consecutive slices (cross-sections) (Figure 2) of VHD are used. First, exact overlapping of all slices is required to ensure the integrity of the model. Therefore, a reference point on every slice is needed to align the slices. The white circle in the top right-hand corner of the VHD cross-section is selected for this purpose. A simple automatic procedure is used to find the centre of this circle and thus the absolute position of each slice. The advanced elastic matching (Dann, *et al.*, 1989) is not applied at this stage. Second, several impurities and other errors in the original photographs have to be discarded manually. Last, different tissues have to be identified on the slices. This is done manually, under the supervision of an expert for anatomy, with image editing software, by painting over the photographs with predefined colours, always the same colour for the same tissue. Only the most prominent tissues are marked, the rest are approximated with the surrounding tissue. An automated procedure is then applied to scale down the slices to 1 mm × 1 mm resolution, convert them from colour images to a format recognized by the computer simulator, and stack them together into a 3-D model. Finally, the 3-D model is checked and edited in custom modelling software to smooth out the jaggedness in z axis appearing as a result of the previous processing done only on x-y cross-sections.

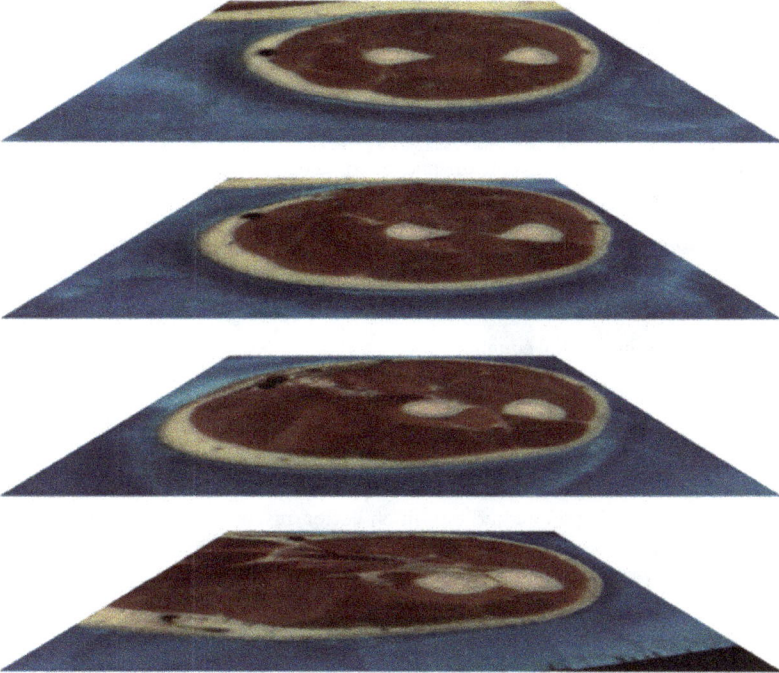

Figure 2. Example of stacked images of VHD for arm.

Each voxel of the geometric model is characterized by thermo-physical properties.

3.2. Discretization

The described physical model does not have a closed form solution and has to be treated by a numerical approach. The present work deals with the transient problems described by parabolic partial differential equations. Such problems are space and time dependent and therefore discretization in space and time is required.

In the present solution procedure, one of the simplest and most intuitive numerical techniques for time dependent simulation is used. Temporal discretization is performed through two-level explicit discretization. The most commonly and widely used unconditionally stable schemes for temporal stepping are implicit and Crank-Nicolson approaches. Besides two-level methods, also multilevel methods can be used to achieve higher-order approximations. The disadvantage of the multilevel methods is the need to store data from several previous time steps. The multilevel data storage issue can be avoided by using Runge-Kutta methods.

For the spatial discretization, the Finite Difference Method (FDM) (Őzisik, 2000, Shashkov, 1996) is used, as it has low calculation complexity, is sufficiently accurate, and is simple to implement. In addition, FDM can directly use the presented discretized domain, where uniform discretization elements are used. There are also several numerical techniques with higher order of accuracy and better stability properties. For example, spatial discretization can be performed through weak form of Finite Element Method (FEM) (Rappaz, et al., 2003) or Boundary Element Method (BEM) (Białecki, et al., 2006, Wrobel, 2002). Furthermore, there are several meshless techniques (Atluri and Shen, 2002, Atluri, 2004, Chen, 2002, Kansa, 1990, Liu, 2003, Šarler, 2007) with convenient properties.

However, the complication in the temporal and/or spatial discretization of the domain gains benefits in accuracy and stability, but loses much in terms of computational time. In the light of this work, the simplest two-level method is used with the lowest CPU complexity.

In general, a partial differential equation

$$\frac{\partial \Theta(\mathbf{p},t)}{\partial t} = L\big(\Theta(\mathbf{p},t)\big) \tag{15}$$

is solved. In equation (15) Θ, L and $\mathbf{p}(x,y,z)$ stand for an arbitrary field/quantity, a second order partial differential operator and a position vector, respectively. In the present work, a transport equation is treated, which is governed by the Laplace $\left(\nabla^2\right)$, the gradient $\left(\nabla\right)$ and the divergence $\left(\nabla\cdot\right)$ operators. First and second spatial derivatives are needed to construct those operators. Both derivatives are evaluated through Taylor's expansion, known as finite differences:

$$\frac{\partial}{\partial p_\varepsilon}\Theta(\mathbf{p}) = \frac{\Theta(\mathbf{p}+\Delta\mathbf{p}_\varepsilon)-\Theta(\mathbf{p}-\Delta\mathbf{p}_\varepsilon)}{2\Delta p_\varepsilon}, \tag{16}$$

$$\frac{\partial^2}{\partial^2 p_\varepsilon}\Theta(\mathbf{p}) = \frac{\Theta(\mathbf{p}+\Delta\mathbf{p}_\varepsilon)-2\Theta(\mathbf{p})+\Theta(\mathbf{p}_{\varepsilon-1}-\Delta\mathbf{p}_\varepsilon)}{\Delta p_\varepsilon^{\ 2}}. \tag{17}$$

In equations (16) and (17), the index ε denotes spatial direction (x,y,z), p_ε stands for the ε component of the position vector, Δp_ε stands for the spatial step in ε direction, and $\Delta\mathbf{p}_\varepsilon$ stands for the offset in ε direction.

The left side of equation (15) is discretised through two-level time discretization with the finite difference representation of the first temporal derivative. It is rewritten in an explicit form as:

$$\frac{\Theta(\mathbf{p}, t + \Delta t) - \Theta(\mathbf{p}, t)}{\Delta t} = L(\Theta(\mathbf{p}, t)) + O(\Delta t). \tag{18}$$

For the finite difference spatial discretization, the Neumann (or Fourier) stability analysis shows that both implicit and Crank-Nicolson are unconditionally stable, while the explicit scheme suffers from stability problems (Smith, 2003). For diffusive-convective problems, as the ones considered in the present work, the differential operator L is generally written as:

$$L = \nabla \cdot \left(\frac{\lambda}{\rho c_p} \nabla(\ldots) \right) - \nabla \cdot (\rho \mathbf{v} \ldots), \tag{19}$$

and for the FDM, the stability criteria derived by Fourier analysis is stated as:

$$\frac{\lambda}{\rho c_p} \frac{\Delta t}{\Delta p_\varepsilon^2} \leq \frac{1}{6} \qquad |v_\varepsilon| \frac{\Delta t}{\Delta p_\varepsilon} \leq 1, \tag{20}$$

where v_ε stands for the ε component of the velocity.

Despite the stability issues, in the present work the explicit method is used because of its straightforward local applicability and simplicity. The main drawback of the implicit and semi-implicit forms is the need for solving the global problem, where systems (usually large) of linear equations have to be considered in order to proceed to the next simulation time-step. On the other hand, the explicit scheme uses series of pre-solved linear systems solutions. In other words, the explicit scheme uses the local information from the current time-step (t_0) to compute spatial derivatives, while the implicit scheme uses the current local time-step information only for the evaluation of time derivative. Finally, the Crank-Nicholson scheme uses linear combination of the current and the next time-step to improve the accuracy of the method.

One of the most convenient properties of the Explicit FDM (EFDM) is its straightforward parallel implementation, as there are no requirements of any kind for global communication.

3.3. Pressure velocity coupling

An important part of the solution procedure represents the solution of the momentum equation. An additional problem lies in the momentum-mass conservation coupling. The pressure is not explicitly included in the continuity equation and therefore, a special treatment is needed to solve the pressure-velocity coupling problem. The most widely known solution for the problem is the Semi-Implicit Method for Pressure Linked Equations SIMPLE (Ferziger and Perić, 2002), where the problem is translated to the Poisson equation problem. However, the SIMPLE algorithm demands the solution of the global system, which is complex from parallel implementation point of view. There are local alternatives, like the Local Pressure Velocity Coupling (LVPC) used in context with local meshless methods (Kosec and Šarler, 2008a), the artificial compressibility method (ACM), which has been recently under intense research (Malan and Lewis, 2011, Rahman and Siikonen, 2008, Traivivatana, *et al.*, 2007), and the framework for the Finite Difference Method in the SOLA approach (Hong, 2004). In this chapter, the more standard and well accepted pressure velocity coupling, based on the SIMPLE algorithm, is used.

The problem domain is discretized to voxels that can be either liquid or solid. For solid voxels only the velocity is assumed to be zero and no pressure computations take place. Otherwise, standard staggered grid approach (Ferziger and Perić, 2002, Shashkov, 1996) is used, with the temperature and the pressure defined in the middle of each voxel, and the velocity components defined on the respective perpendicular voxel edges.

The Navier-Stokes equation is discretized in time by using the explicit time discretization:

$$\frac{\hat{\mathbf{v}} - \mathbf{v}^0}{\Delta t} = \frac{1}{\rho}\left(-\nabla P^0 + \nabla \cdot \left(\mu \nabla \mathbf{v}^0\right) + \mathbf{b}^0 - \nabla \cdot (\rho \mathbf{v}^0 \mathbf{v}^0)\right), \tag{21}$$

where $\hat{\mathbf{v}}$ stands for the intermediate velocity and the superscript 0 denotes that the current time-step values are used. As the explicit time discretization is employed, the new time-step velocity is computed from the velocity values in the previous time-step. However, instead of a new time-step value, the intermediate velocity is introduced, which does not take into account the mass continuity (3). This problem is solved by introducing the velocity and pressure corrections. The added velocity correction drives the intermediate velocity towards divergent free field:

$$\nabla \cdot \left(\hat{\mathbf{v}} + \tilde{\mathbf{v}}\right) = 0, \tag{22}$$

$$P^1 = P^0 + \hat{P}, \tag{23}$$

where $\tilde{\mathbf{v}}$ and $\hat{P}$ stand for the velocity and the pressure correction terms, and the superscript 1 denotes the next time-step. With the corrected velocity and pressure, the momentum equation can be rewritten as:

$$\underbrace{\frac{\hat{\mathbf{v}} - \mathbf{v}^0}{\Delta t}}_{I} + \underbrace{\frac{\tilde{\mathbf{v}}}{\Delta t}}_{II} = \underbrace{\frac{1}{\rho}\left(-\nabla P^0 + \nabla \cdot \left(\mu \nabla \mathbf{v}^0\right) + \mathbf{b}^0 - \nabla \cdot (\rho \mathbf{v}^0 \mathbf{v}^0)\right)}_{I} \underbrace{-\frac{1}{\rho}\nabla \hat{P}}_{II}. \tag{24}$$

From equation (24), directly follows:

$$\underbrace{\tilde{\mathbf{v}}}_{II} = \underbrace{-\frac{\Delta t}{\rho}\nabla \hat{P}}_{II}, \tag{25}$$

thus all other terms subtract to zero. By applying the divergence over equation (25) and taking into account equation (22), the pressure correction equation emerges to:

$$\nabla \hat{v} = \frac{\Delta t}{\rho} \nabla^2 \hat{P} \; . \tag{26}$$

In order to solve the Poisson equation, additional boundary conditions are required. The boundary conditions for pressure field can be formulated by multiplying the momentum equation with the normal boundary vector. The boundary conditions are obtained with a similar approach used to construct the equation for pressure correction, stated as:

$$\frac{\partial \hat{P}}{\partial n} = \frac{\rho}{\Delta t} \left(\hat{v} - v_\Gamma \right) \cdot n \, , \tag{27}$$

where v_Γ stands for the velocity boundary condition and n denotes the normal boundary vector. It is common to use the normal pressure boundary condition when no-slip and no-permeable velocity boundary conditions $\left(v_\Gamma = 0 \right)$ are used:

$$\frac{\partial}{\partial n} \hat{P} = 0 \; . \tag{28}$$

With the pressure corrected according to (23), the velocity can be corrected as:

$$v^1 = \hat{v} - \frac{\Delta t}{\rho} \nabla \hat{P} \, , \tag{29}$$

where v^1 stands for the next time-step velocity. With computed velocity and pressure, the simulation can proceed to the next time-step. The computation cycle is closed.

3.4. Implementation of parallel program

The simulator is written in C++ and compiled with GCC. The parallelization is done by domain decomposition. The domain decomposition is one-dimensional: the domain is split in the z axis into subdomains of approximately equal size. The number of subdomains equals the number of processors participating in the solver execution. The sizes of the subdomains are determined at the program initialization and are constant for the whole execution; therefore, the load-balancing of the processors is static. One-dimensional decomposition was chosen because of its simplicity. To calculate a point on the border of a subdomain, data from one or more points from neighbouring subdomains might be required. In such cases, point-to-point communication between processors of those neighbouring subdomains is performed. MPI library is used for communication. It enables the program to execute in parallel on multicore computers, computer clusters or other distributed computer systems.

The multigrid method, used in our solver, is parallelized using similar principles. Calculations on fine grids are done in parallel using the same domain decomposition, while a single processor does the computation on the coarsest grid, which is negligible compared to the computation on the finest grid.

The developed physical model and solution procedure have been evaluated through comparing some of the simulated results against the measurements. The well-known Pennes (Pennes, 1948) in vivo measurements have been reproduced with the proposed model. The model results have been also compared with real case measurements of temperature distribution within the knee after the ACL surgery. The steady state profiles and the temporal transients have been analysed. For all cases, it is shown that an appropriate modelling of the bio-heat energy transport results in a good agreement with experimental data. The fluid flow solution procedure, however, is designed as a proven approach and tested on several standard benchmark tests, e.g., driven cavity, de Wahl Davis test, etc. The presentation of benchmark tests are beyond the scope of this chapter and are as such omitted. All results related to the bio-heat transfer are presented and elaborated in details in the following sections.

The numerical scheme used to solve the proposed physical model is already well accepted and extensively tested in the past. A detailed convergence of several numerical techniques applied the diffusion problem was published in (Trobec, et al., 2012). It was shown that the FDM has an adequate convergence rate. Based on the obtained results and taking into account that we use the same approach we do not insert the convergence analysis in this chapter.

4. Heat transfer in resting human forearm

The invaluable measurements of the steady-state temperature fields in forearms of subjects that are not anesthetized, published sixty years ago by Pennes (Pennes, 1948), have still not been explored in all details (Wissler, 1998). Pennes' used a thermocouple probe on a wire that was pulled through the forearm in a straight line (referred to as the measurement path in the rest of the section) in steps of less than 1 cm. At each step, the measurement was read after the probe temperature has stabilized. In Figure 7 the original measurements and the author's interpolated temperature curves are reproduced.

Other experimental data are available for the temperature of the forearm during an immersion in water at various temperatures, either as it evolves in time (Barcroft and Edholm, 1946) or near its steady-state. Since then, these measurements have been elaborated by others (Ducharme, 1988). In some recent studies, measured lower resolution temperature profiles in the forearm cross-sections are reported, together with estimated values for the in-vivo thermal conductivities of muscle and subcutaneous tissues (Ducharme, 1988). We recreated the conditions similar to those of the Pennes' measurements and simulated the steady-state temperatures of the proximal forearm. In this section, we consider the effects of blood perfusion, metabolism, position of the arteries, and inhomogeneity of tissues on the steady-state temperature profile of the forearm.

4.1. Geometric model of the forearm

Cross-sections of the VHD male thorax (numbered 1601 – 1800) are used for the forearm model. The right forearm is cropped from those cross-sections as rectangles of 506 (width) × 632 (height) pixels. The VHD right forearm is inclined by 32.5° in the x-z plane and by 43.1°

in the y-z plane. It is more suitable for the simulation if the z axis of the forearm is parallel to the z axis of the model; therefore, the elements of the model are sampled out of the forearm rectangles at an angle, making the z axes parallel. Finally, a regular grid of size $140 \times 116 \times 100$ elements with spatial step of 1 mm is created. For the 2-D simulations, "model slices" are used – individual x-y planes of the grid, numbered by their position on the z axis.

The 3-D domain represents a section of the human forearm, 50 mm distal to the tip of the elbow and 100 mm long (50 mm from the probe insertion point in both directions). For better visibility of the bones and arteries in the model, muscles, skin, and subcutaneous tissues are shown semi-transparent. In Figure 3 model example slice is shown.

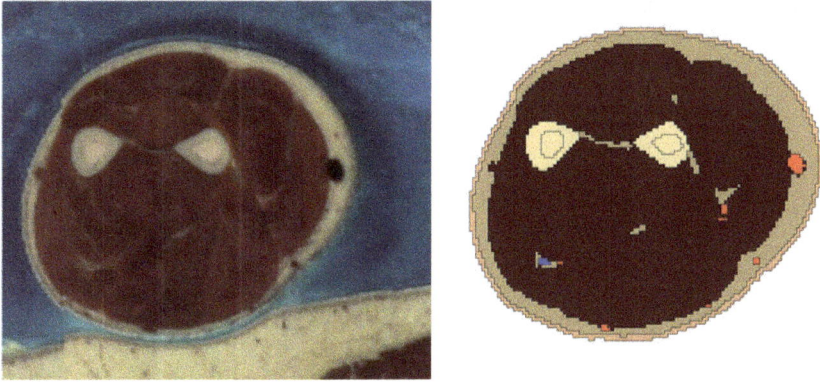

Figure 3. Example of a cropped VHD photographs before the geometric model preparation (left) and corresponding forearm slice after segmentation (right).

The Pennes' measurements (Pennes, 1948) were performed exactly "8 cm distal to tip of ulna olecranon" – or 8 cm from the tip of the elbow, with no respect to the length of the forearm. While the lengths of the forearms are not reported, the diameters of the measured forearms differed greatly (7 – 10 cm). Therefore, the measurements were likely performed at different anatomical positions. In case that the 8 cm were strictly respected, the variation in the position of the measuring probe, regarding the forearm anatomy, could be significant – up to 25% of the distance to the elbow.

The diameter of the described geometric model of the forearm at slice 50, which is in the middle of the model, is 12.4 cm. To obtain a closer match to the measured forearms, which were significantly smaller, we scale down the domain size by linearly scaling down the domain elements. In the rest of the section, we use forearm diameter of 8.3 cm, except when explicitly specifying otherwise.

4.2. Numerical setup

We simulate the steady-state temperatures of a resting forearm near the thermo-neutral conditions. It is known that the thermo-neutral conditions of an unclothed arm with skin

temperature (T_S) of approximately 33 °C are obtained at ambient air temperature of 25.2 ±
1.1 °C. The initial conditions of the simulation comprise the temperatures of the modelled
tissues except arteries, and are set to 36 °C. The arteries acts as heat generators, and ambient
air acts as heat sink. They are set, in accordance to previously measured values, to constant
values of 36.8 °C and 26 °C, respectively. During the simulation, the tissue temperatures
converge with time towards their stationary values. After the simulating three hours of a
resting forearm, the maximal change of temperatures during the last hour is less than 0.01
°C, which confirms that the near-steady state is reached.

Moving air is not simulated because of its significant contribution to the calculation
complexity. We simulate the heat flow between the skin surface of the forearm and the
ambient air using Robin boundary conditions, based on the continuity of the heat flow
perpendicular to the surface of the simulated body part:

$$\lambda\frac{\partial T}{\partial \mathbf{n}}=H(T_{am}-T),\tag{30}$$

where T_{am} stands for the ambient temperature. The convection coefficient $H = 16$ W m²/°C is
determined by numerical experiments, based on simulation of the thermo-neutral
conditions ($T_{am}=26°\text{C}$ and $T_S=33°\text{C}$).

The influence of the parts of the forearm that were not simulated is considered by setting the
heat flow to zero at both "ends" of the modelled arm (slice 1 and 100), indicating a thermal
equilibrium between the modelled and excluded parts of the forearm.

The blood flow is modelled with the parameters set to:

$$\begin{aligned}V_1 &= 5.145\cdot 10^{-5}\\V_2 &= 0.322\\V_3 &= 0.705\end{aligned}\tag{31}$$

The simulation setup is presented in Figure 4.

4.3. Results of numerical integration

The simulated steady-state temperature field of the model slice 50 with diameter of 8 cm,
after three hours of simulation, is shown in Figure 5a. The impact of the two main arteries,
with constant blood temperature, is visible as peaks in the temperature profile at the
locations of the arteries. The peaks are marked by two vertical lines. We can also notice that
the skin temperature depends on the position because its distance from the arteries varies.
The steady-state temperature profile taken from the same slice along the path of the
thermocouple measurements are shown in Figure 5a by a black curve and Figure 5b by a
thick black curve. The dotted curve shows the analytical solution of the Bio-heat equation
(Yamazaki and Sone, 2000) for a cylindrical homogeneous model with blood flow $V = 4.7 \ 10^{-4}$
1 s^{-1}, $h_m = 0.62$ W/kg, skin temperature $T_S = 32.5$ °C, and with the other constants set the

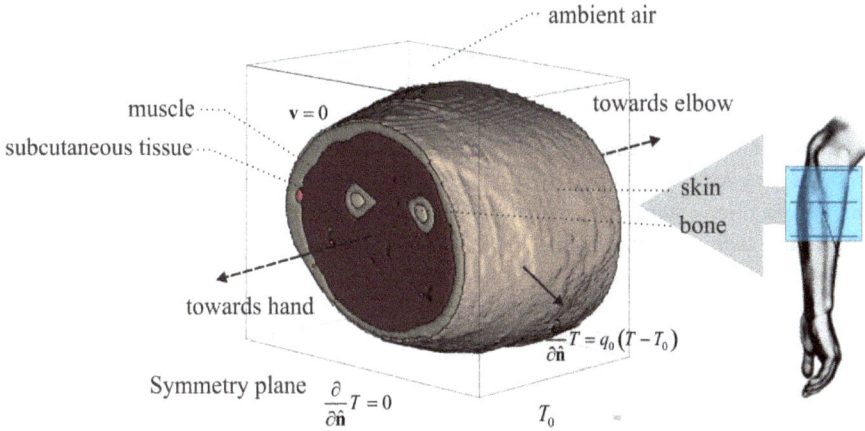

Figure 4. Computational domain

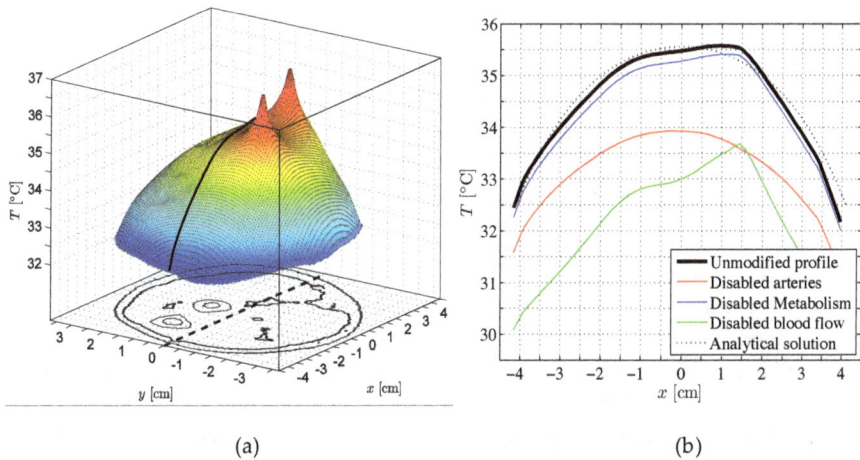

(a)   (b)

Figure 5. (a) Model slice 50 and its steady-state temperature field. The simulated temperature profile along the approximate measurement path (dashed line) is emphasized with a white line. Simulated temperature profiles along the same path as in (a). The unmodified profile (thick curve) is contrasted against the analytical solution of the Bio-heat equation on a homogenous cylindrical model and the profiles acquired with modified simulation: disabled metabolism, arteries not as a heat source, and disabled blood perfusion.

same as in the simulation. We see that the closed form solution of the homogenous model can reproduce the general trends in the temperature gradients; however, it cannot reproduce the measured results because it is symmetric and uniform throughout the model.

By disabling, one at a time, the metabolism, main arteries, and blood perfusion in the simulation, we obtain the last three temperature profiles shown in Figure 5b. The blue curve shows the simulated temperature without metabolism. We can see that the metabolism minimally contributes to the final temperature. The red curve was obtained by not keeping the arteries at a constant temperature in the simulation, thus cancelling their heat generating role. The impact of the arteries not heating up the surrounding tissue makes this temperature profile stand out among the simulated profiles. It shows lower deep muscle temperature than normal, which is more symmetric. The green curve, obtained by the simulation with the arteries at a constant temperature and disabled blood perfusion, confirms that the contribution of the blood perfusion is essential. The average temperature of this profile is significantly lower with pronounced peaks corresponding to the two main arteries.

4.4. Sensitivity analysis

The simulated steady-state solution is evaluated regarding its sensitivity. The input parameters are varied within selected ranges and the variations in the resulting temperature profiles are analysed. In Figure 6 the temperatures of the forearm after three hours of simulated resting are shown.

The temperatures are taken from the model slice 50 along the measurement path. All results are calculated based on the same geometric model with varying forearm diameter, thermal diffusivity, blood flow, and metabolism, all for +20% (dotted line) and -20% (dashed line) of their nominal values (solid line). The nominal values are taken from Table 1. The nominal value for the forearm diameter was determined to be 10 cm.

The most significant impact on the temperature profiles comes from varying the forearm dimension (Figure 6a). Larger arms result in a plateau slightly above 36.3 °C, while smaller arms are cooled more intensively, reducing the maximal temperature to 35.6 °C and with larger temperature gradients in the superficial regions of the forearm. Changes in the thermal diffusivity (Figure 6b) have a much smaller impact on the temperature profiles than the forearm dimension. The effect is also non uniform across the profile, but is rather equalizing the superficial and the deep muscle temperatures. The impact of the blood flow (Figure 6c) is again uniform across the profile, with higher forearm temperatures when the blood flow is larger. The impact of the metabolism (Figure 6d) is analogous to the one of the blood flow – greater metabolism results in higher forearm internal temperatures – but significantly smaller in scale.

4.5. Reproducibility of measurements

We found out that the most important parameter that influences the simulation results is the dimension of the simulated forearm. The impact of the diameter is evident from Figure 6a. Other thermal parameters, like blood perfusion and structural details, have much smaller

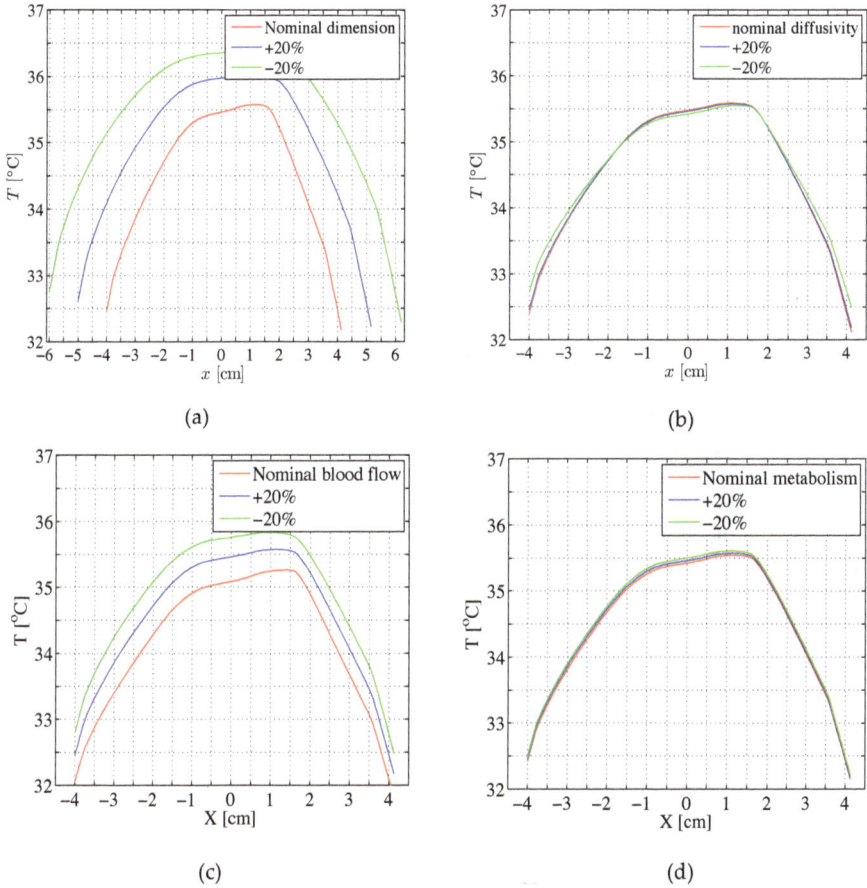

Figure 6. Temperatures from the model slice 50 along the measurement path after three hours of simulated resting forearm at a room temperature of 26 °C for various: (a) forearm diameters, (b) thermal diffusivity, (c) blood flow, and (d) metabolism. All panels feature red, blue, and green lines for the results acquired using the nominal, 20% above nominal, and 20% below nominal values of the given parameter, respectively.

impact on the variations in the simulated temperatures and have been therefore considered in the same way in all individual cases. Only two main vessels have been considered, which are anatomically quite uniform at most subjects. The remaining smaller vessels are encompassed through the blood perfusion as an average temperature regulator. The temperature gradients are very pronounced and therefore the location of the temperature probe provides the second most important contribution in the analysis of the temperature profiles. Because the anatomical probe positions are not known exactly, we try to find them to see if they can explain the unexpected variations in the measured temperature profiles.

We can deduce from Pennes work further sources of measuring errors. The positions of the measuring points and the reading of the temperature values have been with limited accuracy, which could result in errors of 0.1 °C. The dimensions of the thermocouples were up to 0.2 mm and hence, in the areas with steepest temperature gradient, near the forearm surface, an additional error could be introduced.

To reproduce the measurements, we first simulate the resting forearms with the same diameter and under the same conditions as in the real measurements. Using a matching 3-D simulation and a temperature profile measurement, we perform a search for the probe trajectory on which the simulated temperature profile fits the measured temperature profile the most. We do so by defining a fitness function as the root mean square (RMS) error between the simulated temperature profile and the measured referential profile, weighted with Gaussian function to diminish the effect of greater inaccuracies in the measurements near the forearm surface. Then, we use the fitness function in Differential Evolution (Price and Storn, 2008) – a stochastic optimization procedure – to find a probe trajectory that fits the real measurement well. The similarity between the measured and the simulated profiles for all analysed measurements is given by the Pearson correlation coefficient in Table 2. The simulated paths of the measuring probes, which provide the highest correlation between the simulated and the measured profiles, are also determined in Table 2 by two offsets and two angles to the measurement path. They can be obtained by sequentially applying translations in y and x directions and rotations in x-y and x-z planes to the trajectory of the thermocouple probe.

	Subject							
	1	2	3	4	5	7	8	9
Correlation	0.988	0.995	0.993	0.980	0.997	0.989	0.980	0.991
y offset [mm]	-2.6	2.7	4.5	1.8	5.7	-0.4	2.4	1.2
z offset [mm]	-21.4	-32.7	-30.3	-9.3	-28.8	-9.0	-36.3	-4.4
Angle in x-y plane [°]	0	-0.4	-1.6	-0.4	0	0.4	0	3.5
Angle in x-z plane [°]	5.5	5.9	0	0.8	7.8	4.3	3.9	4.3

Table 2. Comparison between the measured forearm temperature profiles and the simulated temperature profiles that fit them best, for all analysed measurements.

Two typical simulated temperature profiles for measurements 4 and 9, together with points from the original measurements, are shown in Figure 7. The central peak in the profile of measurement 4 is expected and can also be predicted by the theory. The temperature profile of measurement 9 with dual maxima can be explained by the probe moving close to and about the same distance from both arteries. The left and right shifted maxima visible in some other measurements can be explained by the path of the temperature probe being closer to one of the arteries. The difference in the forearm dimensions results in significantly different temperatures in the central part of the forearm. As visible from Figure 7, larger forearms induce higher temperatures. We are able to reproduce the original measurements using the same geometric model for all subjects (just scaled to the measured diameter), and by introducing minor changes in the position of the measuring paths.

4.6. Summary

Instead of first simulating and then evaluating the simulated results, our study is designed in reverse direction. First, the well-known in vivo measurements of the steady-state temperatures of the proximal forearm are taken from the famous Pennes' publication; then, the experimental environment is reproduced; next, the bio heat transfer is modelled, and lastly, the measurements are simulated. The primary goal was to explain, by computer simulation, the variations in the Pennes' in vivo measurements. We have known that the measurements vary significantly between subjects and between forearm cross-sections. Several attempts have been made in the past to average the measured steady-state temperatures in order to smooth fluctuations and obtain a standardized response.

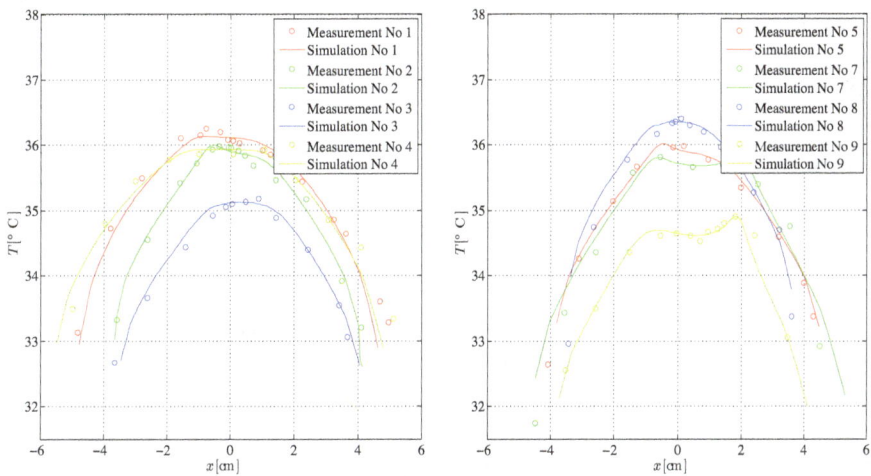

Figure 7. The best fit of the simulated temperature profiles and the corresponding measurements.

The anatomical positioning of the measuring probes is assumed to vary because of the different forearm sizes of the subjects; therefore, their possible paths have been reconstructed by searching for the simulated paths that result in the best agreement between the simulated and the measured temperature fields. Our simulated results suggest that the smoothing approach is inappropriate, and that the fluctuations are natural result and the consequence of a complex interplay between the position of the measuring probe, the anatomical position of the main arteries, the dimensions of the forearm, the blood flow, the inhomogeneity of tissues, and the environmental temperature.

Several innovative approaches have been introduced in the simulation method. A procedure for spatial model generation based on digitized slice data has been developed. A mathematical model and a 3-D computer simulation program have been implemented for the simulation of steady-state temperatures. The heat transfer in the model of non-homogenous tissue was modelled with a well-known Bio-heat equation. The heat production by tissue metabolism was modelled using the Q10 rule. The heat exchange between the blood and the tissue was modelled as a function of the local temperature and the regional blood flow that was modelled as an exponential function fitted to experimental data. The regional blood flow was assumed to depend on the regional temperature in the same way as the mean blood flow rate depends on the skin temperature. Such an approach produced reasonable results. Nevertheless, it needs to be supported by further physiological research. The developed simulation, however, is sufficiently robust to incorporate arbitrary model functions for blood flow and metabolism. The stability of the method was confirmed by varying the simulation parameters, the initial and boundary values, and the model shape, with subsequent analysis of the results.

We have confirmed that blood flow has a significant and complex impact on the steady-state temperatures of the forearm. Lower blood flow results in lower temperature of the central part of the forearm. When the effect of blood flow on the temperature profile is increased by disabling the direct effect of the main arteries, the resulting temperature profiles match very well to the analytically produced profiles from the literature, but not well with the experimental measurements. We have shown that only when the effect of both sources, the heat distributed via blood flow and the heat emitted directly from the arteries, are considered, the measured temperature profiles can be fully explained. Furthermore, we have revealed the dependence of the temperature profile shape on the dimensions of the forearm. Smaller forearms are more likely to feature off-centre peaks in the vicinity of the main arteries than larger forearms. In larger forearms, a central plateau may be formed, masking all other potential peaks. The thermodynamic constants of the tissue and the tissue metabolism were confirmed to have a relatively minor impact on the temperature field.

Our simulation methodology has several limitations. The anatomy of the forearm differs between individuals and changes in time. Consequently, the simulation on one generalized model can differ from the experimental measurements. Minor errors in tissue segmentation and inaccurate thermodynamic constants could produce errors in the simulated results. The possible influence of different mechanisms of thermoregulation on blood perfusion has not been simulated because we assumed that their effects are stable when the forearm temperature is in its steady-state. We confirmed that the above essential findings are in close agreement with in vivo measurements, particularly if the inaccuracy of the position of the measuring path is accepted.

5. Simulation of a human knee exposed to topical cooling after ACL surgery

Topical cooling after surgery or after knee injuries is often performed using gel-packs filled with gel that can be cooled well below 0 °C, yet still remaining soft and flexible; or alternatively, with cryo-cuffs, which provide a constant temperature maintained by circulating cooled liquid (Hubbard and Denegar, 2004). Desired cooling conditions are obtained by differently shaped gel-packs or cryo-cuffs, often influenced by environmental conditions and the temperatures of internal tissues (Barber, 2000). However, different tissues around the knee joint have different physical and thermodynamic properties, and respond diversely to the cooling. The knee temperature is influenced by the muscle metabolism and the circulating blood, both of which depend on the tissue temperature. Therefore, the real cooling efficacy varies in space and time in different parts of the knee. The effects of cooling on the tissue metabolism and the response of inflamed or traumatized tissue have been researched and have shown beneficial effects (Ho, et al., 1994, McGuire and Hendricks, 2006). There is no generally accepted doctrine about topical cooling. Therefore, further investigation of cooling effects and different cooling regimes are needed, but these are difficult to perform by experiments alone.

In this section, we are interested in simulating possible modes of knee cooling after injury or surgery (Grana, 1994). We focused on the comparison of two different methods of topical cooling. First is the cooling with gel-pack filled with refrigerated gel (McMaster, et al., 1978). The gel-pack is exposed to ambient temperature and therefore becomes less and less effective with time. Second is the cooling with a cryo-cuff filled with a liquid. The temperature of the liquid is maintained constant by an external cooling device (Shelbourne and Nitz, 1990).

Lowering the tissue temperature reduces the need for medicaments and shortens the rehabilitation period (Knight, 1985). It is not clear, however, whether different tissue regions in the knee should be cooled uniformly. Some recent findings put the benefit of topical cooling under question (Hubbard and Denegar, 2004, Knight, 1985, McGuire and Hendricks, 2006). The aim of our work is to simulate topical cooling of the knee after injury or surgery, and to calculate and show the development of temperature distribution in all tissues of the knee region.

5.1. Geometric Model of the Knee

The knee area is cropped from the VHD male cross-section photographs number 2200 to 2390 as rectangles of 550 (width) × 610 (height) pixels. Cropped VHD slice 2301 is shown on the left in Figure 8.

The 3-D model is shown in Figure 9. Skin, joint liquid and subcutaneous tissues are not shown. The artero-lateral quadrant is removed for better visibility into the central knee region.

Some surrounding space for the protective bandage, the gel-pack, the blanket, and the ambient air is added around the 3-D knee model. The simulation environment is imitated by an isolated cube composed of $x \times y \times z$ resulting in roughly 10^7 voxels independently characterized by thermodynamic properties and initial temperatures. Each voxel has a volume of 1 mm³. The knee is covered by a 2 mm thick protective bandage and embraced by a 12 mm thick cooling layer (gel-pack or cryo-cuff). Additionally, the cooling layer is covered with a 5 mm thick isolating blanket to reduce the convection and slow down the warming from outside. The protective bandage, the cooling layer, and the isolating blanket are inserted into the model automatically by a computer program.

5.2. Numerical setup

The boundary layers of the simulated box are held at constant initial temperatures to mimic the ambient air. The influence of the rest of the leg, which is not simulated, is managed by setting the temperature flux at the boundaries. The simulation can be carried out in 2-D with the heat flux in the axial direction set to zero. In this way, an infinitely long "knee", with homogeneous structure in the axial dimension, can be simulated. Simulation in 3-D is performed on the described model with all boundaries set to values that are similar to those of the measuring conditions. The simulation setup is presented in Figure 4.

The blood perfusion in bones, cartilage, and ligaments from the knee joint are assumed to be very small. Thus, for these tissues, ten times smaller numerical values for V_0 are taken than for the other tissues. The blood flow is modelled using the following parameters:

$$
\begin{aligned}
V_1 &= 5.722 \cdot 10^{-3} \\
V_2 &= 0.187 \\
V_3 &= 0.000
\end{aligned}
\qquad (32)
$$

As in the case of the blood flow, the metabolism of bones, cartilage and ligaments is assumed to be very small.

The thermo-physical properties of newly introduced materials are stated in Table 3 and initial states of different regions are stated in Table 4.

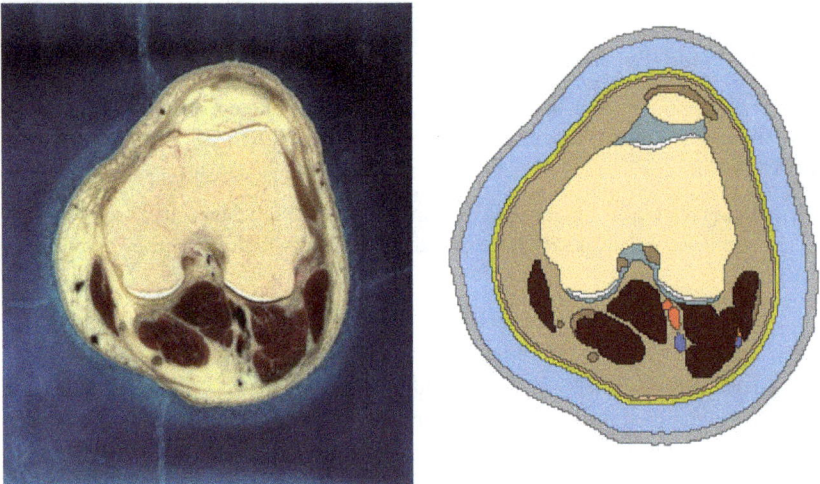

Figure 8. Example of a cropped VHD photographs before the geometric model preparation (left) and corresponding knee slice after segmentation (right).

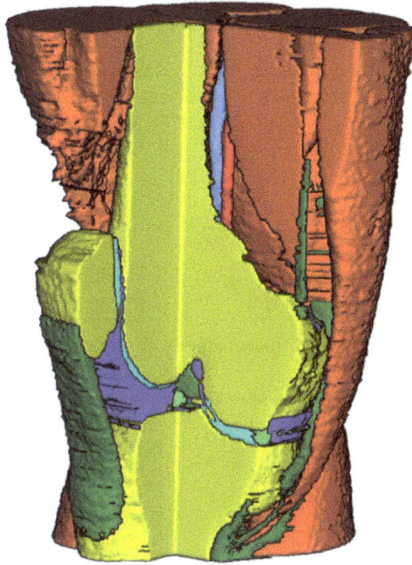

Figure 9. 3-D geometric knee model obtained by stacking 191 consecutive VHD slices from top to bottom. The frontal quadrant, skin and subcutaneous tissue are removed for better visibility.

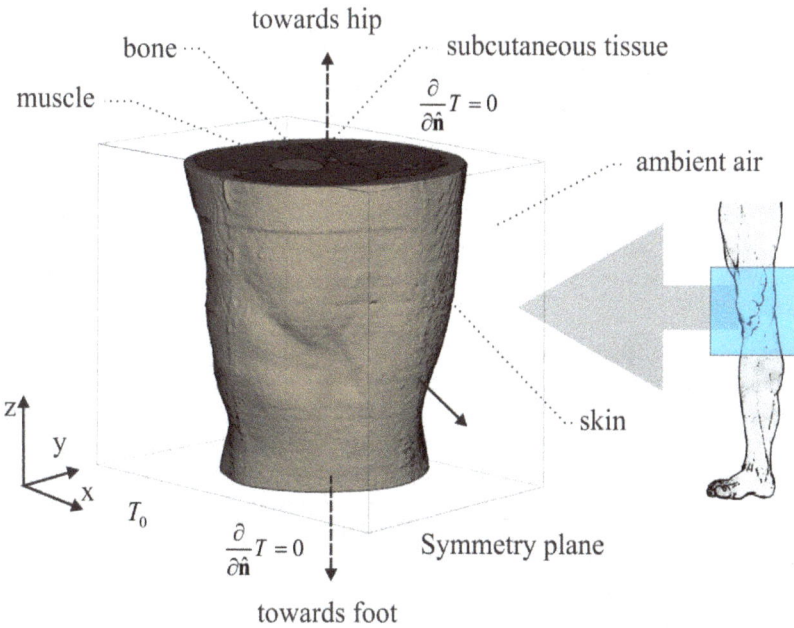

Figure 10. Computational domain.

	$\lambda \left[\dfrac{W}{mK}\right]$	$c_p\left[\dfrac{Jkg}{K}\right]$	$\rho\left[\dfrac{kg}{m^3}\right]$
blanket	0.04	1200	150
cryo-cuff water	0.58	4204	1000
gel-pack	0.1	4000	990
bandage	0.04	1200	150
joint liquid	0.58	4204	1000
venous blood	0.67	3890	1057
arterial blood	0.67	3890	1057
blanket	0.04	1200	150

Table 3. Thermo-physical properties of additional materials.

ambient air	25	muscle	36
blanket	25	bone	36
cryo-cuff water	15*	ligament	36
gel-pack	0	cartilage	36
bandage	30	joint liquid	22
skin	35	nerve	36
subcutaneous tissue	35.6	venous blood	36
arterial blood	36.8*		

* Constant temperatures are denoted by asterisks.

Table 4. Initial temperatures in °C.

5.3. Results of numerical integration

First, we simulate the steady-state temperatures of a resting knee in thermo-neutral conditions (Ducharme, *et al.*, 1991) at ambient air temperature of 27°C, which is equivalent to the ambient water, e.g., a water bath at 33°C. A steady-state is reached after three hours of simulation, with the maximal change in the temperature near the end of the third hour being less than 0.01 °C.

At that state, the simulated temperatures can be recorded at arbitrary positions. For example, the temperature field over the model slice $z = 102$ is shown with a 2-D surface in Figure 11. For better visibility of the temperature field, the point (0,0) is on the right and the frontal part of the knee with the patella on the left. The temperature along the transverse axis Y for $x = 106$ is used in a later analysis. The obtained steady-state is used as an initial condition in all our subsequent simulations. The tissues nearer the patella, toward the surface, are seen to be colder than the internal part. The impact of the main knee artery with its constant blood temperature is visible as a peak in the temperature field. The location of the bones is also evident as shallow depressions in the temperature field, mainly as a result of the lower blood flow in the bones.

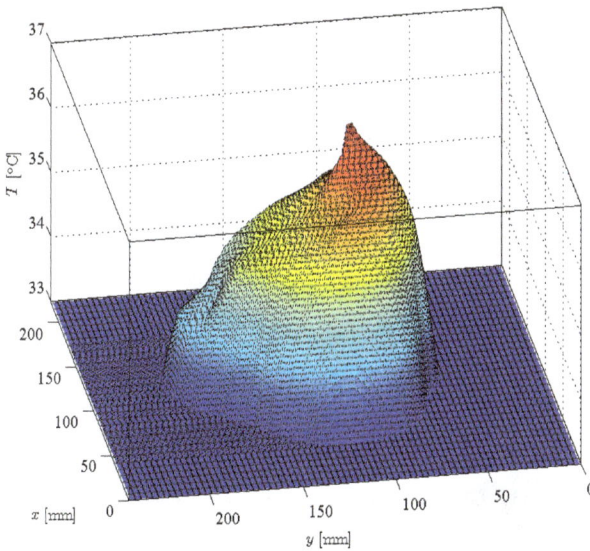

Figure 11. Steady-state temperature field over model slice z = 102. Point (0,0) is in the right corner for better visibility.

In Figure 12 the solid curve shows the simulated steady-state temperature obtained with the 3-D simulation. The dotted curve is obtained under the same conditions with a 2-D simulation and zero flux in the third dimension (infinite knee). The difference between the simulations is in the range of 0.5 °C. It can be explained by underestimating the amount of muscle tissue in the whole knee region. Slice 102 is taken, namely, from the central part of the knee with less muscle, which could contribute additional heat through the blood flow. However, the shapes of the profiles are similar. Therefore, for the initial analysis we use 2-D simulation to reduce the simulation time.

We can observe and analyse arbitrary simulated points or regions of the domain. For example, in Figure 13, the steady-state temperature profiles of the knee are shown for temperatures on the transverse knee axis y and model slice $z = 102$ for $x = 96$, $x = 106$ and $x = 116$. It is clear that there can be significant differences in the temperature, even for analysed points as close as 10 mm. The effect of altering the observed position in other directions is similar to that seen in Figure 13.

The stability of the simulated steady-state solution is evaluated by varying the simulation parameters. Input parameters are varied within selected ranges and the variations in the solution were analysed. In Figure 14 the simulated steady-state temperatures from model slice 102 on the transversal axis y for $x = 106$ are shown for various knee dimensions, diffusion constants, blood flow and metabolism, all differing for +20 % and -20 % of their nominal values.

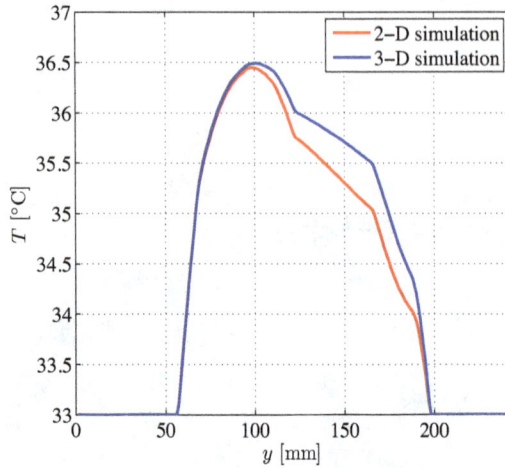

Figure 12. Simulated steady-state temperature profiles by thermo-neutral conditions at the slice $z = 102$ along the transverse knee axis y for $x = 106$, for 2-D and 3-D simulation.

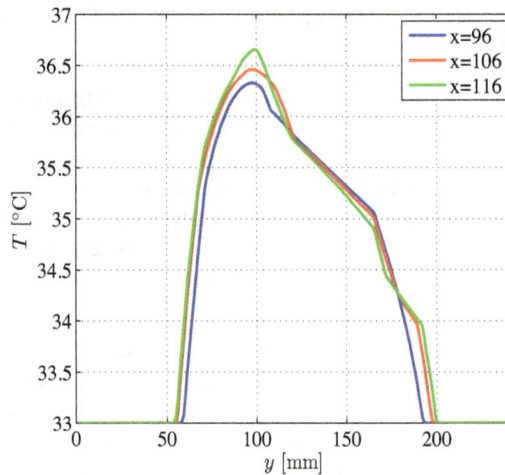

Figure 13. Simulated steady-state temperature profiles by thermo-neutral conditions on the transverse knee axis Y and model slice $z = 102$ for $x = 96$, $x = 106$ and $x = 116$.

The most important impact on the temperature profiles seen in Figure 14 arises from varying the knee dimension (Figure 14a). Larger knees result in a temperature plateau slightly above 36.7 °C, while smaller knees are cooled more intensively to the central temperature of about 36 °C, with larger temperature gradients in the superficial regions of the knee.

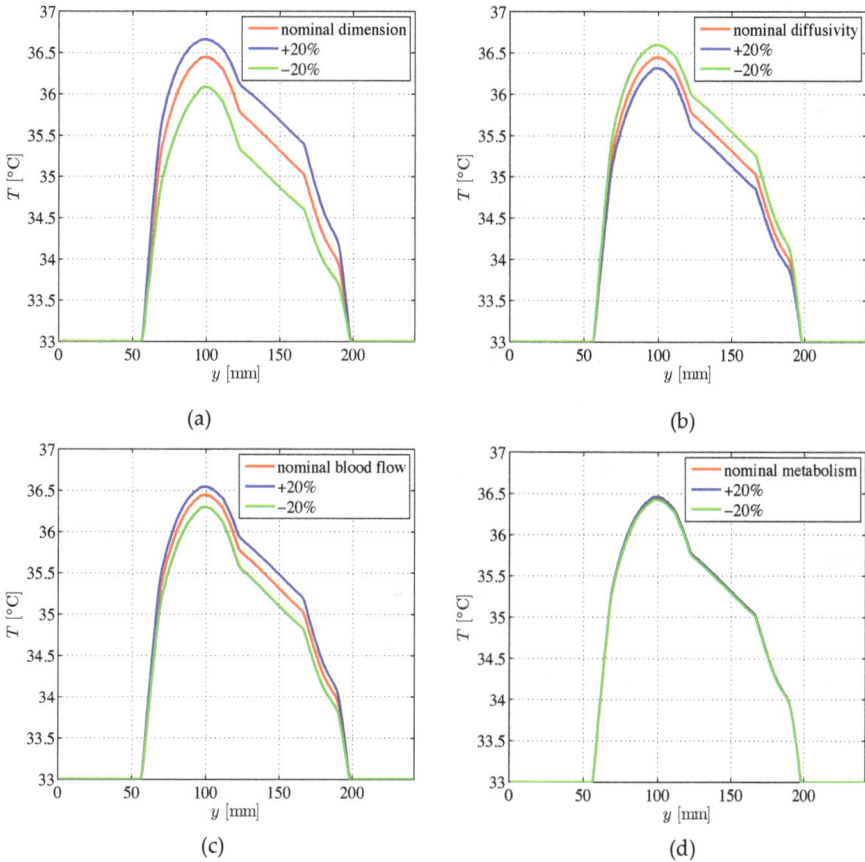

Figure 14. Simulated steady-state temperatures from model slice $z = 102$ and the transversal axis Y for $x = 106$ for various: (a) knee dimension, (b) thermal diffusivity, (c) blood flow and (d) metabolism. In all Figure-panels nominal values are red, 20 % above nominal values are blue, and 20 % below nominal values are green.

Changes in thermal diffusivity (Figure 14b) have a smaller impact on the temperature profiles than changes in the knee dimensions, and in the opposite direction, i.e., larger diffusivity constants result in lower central temperatures.

The impact of blood flow (Figure 14c) is similar to that of knee dimensions, with higher internal temperatures arising from greater blood flow. However, an important difference is that the shapes of the temperature profiles remain unchanged with variations of the blood flow.

The impact of metabolism (Figure 14d) is analogous to that of the blood flow – increased metabolism results in higher internal temperatures, but its impact is so small that it can be neglected in our experiments.

Simulated steady-state temperatures of a resting knee under thermo-neutral conditions (TN) are used as an initial condition for all further simulations. First, we simulated a naked knee in steady-state at ambient air temperature of 25 °C. Then, we simulate a two-hour period of an arthroscopic operation, during which the knee joint is washed out by sterilized water at 22 °C, and therefore cooled. During the following two-hour period we simulate the temperature evolution in the operated knee while it is resting and is covered by a blanket, and therefore warming. Finally, we simulate a subsequent two-hour postoperative topical cooling.

5.4. Washing out during arthroscopy

During arthroscopic reconstruction of ligaments, the central part of the knee is washed by sterilized water at 22 °C. The water circulates in the space around the femoral intercondylar notch, normally filled by the joint liquid. The initial temperatures are taken from the steady-state of a naked knee at an ambient temperature of 25 °C. The temperature of the joint liquid is fixed at 22 °C. The temperature profile after two hours of washing out are shown in Figure 15 for the same plane as before; that is, along the Y axis for $x = 106$, but for the 7 mm higher slice at $z = 95$, because this is nearer the actual position of our measuring probes for validation of the simulated results. The internal knee temperature decreases significantly, maximally by more than 14 °C in places that have direct contact with the washing water, as can be seen from the temperature profiles in Figure 15.

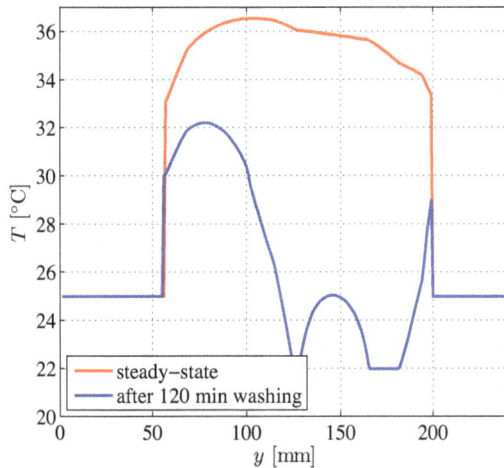

Figure 15. Steady-state temperature profiles for a naked knee and after a two hour washing out during surgery, for the model slice $z = 95$ and the transversal axis y for $x = 106$.

5.5. Resting after arthroscopy

Next, we simulate the two hour period immediately after surgery. The knee is resting and is covered with a blanket in a room temperature of 25 °C. The evolution of the temperatures in voxel (106,128,95), which is at the level of the femoral intercondylar notch in the central part of the knee, and in voxel (52,126,95), which is nearer to the knee surface, 1 cm below the skin in the subcutaneous tissue, are shown in Figure 16. Note that arbitrary voxels can be selected for the analysis. The knee is initially colder in the central region because of the previous washing out with cold water. During resting, its temperature increases and approaches the steady-state, with colder regions nearer the skin.

5.6. Postoperative topical cooling

Finally, we simulate postoperative topical cooling with two different cooling methods, gel-pack and cryo-cuff. In both cases, the knee is bound with a protective blanket surrounded by ambient air at 25 °C.

In Figure 17 the temperatures of voxels (106,128,95) and (52,126,95) are shown for the two-hour simulated period, for cooling with a gel-pack (initial temperature 0°C) and a cryo-cuff (water with constant temperature of 15 °C). The effects of the two methods on the knee temperatures are quite different.

When cooling with gel-pack, the temperature of the voxel (106,128,95) in the central knee region initially slightly increases because of the arterial blood perfusion and metabolism, and the weak influence of the initial cooling. After 5 minutes, the temperature of the voxel starts to decrease. However, after 40 minutes, the gel-pack has received enough heat from

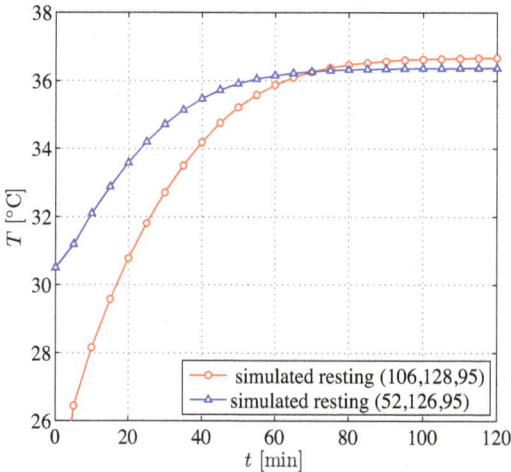

Figure 16. Simulated temperature evolution in two resting hours after surgery for voxels (106,128,95) (the central part of the knee) and (52,126,95) (in the subcutaneous tissue).

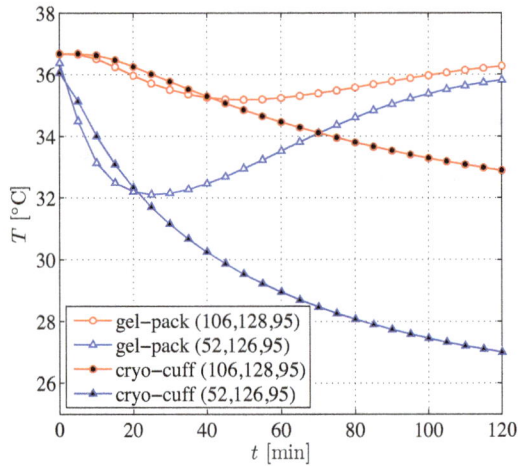

Figure 17. Simulated temperature evolution in voxels (106,128,95) and (52,126,95), as a function of time for cooling with gel-pack and cryo-cuff.

the knee surface and the ambient air, allowing the inner knee temperature to increase during the second part of the cooling period. For the voxel (52,126,95) in the subcutaneous tissue, 10 mm below the skin, the temperature first sharply decreases; the effectiveness of the gel-pack then becomes weaker, the voxel starts warming up, and after 120 minutes reaches almost 36.0 °C.

The cooling with a cryo-cuff is found to be more effective. It induces lower tissue temperatures, even if the temperature of the cooling liquid is as high as 15 °C. In the initial phase, both voxels experience the same cooling rate as the one with the gel-pack. However, there is no subsequent increase in temperature because the cryo-cuff is a constant heat sink which gradually cools the knee. After two hours of cooling, the near surface voxel reaches a temperature of 27 °C. In the same way, but with smaller intensity, the inner voxel is cooled to 33 °C.

In the case of topical cooling with gel-pack, the heat of fusion should also be simulated, which is necessary for the transition between aggregate states. Obviously, for a crushed gel-pack, a significant part of the heat is needed for such a transition, which would prolong the effective cooling time. This phenomenon has not been incorporated in our mathematical model. Instead, we increase the heat capacity c of the gel-pack to recognize and account for this behaviour. The simulated results show that the topical cooling with a cryo-cuff provides more constant lowering of the temperatures in the whole region of the knee. Cooling with gel-packs is less stable; consequently, they should be changed every half an hour in order to be effective.

In Figure 18 temperature profiles after one hour of simulated cooling with gel-pack and cryo-cuff are shown for a cross-section from the patella to the lateral side of the knee on our standard axis, i.e., along the Y (anteroposterior) axis for $x = 106$ on the model slice $z = 95$.

After one hour of simulated cooling, the gradients in the temperature profiles were much more pronounced than in the initial state. When cooling with gel-pack, the temperature of the outer knee layers at the skin level remained cooled to 32 °C, and in the centre of the knee to 36 °C. The peaks in the tissue temperatures around $y = 100$ result from the simulated heat conduction from the middle popliteal artery. Significantly lower temperatures are observed for the cooling with cryo-cuff, even though its constant temperature was as high as 15 °C.

Given the above results, it would be interesting to test the effectiveness of a simple cooling with ambient air. The knee would remain uncovered and exposed to the ambient air temperature of approximately 20 °C. We expect, from the results under thermo-neutral conditions, that the skin temperature would be about 25 °C, which could lower the temperatures inside the knee. This could be tested preliminarily using the proposed simulation method.

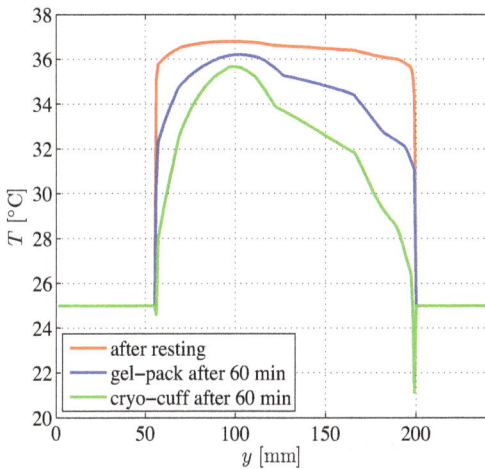

Figure 18. Simulated initial knee temperature profile after resting and temperatures after one-hour period of cooling with the gel-pack and cryo-cuff along the Y axis for $x = 106$ and $z = 95$.

5.7. Validation of results

To evaluate the simulated results, we made two control measurements of the knee temperature following surgery in a room with constant ambient temperature of 25 °C. We measured the temperature of the knee covered with a blanket during the two-hour resting period immediately after surgery; then, during the next two-hour period in which topical cooling with a gel-pack was applied.

Two small thermistors were placed into the knee in thin sterile tubes (Foley-catheter with temperature sensor, 3 mm, Ch8-thermistor; Curity, Degania Silicone Ltd., Degania Bet, Israel). Similar tubes, without thermistors, are ordinarily inserted for wound drainage following surgery. The thermistors were connected to a registration device for continuous measurement with a sample rate of 0.1 Hz and resolution of 0.01 °C.

The first thermistor was placed in the centre of the knee near the voxel (106,128,95), and the second approximately 1 cm below the skin in the subcutaneous tissue, near the voxel (52,126,95). The measurements were approved by the Slovenian State Medical Ethics Committee and the patient gave written informed consent prior to participation.

In Figure 19 the simulated and measured temperatures are shown for the two-hour resting period after washing out during arthroscopic surgery. The knee was wrapped in a protective blanket at ambient temperature of 25 °C under the same conditions as in the simulation.

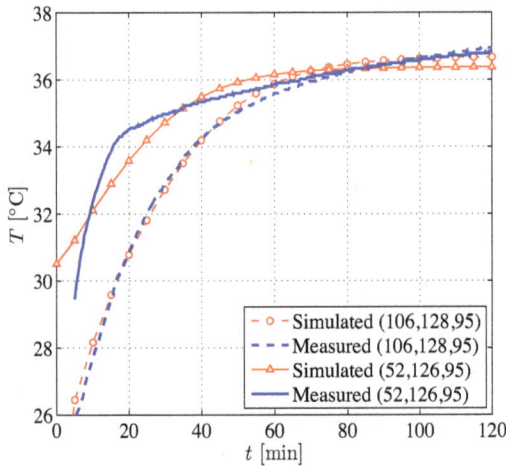

Figure 19. Measured and simulated temperature evolution of voxels (106,128,95) and (52,126,95) in a two hour resting period after arthroscopic surgery.

The simulated temperature evolution for the resting period shows very good agreement with the measured values for the test point (106,128,95) in the central part of the knee. However, the simulated rate of cooling in point (52,126,95) in the subcutaneous tissue was much smaller in the initial phase than the one obtained by the measurements. One of the possible reasons is the fact that we did not simulate the cold washing out inlet that also cooled the surrounding tissue from the skin to the central part of the knee. The subcutaneous thermistor was placed in such cooled environment which could then exhibit faster warming than in our simulation. In Figure 20 the simulated and measured temperatures are shown for two hours of cooling with a gel-pack, which follows immediately after the resting period. The initial temperature of the gel-pack was 0 °C and the ambient temperature 25 °C. The knee was bound with elastic bandages approximately 2 mm thick, surrounded with fixed gel-packs and wrapped in a protective blanket.

The simulated evolution of temperature for the cooling period shows good agreement with the measured values for both test points. The simulated rates of warming in the second hour are slightly greater than by measurement. One of the possible reasons is the incomplete mathematical model that does not include the heat of fusion for the gel-pack. In fact, all the thermodynamic characteristics of the gel were not available, and we just took some approximate values provided by the supplier. Another possible reason is inaccurate measurement. We did not collect detailed data during the measurements, e.g., the wetness of the protective bandage, which could have a significant impact on the cooling intensity

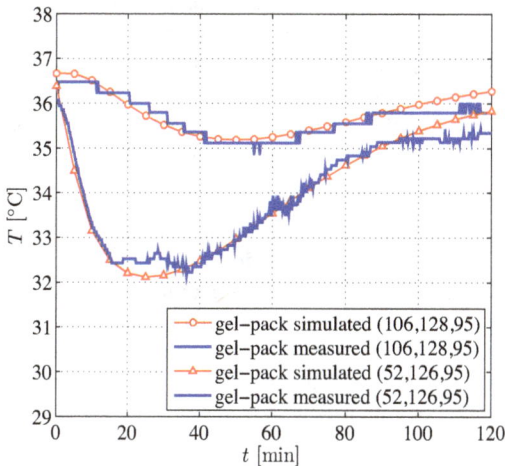

Figure 20. Measured and simulated temperature evolution of voxels (106,128,95) and (52,126,95) for cooling with a gel-pack.

5.8. Summary

We demonstrated a practical application of the simulation program for two different methods of topical postoperative cooling of a knee. We simulated topical cooling of a knee with a gel-pack. The inner knee tissues reached their lowest temperature in 40 minutes. For continued effective cooling, the gel-pack has to be replaced. The topical cooling with a cryo-cuff was more effective. We simulated situations with relatively small cooling rates to analyse the small influences of the blood flow and metabolism. The thickness of a protective bandage and isolating blanket, or their thermal conductivity, together with the cooling temperature, can be used to regulate the cooling intensity. The results have been validated by experimental measurements of knee temperatures. The simulation results confirm that the model and methodology used are appropriate for the thermal simulation of bio-tissues.

We have shown that the blood flow has a significant and complex impact on the stationary state temperatures, and the gradient of temperature change in the subcutaneous tissue and in the tissues nearer the central part of the knee. Lower blood flow results in linear temperature profiles with larger gradients, for example in bones. The dimensions of the knee are also important factors influencing the temperature distributions and gradients. Temperatures in smaller knees will differ more from the arterial blood temperature than those in larger knees. At the same time, the temperature gradient will be much greater in smaller knees. Thermal constants and metabolism have a relatively minor impact on the temperature field.

The simulated results were visualized and compared with measured values. Good agreement was obtained, leading to the conclusion that the model and method used in our simulation are appropriate for such medical simulations. Although there are not many studies of knee temperatures measured in vivo, and the measuring conditions are often not described in sufficient detail, we have also run our simulation software with the initial conditions described in some published measurements. In (Martin, et al., 2001) the temperature in the lateral gutter of the knee decreased by 4 °C, one hour after knee arthroscopy. Similar values have been obtained by our simulation program.

The method described has several limitations. The 3-D knee model used in our simulation is not complete, as only a small part of the leg above and below the knee was included. The remaining part of the leg that was not included was compensated by a constant flux in boundary conditions. Spatial models differ with different persons and with time, and consequently, the simulated results can differ. Minor errors in tissue segmentation and inaccurate thermodynamic constants could also produce small errors in the simulated results. Moving air was compensated by an artificial thin layer of air with non-constant temperature. Incorporation of a fluid-flow model would be needed for even more accurate results. The possible influence of blood perfusion by different regulatory mechanisms has not been simulated. Personal regulatory mechanisms have not been included in the simulation model, but could easily be incorporated. Listed limitations could have some impact on the simulated temperatures, but the essential findings agree remarkably well with existing experimental in vivo measurements.

6. Simulation of human heart cooling during surgery

During longer surgery the human body and the heart have to be managed appropriately in order to slow down their vital functions. One of the options for lowering metabolic requirements is the cooling of the body and the heart, e.g., by whole body hypothermia or infusion of a cold solution in coronary arteries (cardioplegia). For even better cardiac cooling, a method of topical cooling of the heart is sometimes applied (Olin and Huljebrant, 1993) by using packs filled with cooled gel or ice slush.

The efficiency of cooling is reflected in the temperature distribution within the heart, which is hard to determine. In vivo temperature measurements are invasive and, if applicable, limited to a few test points. These issues can be circumvented to some extent by physical modelling and computer simulation. Appropriate simulation tools provide insight in the temperature distribution through the whole heart volume and its neighbourhood, and therefore, various preliminary evaluations and analyses of different cooling regimes can be assessed.

Initial results on the simulation of heart cooling based only on diffusion have been reported in (Trunk, *et al.*, 2003). In this section, we present simulation results obtained through much more complex physical model, incorporating, beside diffusion, also advection, bio-heat sources and fluid flow. Such a model stands for more realistic variant of already developed simulation tools (Šterk and Trobec, 2005, Trobec, *et al.*, 1998).

Our final goal is to implement an accurate and reliable simulation framework that enables simulations of various diagnostics or therapeutic medical procedures. Simulation of different protocols of myocardial protection is taken as a test-case because they incorporate most of the physical phenomena during surgery that are related to the bio-heat transport. In this final section we focus on topical cooling as an adjunct to cardioplegia in classical non-beating heart operation (Allen, et al., 1992). During cardiac surgery, local hypothermia is reported to be successful in slowing down the tissue metabolism (Birdi, et al., 1999). This is achieved by infusing a cold cardioplegic solution into the coronary vessel system. Alternatively, topical cooling is applied with a cold saline solution, sometimes mixed with ice slush, which is distributed around the heart to further lower the temperature of the heart tissue. In non-perfused tissues, ischemia could provide deleterious effects on the cells, finally producing necrosis. Lowering of tissue temperature lengthens the time in which the ischaemic damage to the cells is reversible, which can prolong the safe time available for cardiac surgery. To provide adequate protection to the whole heart tissues, the heart has to be cooled down as uniformly as possible (Gonzales, *et al.*, 1985). The anterior part of the heart, especially the right ventricular wall is likely to remain outside the topical cooling solution during the operation, because it represents the uppermost part of the heart in supine patient. The importance of uniform temperature distribution in heart tissues comes also from the fact that areas with different temperatures could change myocardial contraction velocity (Riishede and Nielsen-Kudsk, 1990) and trigger cardiac arrhythmias. Some previous computer simulations have already shown irregular temperature distribution in the human heart during topical cooling (Geršak, *et al.*, 1997). Opinions about the cryo-therapeutic effects on the heart during cardiac surgery and during post-surgery rehabilitation period are still controversial. Furthermore, recent findings put the real benefit of topical cooling under question (Cordell, 1995). The optimal temperature at which the heart tissues should be cooled down and the time intervals of optimal cooling are still not clearly defined, probably because of the lack of in vivo measurements and difficult detailed post-surgery patients screening.

The aim of our work is to accurately simulate topical cooling during induced cardiac arrest in a typical heart operation. We want to use 3-D geometric model with all details, including coronary vessels. The desired result is the temperature distribution in the heart tissues as evolving in the time of surgery. The topical cooling is mimicked by submerging the heart in a cooling liquid. We are particularly interested in the influence of the liquid dynamics on the cooling performance.

Detailed quantitative validation of the results is not the main goal of our investigation. We are more concerned about the adequacy analysis of the presented physical model for simulation of heat transfer phenomenon during open heart surgery. For the sake of simplicity and transparency of results, we assume some simplifications, which can be declared as limitations of our study.

First, we simulate only the topical cooling and do not take into account the heart cardioplegia and the whole body hypothermia. It allows us to be concentrated only on a single cooling modality. The impacts of the remaining cooling methods remain to be studied in the future. The whole body hypothermia could be simulated by setting the domain boundary conditions to lower temperatures. The cardioplegia could be approximated by setting the temperature of blood in the coronary vessels to a constant value with Dirichlet boundary conditions on the vessel walls, assuming well mixed blood with a significant flow. Next, we simplify the domain boundary conditions to the Dirichlet, although the cooling would affect the body as well; however, the body temperature is assumed to remain constant through physiological regulatory loops. Next, because we assume well mixed air we neglect the natural convection of air in the operating room. The ambient temperature is set to constant value. Finally, blood in heart chambers is assumed to be steady.

6.1. Geometric model of a heart

VHD male cross-section photographs, numbered 1350 to 1505, are selected for the construction of the geometric model. Z axis of the VHD is parallel to the heart axis, pointing approximately from the apex towards the atria and the base of the heart. The selected cross-sections are cropped to the size of 512 (width)$\times$512 (height) pixels. Different tissues are classified (segmented) manually with an assistance of expert anatomist. The space occupied by voxels not belonging to the heart tissues or major vessels (aorta, vena cava, pulmonary vessels) are replaced either by voxels of air, water or the box wall. The final 3-D model used in the simulation is shown in Figure 22.

During the tissue segmentation, a few problems that could deteriorate the desired spatial accuracy of the heart model have to be resolved. The details of the heart structures have to be preserved as much as possible. The accuracy of the simulations relies on the resolution of the geometric model, which serves as a substrate for the simulation. For example, it is impossible to distinguish between the epicardium and pericardium on the VHD photographs; therefore, these two tissues are treated as a single tissue. The wall of the large vessels originating in the heart chambers is also not distinguishable from the neighbouring pericardium. The myocardial tissue contains several layers of muscular fibres that run in different directions. This could influence the heat propagation through the tissue, if different directions are examined; however, these layers could not be distinguished on the original cross-sections. Thus, the myocardium is classified as a homogenous mass of muscular tissue. The left ventricle, usually much larger in a heart in function, is quite small in the VHD dataset, evidently contracted and being in a systole like phase. On the contrary, the right ventricle is huge and elliptically shaped with an extremely thin wall. The endocardium, which normally lines the entire interior of the heart chambers, is microscopically thin, so it could not be differentiated in our model.

The border between two tissues was sometimes difficult to be determined from the VHD photographs because of the similar pixel colours. In such cases, the borders were drawn subjectively according to the anatomical knowledge. The majority of problems are detected in the area of the pulmonary and other large vessels at the base of the heart. These vessels can be identified only with concurrent examination of neighbouring slices and subsequent merging of multiple-slice information. Also, the pericardium is discontinued in some places, particularly on the posterior surface of the heart, because it is not distinguishable from the surrounding tissues on the posterior side of the heart. In such cases, the borders are drawn subjectively with some variations, taking into account a few consecutive neighbouring cross-sections. The small coronary vessels are also hard to be identified. The left and right coronary arteries, the circumflex artery and the largest veins are visible and one can follow such vessels from their origin to the periphery. On the periphery, coronary arteries and veins sometimes lie in close proximity and look like a single large vessel. Some vessels do not have a uniform course in the 3-D view. There are discontinuations that occurred because the vessels are not recognized correctly on every cross-section. Some further artefacts are also detected, like spots painted on some cross-sections as vessels, but actually not corresponding to vessel tissues. This results in a single pixel or a group of pixels apart from the coronary vessels. Therefore, the coronary vessel system was re-segmented again in a separate phase.

The segmentation errors of the generated spatial heart model have to be corrected, either manually or automatically by a computer program. A spatial editor program is implemented to allow the user to easily spot and correct the smaller failures in the generated 3-D models. The spatial editor is also used for several automatic modifications of the 3D-model, e.g., a uniform reduction of the model size by a user supplied factor n. Another function is designed for finding small openings or discontinuities in a selected tissue (exposing other tissues) and repairing them by adding small portions of missing tissue onto adequate places of the cross-sections. This function was mostly used to fill in small holes in the pericardium.

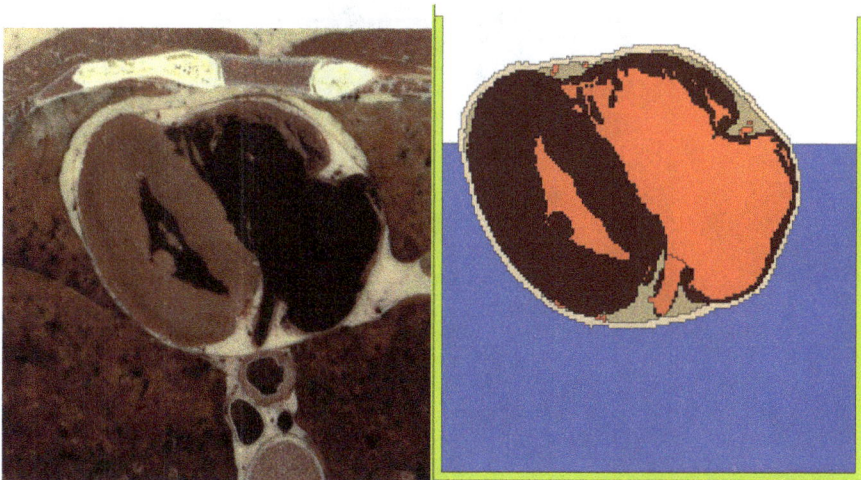

Figure 21. Example of a cropped VHD photographs before the geometric model preparation (left) and corresponding heart slice after segmentation (right).

6.2. Numerical setup

The thermal conditions during cardiac surgery are imitated with an isolated cube large enough to contain the simulated heart model. The computational domain is shown Figure 22. The lower part of the cube represents a cooling container filled with a cooling liquid that is initially set to a temperature of 0.2 °C, assumed to be cooled via ice slush. The container walls are kept at a constant temperature of 36.5 °C, simulating the body temperature. The initial temperature of the heart is also set to 36.5 °C. The upper part of the cube represents the open thorax surrounded by air at room temperature. The surrounding air is assumed to be well mixed and, as such, is kept on a constant temperature of 20 °C. The cube is composed of $163 \times 169 \times 163$ voxels of 1 mm³ representing a 3-D box with size $x = 163$ mm, $y = 196$ mm, $z = 163$ mm independently characterized by specific thermodynamic constants, defined earlier in Table 1 (Section 2) and Table 3 (Section 5).

The blood flow in a Bio-heat equation (7) is modelled using the following parameters:

$$V_1 = 5.145 \cdot 10^{-5}$$
$$V_2 = 0.322 \quad\quad\quad (33)$$
$$V_3 = 0.705.$$

The cooling liquid is assumed to be water with thermal expansion coefficient $\beta = 4 \cdot 10^{-4}$ and viscosity $\mu = 1.2 \cdot 10^{-3} \dfrac{Pa}{s}$. The heat transfer in water is simulated without bio heat sources (10).

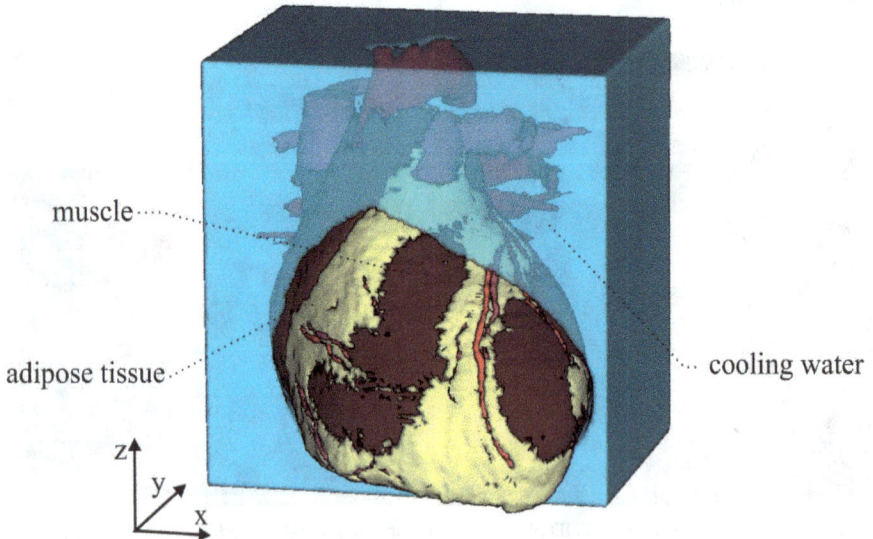

Figure 22. Computational domain of simulated heart.

6.3. Results

The present analysis is focused on the detailed assessment of the impact of fluid flow on the cooling process of the heart. Note that in cases described in previous sections the cooling is modelled only as a diffusive process.

The heat flux from the boundaries and the heart warms the surrounding water. The temperature dependent density governed by the Boussinesq approximation induces non-uniformly distributed volumetric body force in the Navier Stokes momentum equation, thus, a velocity field develops. The phenomenon is also known as natural convection (Amimul, 2011). The moving fluid extensively affects the cooling process through the advection (second term in the right side of equation (7)) term. In this section, we show the importance of heat advection. Simulation time of 10 minutes is considered, as the characteristics relevant to our discussions already develop in that time. The results are presented in terms of velocity profiles and temperature contour plots on x-y cross-section at $z = 80$ mm in four characteristic time instants t = [60, 240, 420, 600] s, and in terms of cross-lines in x-y plane for four different cross-sections z = [30, 60, 90, 120] mm at the end of simulation (600 s).

We compare two different scenarios; first, with fluid flow and second, without fluid flow. The main difference is in the treatment of the cooling liquid. The first case represents complex simulation with diffusion, advection and bio-heat sources, while the second case is similar to previously presented simulations of arm and knee. Such a cooling method with the "still" water can be approximated by cold gel packs or ice slash packs, where the flow is not possible.

In Figure 23 the temporal development of the setup with fluid flow is presented. The effects of advection, due to the natural convection, are clearly visible. Near the boundaries, the water is warmer as it is constantly heated through the domain walls and the heart. The heat flux generates upward fluid flow that eventually turns inwards as it reaches the boundary of the free liquid media. The consequent fluid flow significantly increases the heat transport from the domain boundaries, as well as from the heart into the cooling water. Note that in the bottom of the domain, local flow instabilities can be identified. The local bursts of upward flow result from instable buoyant force distribution. The warmer fluid pressures upward and only a small perturbation is needed to start a local upward burst of fluid, resulting in a local vortex. The induced circulation significantly increases the heat transfer between the walls and the heart.

In Figure 24 the simulation without fluid flow is presented. There is no advection and the heat extracted from the heart is transported only by means of diffusion, which is much slower in comparison with the advection. There is significantly less heat transported from the walls and, therefore, more heat is extracted from the heart before water reaches steady state temperature. The steady state, although not simulated in the present work, is also reached considerably later if the advection is not in place, as the heat flow is much slower (see water temperature on the last figure panel - 600 s). In the case with diffusion only, the water plays an additional role of an insulator between the heart and the boundary walls.

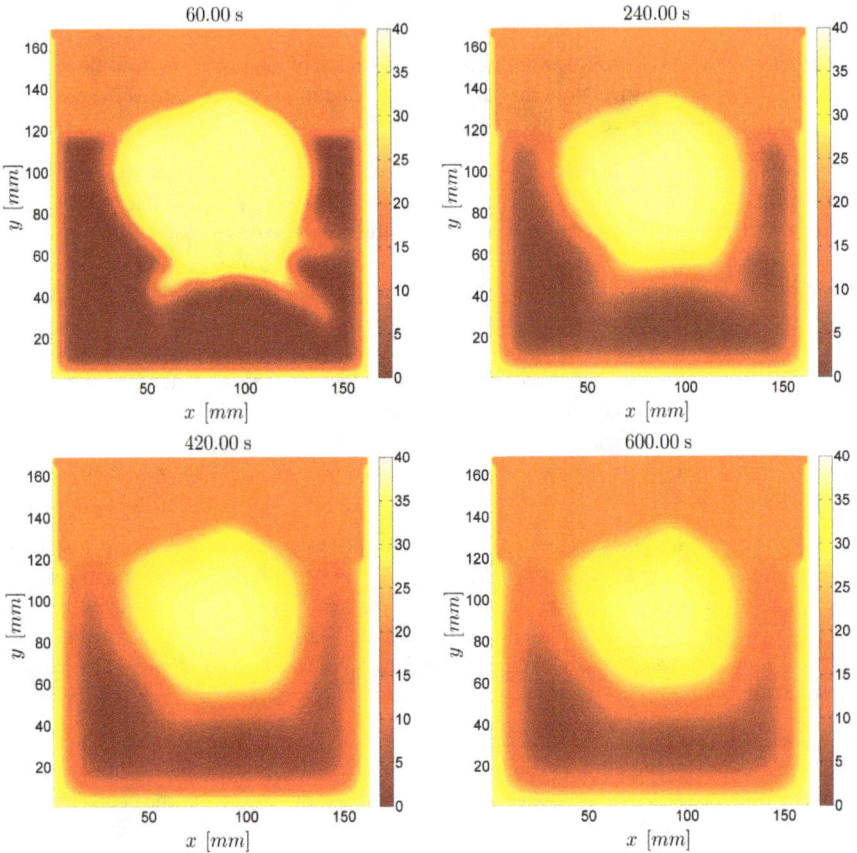

Figure 23. Temperature and velocity fields with natural convection shown on the cross-section at $z = 80$ mm at four time instants [60, 240, 420, 600] s.

Finally, the whole domain is examined by considering temperature cross-section plots after the simulated 600 s. Three cross lines in x-y plane at z = [30, 60, 90, 120] mm for both simulated cases (with and without natural convection) are shown in Figure 25. These additional representations confirm the earlier conclusions. The water retains its initial temperature for much longer period of time in the case without flow. The immediate effect, visible after ten minutes of simulation, is lower heart temperature in the case without advection (curves marked by circles).

6.4. Summary

We confirmed that the fluid flow plays a crucial role in the process of topical cooling of the heart. The natural convection in the cooling liquid increases the dynamics of the heat transport. The advection increases the flux from the domain boundaries, as well as from the

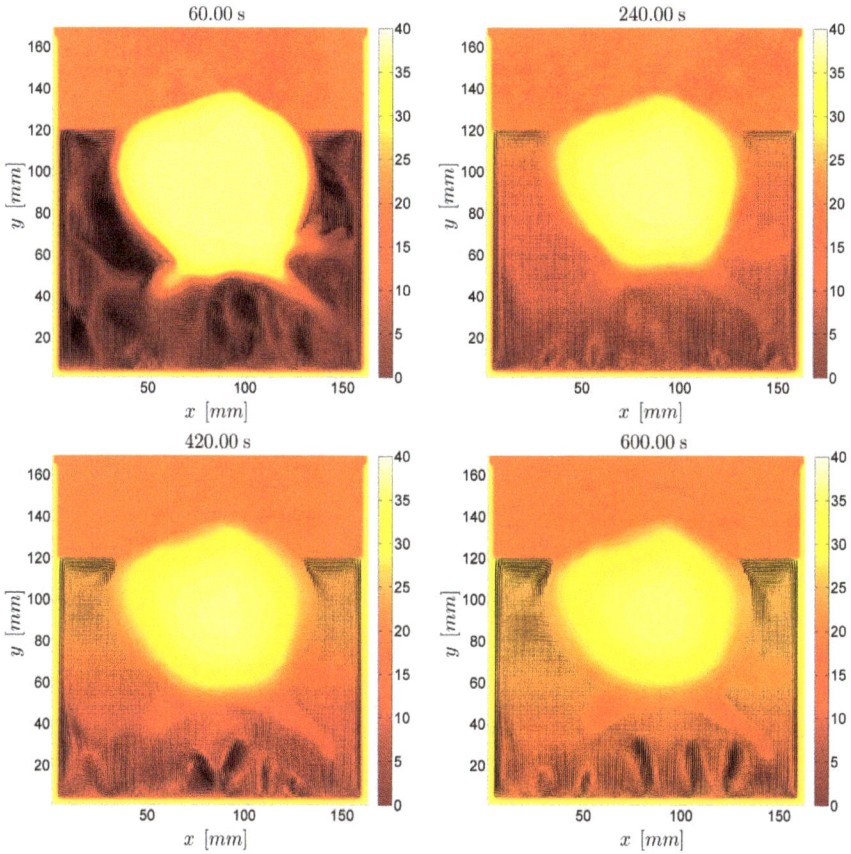

Figure 24. Temperature fields without natural convection shown on the cross-section at $z = 80$ mm at four time instants [60, 240, 420, 600] s.

heart walls. In the presented cases, the walls are kept at a constant temperature. Hence, the increased flux primarily affects the heating of the water. Consequently, the cooling process is less effective. If the domain boundaries would be kept at lower temperatures or kept isolated, e.g., by heart transplantation, the advection induced cooling would be much more pronounced. Alternatively, the water could be externally cooled or completely replaced after certain period of time to keep the cooling process more effective; however. The above cooling options are beyond the scope of our work and remain to be further investigated. In the case of omitted advection, the impact of cold "still water" lasts longer, resulting in lower simulated temperatures of the heart. We expect that new standards for the cooling protocols have to be developed in the future, supported also with the results and methodology presented in our work.

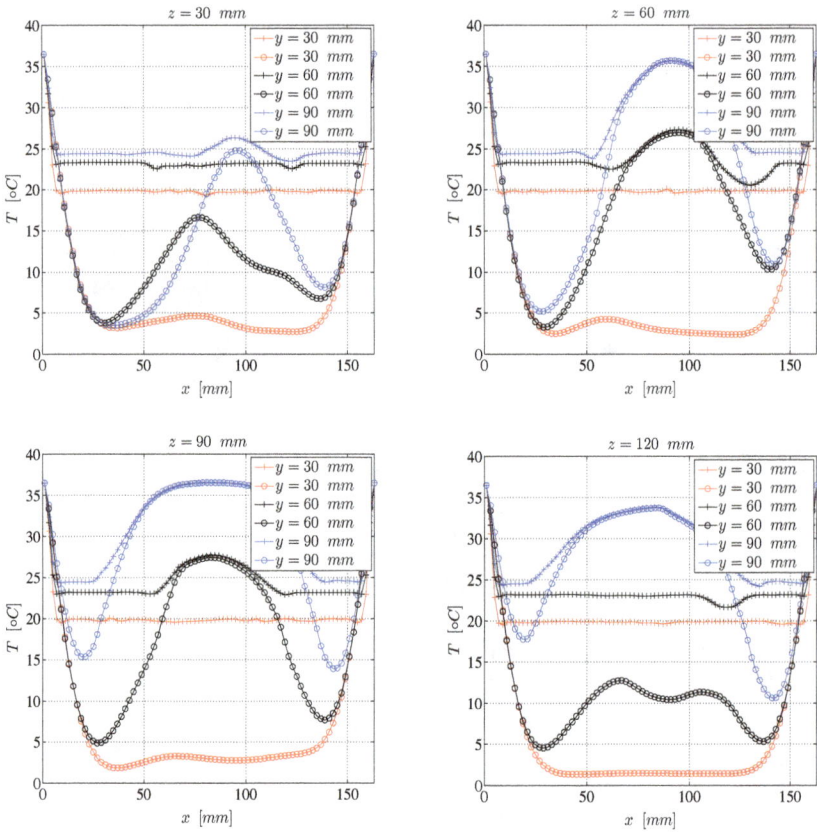

Figure 25. Comparison of temperature distribution at different cross-sections after 10 minutes of simulation: complete simulations (including advection) are marked with plus signs, simulations without advection are marked by circles.

7. Discussion and conclusions

We developed and implemented a physical model for the solution of the Bio-heat equation described in Section 2. The model incorporates heat diffusion, heat advection, heat transfer through blood perfusion and heat generation by tissue metabolism or other bio sources. A corresponding mathematical model and a computer simulation program have been implemented. Some important extensions have been introduced, in particular, an inhomogeneous spatial model composed of tissues with different characteristics, and modelling of the blood perfusion and heat sources as functions of the surrounding tissue temperature.

The heat produced by tissue metabolism was modelled using the Q10 rule, while the heat exchange between the blood and the tissue was modelled as a function of the regional blood flow and the local temperature. The regional blood flow was modelled as an exponential function fitted to experimental data. It was assumed to depend on the regional temperature in the same way as the mean blood flow rate depends on the skin temperature. Such an approach produced reasonable results. Nevertheless, it needs to be supported by further physiological research. The developed simulation, however, is sufficiently robust to incorporate arbitrary model functions for blood flow and metabolism. The stability of the method was confirmed by varying the simulation parameters, the initial and boundary values, and the model shape, with subsequent analysis of the results.

The construction of geometrical model, described in Section 3, was based on VHD slices. The models were composed of equal voxels, which results in a regular spatial discretization. Consequently, an explicit finite difference scheme has been developed and optimized for this purpose. A general methodology for the development of 3-D models, based on digitalized slice data, is described and applied for generating our spatial geometric models.

We have shown that the computational complexity of the simulation is proportional to the geometric resolution. The long run times of the presented 3-D simulation, which was in the range of several hours per run, constitute a technical limitation. The initial simulations were done in 2-D on a single model slice, which gives first merits about the temperature of deep tissue. Although the 2-D simulation times were shorter by several orders of magnitude, there were only minor differences in the steady-state temperature profiles. The final results presented in this chapter have been obtained by the 3-D simulation. Simpler models with lower geometric resolution could not reproduce vessels, which significantly deteriorate the accuracy of the simulation results. Regarding the available measuring equipment and environmental clinical conditions (e.g., variations in room temperature), the simulation accuracy should be approximately 0.1 °C.

We also implemented a parallel version of the proposed method, which runs efficiently on 16 or more connected computers in a computing cluster. In this way, the computation time can be shortened significantly. This approach enables the solution of several millions of equations for each time-step. Besides the simulated results, the computer simulations also support the development of new ideas and theories because a lot of "unexpected" situations can be explained. The proposed method can be applied in investigations of a variety of living tissues.

The solution of the Bio-heat equation was used in three different cases in the field of medicine. In the first case, described in Section 4, our primary goal was to explain, by computer simulation, the variations in the Pennes' in vivo measurements of the steady-state temperatures along the transverse axis of the proximal forearm. Assuming that the anatomical positioning of the measuring probes varied because of the different forearm sizes of the subjects, we have reconstructed their possible paths by searching for the simulated paths that result in the best agreement between the simulated and the measured temperature fields. Several attempts have been made in the past to average the measured steady-state temperatures in order to smooth fluctuations and obtain a standardized response. Our simulated results suggest that this approach is inappropriate, and that the fluctuations are natural result and the consequence of a complex interplay between the position of the measuring probe, the anatomical position of the main arteries, the dimensions of the forearm, the blood flow, the inhomogeneity of tissues, and the environmental temperature.

We have confirmed that blood flow has a significant and complex impact on the steady-state temperatures of the forearm. Lower blood flow results in lower temperature of the central part of the forearm. When the effect of blood flow on the temperature profile is increased, by disabling the direct effect of the main arteries, the resulting temperature profiles match very well to the analytically produced profiles from the literature, but not well with the experimental measurements. We have shown that only when the effect of both sources, the heat distributed via blood flow and the heat emitted directly from the arteries, are considered, the measured temperature profiles can be fully explained. Furthermore, we have revealed the dependence of the temperature profile shape on the dimensions of the forearm. Smaller forearms are more likely to feature off-centre peaks in the vicinity of the main arteries than larger forearms. In larger forearms, a central plateau may be formed, masking all other potential peaks. The thermodynamic constants of the tissue and the tissue metabolism were confirmed to have a relatively minor impact on the temperature field.

In the second test case, described in Section 5, we demonstrated a practical application of our simulation program in topical postoperative cooling of a knee with two different cryotherapeutic methods. First, we simulated the topical cooling of a knee with a gel-pack. The inner knee tissues reached their lowest temperature in 40 minutes. For continued effective cooling, the gel-pack has to be replaced. The topical cooling with a cryo-cuff was more effective. We simulated situations with relatively small cooling rates in order to be able to analyse the small influences of the blood flow and metabolism. We observed that various thicknesses of a protective bandage and isolating blanket, or variations in their thermal conductivity and cooling temperature, can be used to regulate the cooling intensity.

The results have been validated by experimental measurements of knee temperatures. The simulation results confirm that the model and methodology used are appropriate for the thermal simulation of bio-tissues. We have shown that the blood flow has a significant and complex impact on the stationary state temperatures, and on the gradient of temperature change in the subcutaneous tissue and in the tissues nearer the central part of the knee. Lower blood flow results in linear temperature profiles with larger gradients, for example, in bones. The dimensions of the knee are a very important factor influencing its temperature distribution and gradient. Temperatures in smaller knees will differ more from the arterial blood temperature than those in larger knees. At the same time, the temperature gradient will be much greater in smaller knees. Thermal constants and metabolism have a relatively minor impact on the temperature field.

We confirmed that fluctuations in temperature profiles are natural results and a consequence of the complex interplay between the positions of the measuring probes, the dimensions of the investigated body part, the anatomical positions of the main arteries and bones, and the environmental temperature. The fluctuations and augmented peaks in the temperature profiles can be explained by our simulation results, since the temperature changes in the centre of the knee are influenced by the vicinity of the main leg artery with constant blood temperature. The steepest, almost linear, gradients in the area of the subcutaneous tissues are observed in the outer parts of the simulated temperature profiles, particularly in cases with lower cooling temperatures. The temperature plateaus, measured earlier by other investigators, in the inner knee region with increased dimensions of the knee, were also demonstrated by our simulation.

The simulated results were visualized and compared with measured values. Good agreement was obtained, leading to the conclusion that the model and method used in our simulation are appropriate for such medical simulations. Although there are not many studies of knee temperatures measured in vivo, and the measuring conditions are often not described in sufficient detail, we have also run our simulation software with the initial conditions described in some published measurements. In (Martin, et al., 2001) the temperature in the lateral gutter of the knee decreased by 4 °C, one hour after knee arthroscopy. Similar values have been obtained by our simulation program. More detailed control studies should be done to compare experimentally measured and simulated results in order to fine-tune the simulation method.

The last test case, described in Section 6, stands for the most complex computation in the present chapter. It incorporates all physical model elements, including incompressible fluid flow. We confirmed that the fluid flow plays a crucial role in the process of topical cooling of the heart. We showed that, under simulation assumptions, the cooling process is less effective if natural convection is considered. In the case with no advection, the impact of rigid cold substance lasts longer, resulting in lower simulated temperatures of the heart. However, if the cooling liquid would be externally regulated, the results might be different. The advection increases the heat transfer and if appropriate cooling water management would be chosen, the cooling efficiency might be improved.

We are aware that our simulation methodology still has several limitations. A separate study should be devoted to the validation of simulated results; however, in many cases, in vivo measurements are not applicable or significantly limited. The anatomy of the human body differs between individuals and changes in time; consequently, the simulation on a single generalized model can differ from the experimental measurements. Some kind of personalized approach is foreseen in the future, also because of possible influences of different mechanisms of thermoregulated blood perfusion. When known, these mechanisms could be easily incorporated into the simulation model. Finally, the moving air and the blood in heart chambers were approximated without fluid flow in order to simplify the simulation. Although all these limitations could have some impact on the simulated temperatures, the essential findings are in close agreement with in vivo measurements.

Future work includes the simulation of the temperature field evolution in time and analysis of the possible impact of the measuring probes, surgical instruments and other medical equipment on the temperature fields. Using an approach analogous to the presented one, other parts of the human body could also be simulated and their temperature fields could be analysed. The proposed methodology could be useful for researchers in developing methods for accurate prediction of the temperature distribution in inhomogeneous tissues. Computer simulations could help in the study of various medical applications. It has been proved that the techniques described can be used to predict the temperature distribution in deep tissues, at any body point and time of interest, and for different environmental conditions.

For future development of the numerical approach we will consider the promising local meshless methods, like Radial Basis Function Collocation Method (LRBFCM) (Šarler, 2007) or Diffuse Approximate Method (DAM) (Prax, et al., 1996). The meshless methods are not limited to the equidistant uniform nodal distribution as the FDM is. The meshless approach would improve the accuracy near irregular geometric model boundaries. The DAM shows good stability and accuracy behaviour on non-regular nodal distributions (Trobec, et al., 2012), and consequently, is convenient for treatment of the presented bio-medical computations. The adaptive nodal distribution (Kosec and Šarler, 2011) would also improve the treatment of areas with high gradients, e.g., the interfaces between different materials with discontinues in thermo-physical properties. The meshless methods have been recently under intense research in fluid flow context, as well. The local pressure velocity coupling procedure in combination with LRBFCM has been proposed (Kosec and Šarler, 2008a, Kosec and Šarler, 2008b). The local fluid flow solution procedure is simpler to paralelize as it does not require anykind of global treatment.

The spatial models and the simulation programs, presented in this work, are available from the authors for research purposes that could contribute to any kind of knowledge in the area of biomedical simulation.

Nomenclature

ρ	density	t	time
P	pressure	τ	stress tensor
S	heat source	c_p	specific heat
j	heat flux	λ	thermal conductivity tensor
T	temperature	λ	thermal conductivity
v	velocity	b	body force
I	identity matrix	ρ^b	density of the blood
c_p^b	specific heat of the blood	h_r, v_r	spatial/material indicators

V_r	blood flow rate	h_m	metabolism response
T_a	reference temperature	V_0	reference blood flow rate
$V_{1,2,3}$	blood flow rate parameters	h_0	reference metabolism heat production
T_r	tissue reference temperature	μ	dynamic viscosity
ρ_{ref}	Boussinesq reference density	T_{ref}	Boussinesq reference temperature
g	gravitational acceleration	ε	coordinate (x, y, z)
L	differential operator	p	position vector
Δp_ε	spatial step in ε direction	p_ε	ε component of the position vector
v_ε	ε component of the velocity	Θ	arbitrary field
t_0	current time step	t_1	next time step
$\hat{v}$	intermediate velocity	$\hat{\hat{v}}$	velocity correction
$\hat{P}$	pressure correction	v_Γ	velocity boundary condition
n	normal boundary vector	T_S	temperature of skin
T_{am}	ambient temperature	q_0	convection coefficient

Author details

Aleksandra Rashkovska, Roman Trobec, Matjaž Depolli, Gregor Kosec
Jožef Stefan Institute, Slovenia

Acknowledgement

We thank the staff of the University Medical Centre Ljubljana, Department of Cardiovascular Surgery and Department of Traumatology, Slovenia. We also acknowledge the financial support from the state budget by the Slovenian Research Agency under the grant P2-0095.

8. References

Ackerman, M. J. (1998), The visible human project, *Proc IEEE*, Vol. 86, pp. 505–511.
Akl, S. G. (1997), *Parallel Computation: Models and Methods*, Prentice Hall, New Jersey.
Allen, B. S.; Buckberg, G. D.; Rosenkranz, E. R.; Plested, W.; Skow, J.; Mazzei, E. and Scanlan, R. (1992), Topical cardiac hypothermia in patients with coronary disease, *J Cardiovasc Surg*, Vol. 104, pp. 626–631.
Amimul, A. (2011), *Convection and Conduction Heat Transfer*, Intech, Rijeka.
Atluri, S. N. and Shen, S. (2002), *The Meshless Method*, Tech Science Press, Encino.

Atluri, S. N. (2004), *The Meshless Method (MLPG) for Domain and BIE Discretization*, Tech Science Press, Forsyth.

Barber, F.A. (2000), A comparison of crushed ice and continuous flow cold therapy, *Am. J. Knee Surg.*, Vol. 13, pp. 97-101.

Barcroft, H. and Edholm, O. G. (1942), The effect of temperature on blood flow and deep temperature in the human forearm, *J. Physiol.*, Vol. 102, pp. 5–20.

Barcroft, H. and Edholm, O. G. (1946), Temperature and blood flow in the human forearm, *J. Physiol. (Lond.)*, Vol. 104, pp. 366–376.

Bernardi, P.; Cavagnaro, M.; Pisa, S. and Piuzzi, E. (2003), Specific Absorption Rate and Temperature Elevation in a Subject Exposed in the Far-Field of Radio-Frequency Sources Operating in the 10-900-MHz Range, *IEEE Trans Biomed Eng*, Vol. 50, pp. 295–304.

Białecki, R.; Divo, E. and Kassab, A. (2006), Reconstruction of time-dependent boundary heat flux by a BEM-based inverse algorithm, *Engineering Analysis with Boundary Elements*, Vol. 30, pp. 767-773.

Birdi, I.; Caputo, M.; Underwood, G.; Angelini, D. and Bryan, A.J. (1999), Influence of normothermic systemic perfusion temperature on cold myocardial protection during coronary artery bypass surgery, *Cardiovasc Surg*, Vol. 7, pp. 369-374.

Bowman, H. F.; Cravalho, E. G. and Woods, M. (1975), Theory, measurement, and application of thermal properties of biomaterials, *Annual Review Biophysics and Bioengineering*, Vol. 4, pp. 43–80.

Bullock, B. L. and Rosendahl, P. P. (1992), *Pathophysiology, Adaptations and Alteration in Function*, Lippincott,

Charny, C. K. (1992), *Mathematical models of bioheat transfer, in: Y.I. Cho (Ed.), Advances in Heat Transfer*, Academic Press, New York.

Chen, W. (2002), New RBF collocation schemes and kernel RBFs with applications, *Lecture Notes in Computational Science and Engineering*, Vol. 26, pp. 75-86.

Cooper, K. E.; Edholm, O. G. and Mottram, R. F. (1955), The blood flow in skin and muscle of the human forearm, *J. Physiol. (Lond.)*, Vol. 128, pp. 258–267.

Cordell, A.R. (1995), Milestones in the developement of cardioplegia, *Ann Thorac Surg*, Vol. 60, pp. 793–806.

Dann, R.; Hoford, J.; Kovacic, S.; Reivich, M. and Bajcsy, R. (1989), Evaluation of elastic matching system for anatomic (CT, MR) and functinal (PET) cerebral images, *Journal of Computer Assisted Tomography* Vol. 13, pp. 603-611.

Detry, J-M. R.; Brengelmann, G. L.; Rowell, L. B. and Wyss, C. (1972), Skin and muscle components of forearm blood flow in directly heated man, *J. Appl. Physiol.*, Vol. 32, pp. 506–511.

Ducharme, M. B. and Frim, J. (1988), A multicouple probe for temperature gradient measurements in biological materials, *J. Appl. Physiol.*, Vol. 65, pp. 2337–2342.

Ducharme, M. B.; VanHelder, W. P. and Radomski, M. W. (1991), Tissue temperature profile in the human forearm during thermal stress at thermal stability, *J. Appl. Physiol.*, Vol. 71, pp. 1973–78.

Ducharme, M. B. and Tikuisis, P. (1994), Role of blood as heat source or sink in human limbs during local cooling and heating, *J. Appl. Physiol.*, Vol. 76, pp. 2084–2094.

Ducharme, M. B. ; Frim, J. (1988), A multicouple probe for temperature gradient measurements in biological materials, *J. Appl. Physiol.*, Vol. 65, pp. 2337-2342.

Ferziger, J. H. and Perić, M. (2002), *Computational Methods for Fluid Dynamics*, Springer, Berlin.

Geršak, B.; Gabrijelčič, T.; Trobec, R. and Slivnik, B. (1997), Temperature distribution in human heart during hypothermic cardioplegic arrest, *Cor Europeum*, Vol. 6, pp. 172–176.

Gonzales, A.C.; Brandon, T.A.; Fortune, R.L.; Casano, S.F.; Martin, M. and Benneson, D.L. (1985), Acute right ventricular failure is caused by inadequate right ventricular hypothermia, *J. Thorac Cardiovasc Surg*, Vol. 89, pp. 386-99.

Grana, W. A. (1994), *Cold modalities, in: J. C. DeLee and D. Drez (Eds.), Orthopaedic Sports Medicine, Principles and Practice*, WB Saunders, Philadelphia.

Harris, J. and Benedict, F. (1918), A biometric study of human basal metabolism, *Proc Natl Acad Sci USA*, Vol. 4, pp. 370–373.

Ho, S.S.; Coel, M.N.; Kagawa, R. and Richardson, A.B. (1994), The effects of ice on blood flow and bone metabolism in knees, *Am. J. Sports Med.*, Vol. 22, pp. 537-540.

Hong, C. P. (2004), *Computer Modelling of Heat and Fluid Flow Materials Processing*, Institute of Physics Publishing Bristol.

Hubbard, T.J. and Denegar, C.R. (2004), Does cryotherapy improve outcomes with soft tissue injury, *J. Athl. Train*, Vol. 39, pp. 278-279.

Kansa, E. J. (1990), Multiquadrics - a scattered data approximation scheme with application to computational fluid dynamics, part I, *Computers and Mathematics with Applications*, Vol. 19, pp. 127-145.

Knight, K.L. (1985), *Cryotherapy: theory, technique and physiology*, Chatanooga Corporation, Chattanooga.

Kosec, G. and Šarler, B. (2008a), Solution of thermo-fluid problems by collocation with local pressure correction, *International Journal of Numerical Methods for Heat and Fluid Flow*, Vol. 18, pp. 868-882.

Kosec, G. and Šarler, B. (2008b), Local RBF collocation method for Darcy flow, *CMES: Computer Modeling in Engineering & Sciences*, Vol. 25, pp. 197-208.

Kosec, G. and Šarler, B. (2011), H-Adaptive Local Radial Basis Function Collocation Meshless Method, *CMC: Computers, Materials & Continua*, Vol. 26, pp. 227–254.

Liu, G.R. (2003), *Mesh Free Methods*, CRC Press, Boca Raton.

Malan, A. G. and Lewis, R. W. (2011), An artificial compressibility CBS method for modelling heat transfer and fluid flow in heterogeneous porous materials, *International Journal for Numerical Methods in Engineering*, Vol. 87, pp. 412-423.

Martin, S. S.; Spindler, K. P.; Tarter, J. W.; Detwiler, K. and Petersen, H. A. (2001), Cryotherapy: an effective modality for decreasing intraarticular temperature after knee arthroscopy, *Am. J. Sports Med.*, Vol. 29, pp. 288–291.

Martino, R. L.; Johnson, C. A. and Suh, E. B. (1994), Parallel Computing In Biomedical-Research, *Science*, Vol. 265, pp. 902–908.

McElfresh, E. C. and Kelly, P. J. (1974), Simultaneous determination of blood flow in cortical bone, marrow, and muscle in canine hind leg by femoral artery catheterization, *Calcified Tissue International*, Vol. 14, pp. 301–307.

McGuire, D.A. and Hendricks, S.D. (2006), Incidences of frostbite in arthroscopic knee surgery postoperative cryotherapy rehabilitation, *Arthroscopy*, Vol. 22, pp. 1141e1-e6.

McMaster, W.C.; Liddle, S. and Waugh, T.R. (1978), Laboratory evaluation of various cold therapy modalities, *Am. J. Sports Med.*, Vol. 6, pp. 291-294.

Moses, W. M. and Geddes, G. L. (1990), Measurement of the thermal conductivity of equine cortical bone, *American Society of Mechanical Engineers*, Vol. 17, pp. 185–188.

Olin, C. L. and Huljebrant, I. E. (1993), Topical Cooling of the Heart - A Valuable Adjunct to Cold Cardioplegia, *Scand. J. Thorac. Card.*, Vol. 41, pp. 55–58.

Őzisik, M. N. (2000), *Finite Difference Methods in Heat Transfer*, CRC Press, Boca Raton.

Pennes, H. H. (1948), Analysis of tissue and arterial blood temperature in the resting human forearm, *J. Appl. Physiol.*, Vol. 1, pp. 93–122.

Ponder, E. (1962), The coefficient of thermal conductivity of blood and various tissues, *J Gen Phisiol*, Vol. 45, pp. 545-551.

Poppendiek, H. F.; Randall, R.; Breeden, J. A.; Chambers, J. E. and Murphy, J. R. (1967), Thermal conductivity measurements and predictions for biological fluids and tissues, *Cryobiology*, Vol. 3, pp. 318–327.

Prax, C.; Sadat, H. and Salagnac, P. (1996), Diffuse approximation method for solving natural convection in porous Media, *Transport in Porous Media*, Vol. 22, pp. 215-223.

Price, K. V. and Storn, R. (2008), Differential evolution: A simple evolution strategy for fast optimization, *Dr. Dobb's Journal*, Vol. 22, pp. 18-24.

Proulx, C. I.; Ducharme, M. B. and Kenny, G. P. (2003), Effect of water temperature on cooling efficiency during hyperthermia in humans, *J. Appl. Physiol.*, Vol. 94, pp. 1317–1323.

Rahman, M. M. and Siikonen, T. (2008), An artificial compressibility method for viscous incompressible and low Mach number flows, *Internation Journal of Numerical Methods in Engineering*, Vol. 75, pp. 1320-1340.

Rappaz, M.; Bellet, M. and Deville, M. (2003), *Numerical Modelling in Materials Science and Engineering*, Springer-Verlag, Berlin.

Riishede, L. and Nielsen-Kudsk, F. (1990), Myocardial effects of adrenaline, isoprenaline and dobutamine at hypothermic conditions, *Pharmacology and Toxicology*, Vol. 66, pp. 354-360.

Shashkov, M. (1996), *Conservative Finite-Difference Methods on General Grids*, CRC, Boca Raton.

Shelbourne, K. and Nitz, P. (1990), Accelerated rehabilitation after anterior cruciate ligament reconstruction, *Am. J. Sports Med.*, Vol. 18, pp. 292-299.

Silverthorn, D. H. (2001), *Human Physiology, An Integrated Approach*, Prentice-Hall, New Jersey.

Smith, G. D. (2003), *Numerical Solution of Partial Differential Equation - Finite Difference Method*, Oxford University Press, Oxford.

Sugimoto, H. and Monafo, W.W. (1987), Regional Blood Flow in Sciatic Nerve, Biceps Femoris Muscle, and Truncal Skin in Response to Hemorrhagic Hypotension, *J. Trauma*, Vol. 27, pp. 1025–1030.

Šarler, B. (2007), *From global to local radial basis function collocation method for transport phenomena*, Springer, Berlin, pp. 257-282.

Šterk, M. and Trobec, R. (2005), Biomedical Simulation of Heat Transfer in a Human Heart, *J. Chem. Inf. Mod.*, Vol. 45, pp. 1558–1563.

Tikuisis, P. and Ducharme, M. B. (1991), Finite-element solution of thermal conductivity of muscle during cold water immersion, *J. Appl. Physiol.*, Vol. 70, pp. 2673–2681.

Traivivatana, S.; Boonmarlert, P.; Thee, P.; Phongthanapanich, S. and Dechaumphai, P. (2007), Combined adaptive meshing technique and characteristic-based split algorithm for viscous incompressible flow analysis, *Applied Mathematics and Mechanics*, Vol. 28, pp. 1163-1172.

Trobec, R.; Slivnik, B.; Geršak, B. and c, T. Gabrijelčič (1998), Computer simulation and spatial modelling in heart surgery, *Computers in Biology and Medicine*, Vol. 28, pp. 393–403.

Trobec, R.; Šterk, M.; Almawed, S. and Veselko, M. (2008), Computer Simulation of Topical Knee Cooling, *Computers in Biology and Medicine*, Vol. 38, pp. 1076–1083.

Trobec, R.; Šterk, M. and Robič, B. (2009), Computational complexity and parallelization of the meshless local Petrov-Galerkin method, *Computers & Structures*, Vol. 87, pp. 81-90.

Trobec, R.; Kosec, G.; Šterk, M. and Šarler, B. (2012), Comparison of local weak and strong form meshless methods for 2-D diffusion equation, *Engineering Analysis with Boundary Elements*, Vol. 36, pp. 310-321.

Trunk, P.; Geršak, B. and Trobec, R. (2003), Topical Cardiac Cooling - Computer Simulation of Myocardial Temperature Changes, *Computers in Biology and Medicine*, Vol. 33, pp. 203–214.

Wenger, C. B.; Stephenson, L. A. and Durkin, M. A. (1986), Effect of nerve block on response of forearm blood flow to local temperature, *J. Appl. Physiol.*, Vol. 61, pp. 227–232.

Wissler, E. H. (1998), Pennes' 1948 paper revisited, *J. Appl. Physiol.*, Vol. 85, pp. 35–41.

Wrobel, L. C. (2002), *The Boundary Element Method: Applications in Thermo-Fluids and Acoustics*, John Wiley & Sons, West Sussex.

Yamazaki, F. and Sone, R. (2000), Modulation of arterial baroreflex control of heart rate by skin cooling and heating in humans, *J. Appl. Physiol.*, Vol. 88, pp. 393–400.

Yue, K; Zhang, X and Yu, F (2007), Simultaneous estimation of thermal properties of living tissue using noninvasive method, *Int J Thermophysics*, Vol. 28, pp. 1470–1489.

Chapter 5

Numerical Analysis and Experimental Investigation of Energy Partition and Heat Transfer in Grinding

Lei Zhang

Additional information is available at the end of the chapter

1. Introduction

In industry, structural parts have to be subjected to a heat treatment to increase fatigue strength or wear resistance of the surface layers and to adjust special combinations of material properties, which strengthen the surface layer of components and consequently improve the load capacity and lifetime. The heat treatment processes of the surface layer are some typical advantages compared to the full hardening heat treatments (Zhang 2004). In industry various heat treatment methods are used for the production of required surface layer properties. But these processes cannot simply be integrated into the production line because of economical disadvantage. A new technology called grind-hardening can utilize the grinding heat for martensitic phase transformation of the work piece surface layer (Brinksmeier and Brockhoff 1996; Brockhoff 1999). In contrast to conventional grinding processes, grind-hardening is based on extensive heat generation in the contact zone between grinding wheel and workpiece causing the austenitizing the surface layer material. The martensitic hardening is usually achieved by self-quenching mechanisms. Metallurgic investigations as well as hardness and residual stress measurements confirmed the possibility and the potential of this new heat treatment method, which can be integrated into existing production lines to decrease manufacturing cost, improve machining efficiency and reduce production time (Zarudi and Zhang 2002).

Fundamental investigations were carried out to determine the basic mechanisms of the short-time metallurgical processes, as well as the influence of the process parameters on the hardening result. The properties of the grind-hardened surface layers were evaluated by residual stress measurements and metallographic analyses as well as hardness measurement (Zarudi and Zhang 2002; Liu and Wang 2004; Wang and Liu 2005). Due to the process kinematics, grind-hardening is classified as a short-time heat treatment. Particularly in short-time heat treatments, composition and initial structural state of the material have an important influence on the hardening result. For grind-hardening the same mechanisms as for conventional heat treatments were verified. The hardening result depends on the content of carbon and alloying elements, as well as on the distribution of carbon at the initial materials state. High hardness penetration depths were obtained with tempered alloyed hyper-and hypo-eutectoid steels (Liu and Wang 2004).

Besides the material composition and initial state, as well as the grinding wheel specification, the parameters of the grind-hardening process also significantly influence the hardening result. The cut depth has the most important influence on the result of the grind-hardening operation. With increasing cut depth the cutting power and the generated heat increase. Thus the hardness penetration depth rises with increasing cut depth (Liu and Wang 2005). The interrelation of the table speed and the hardening result is more complex, because the table speed influences two factors that are highly important for the material phase transformation. With increasing table speed the cutting power and the heat quantity increases. Also the duration of the heat impact decreases with increasing table speed. Thus the table speed has to be adjusted to generate high heat quantities and sufficient heating durations. Concerning the cutting speed basic investigations did not prove a definite influence on the hardness penetration depth (Brinksmeier and Brockhoff 1996).

Most of analytical thermal models are based on the early work by Jaeger, where the grinding wheel is represented by a heat source, equally distributed over the contact length between the workpiece and the grinding wheel, moving along the surface of the workpiece with a speed equal to the speed of workpiece (Jaeger 1942; Rowe and Jin 2001; Ramesh et al. 1999; Lavine 2000). In grinding the total grinding energy is distributed not only in the workpiece but also in the grinding wheel, the chip, and the coolant. A prediction of these energy components and the effect of cooling, along with the various assumptions made, related to the thermal prosperities of the workpiece material and the exact heat source profile have been made in order to improve the simulation model (Zhang 2004; Zhang 2005).

The finite element method has been used for modelling the grinding process, which allows for the calculation of the grinding temperatures and their distribution within the workpiece, in order to achieve a greater accuracy and more reliable results. Experiment was performed with the different conditions. Semi-natural thermocouple is used to measure the temperature field of surface layer in grinding. The maximal surface temperature and the temperature field in the surface layer of the workpiece during grinding can be theoretically predicted by using the finite element model.

In grinding process almost all the grinding energy is converted into heat within a small grinding zone. A large part of the generated heat flows into the workpiece, which results in extremely high temperature at the interface between the wheel and the workpiece. This can cause elevated temperatures and harden layer of the work piece. Analysis of the grinding temperature requires detailed knowledge about the grinding energy, heat flux distribution along the grinding zone, the fraction transported as heat to the workpiece, and cooling by the applied fluid. In order to calculate the temperature, it is necessary to specify the energy partition within the workpiece.

One approach to estimate the energy partition within the workpiece is fitting the subsurface temperature respond to analytically calculated values. The energy partition is obtained using temperature matching and inverse heat transfer methods. The temperature response in the workpiece subsurface can be measured during straight surface plunge using an embedded thermocouple. With each grinding pass, the temperature response is measured closer to the surface being ground. Another approach for analyzing the energy partition is the single grain model. This model includes the effects of heat transfer to the abrasive grains, fluid, and work piece by considering a single grain surrounded by fluid interacting with the workpiece. Each single active grain is modeled as a truncated cone moving along the workpiece surface at the wheel speed with all of the grinding energy uniformly dissipated at the grain and workpiece interface area. Cooling by the fluid is then taken into account by considering the temperature at the fluid-workpiece interface within the grinding zone, but the convective coefficient is difficult to calculate. In order to solve the problem, a composite model is analyzed. It assumes that the surface porosity is completely filled with grinding fluid and that the thermal properties of the composite can be approximated by a weighted volumetric average of the thermal properties of the grain and grinding fluid. Some energy would be carried away with the grinding chips, but this is usually negligible in energy partition model.

In order to estimate the energy partition within the workpiece, an integral approximation solution of energy partition is studied to calculate the energy partition within the work piece, wheel, chip and fluid in the grinding process. Heat transfer models of the abrasive grain, workpiece, chips and fluid were analyzed by using the integral approximation method. The present model can calculate the energy partition with and without film boiling in the grinding zone. Experiments were performed with the different grinding conditions. Semi-natural thermocouple is used to measure the temperature field of steel surface layer in grinding. Temperature field of alloy steel is simulated by finite element method. The workpiece background temperature calculated by the present model agreed very well with the experimental results. Energy partition model can be used to calculate the grinding temperature for controlling thermal damage and predicting the harden layer depth in grinding.

2. Thermal principle

The following describes developments achieved over more than twenty years of research. The approach developed by the author and his colleagues has been influenced by many previous workers whose achievements are acknowledged. The most recent revised approach is the result of many refinements and experiments undertaken to achieve generality for all grinding processes.

2.1. Heat flows

In grinding, there are four kinds of heat flows. Heat flows to the workpiece, to the abrasive grains, to the grinding fluid and into the chips. If it assumed that all heat goes into the workpiece, temperatures predicted are much too high. In many cases the workpiece would completely melt. However, if heat flows to the wheel, chips and fluid are subtracted from the total heat, the maximum workpiece temperature can be simply estimated. There is a special case where ignoring heat to the wheel, chips and fluid is reasonable but over-estimates workpiece temperatures. The over-estimate is typically about a third. This special case is for dry shallow-cut grinding steels and cast irons with conventional abrasives at high values of specific energy. However, simple techniques have been developed to take account of the four kinds of heat flows. The approach becomes very general and applies for all grinding situations including low specific energies, creep-feed grinding and HEDG. This has been demonstrated by numerous case studies. Super-abrasives, easy-to-grind materials and deep cuts make the general approach absolutely essential.

2.2. Workpiece heat transfer

An early assumption was that heat was generated at the shear plane. However, the main source of heat in grinding was shown to be the grain-workpiece rubbing surface. In either case, temperatures must be solved using the theory of moving heat sources. A sliding heat source solution applies for shallow-cut grinding. An oblique heat source solution applies for both shallow-cut and deep grinding [Rowe 2001]. Although the oblique heat source solution is a large improvement on the Jaeger sliding heat source and is accurate for shallow cuts, for deep cuts, it slightly over-estimates maximum contact-surface temperatures. A circular arc heat source solution extended the oblique heat source approach [Rowe and Jin 2001]. The circular arc heat source data presented in this chapter has been re-computed and provides better accuracy than the oblique heat source.

2.3. Fluid convection

For shallow grinding, convective cooling occurs mainly outside the contact region. However, it is cooling within the contact region that prevents thermal damage and therefore it is necessary to make this distinction as in the approach outlined below. Much greater fluid cooling takes place inside the grinding contact with deep cuts. This is due to the large contact length in deep grinding. Usually in creep-feed grinding, most of the heat goes to the fluid. The energy which may be extracted is limited by fluid boiling. This was confirmed for shallow grinding. Measurements show that effective cooling techniques can produce very high fluid convection factors within the grinding contact area.

2.4. Chip energy

The energy q_{ch} carried away by the chips is strictly limited but can easily be estimated. The limit is the energy that causes melting. There is also a small amount of kinetic energy that can easily be shown to be negligible. It is known that chips do not usually melt before being detached. For ferrous materials, the maximum energy carried within the chips is approximately 6 J/mm^3 of material removed.

2.5. Heat partitioning

Heat partition is the process of sharing out the four kinds of heat flows to determine the heat into the workpiece.

- **The work partition ratio R_w:** R_w defines the net heat entering the workpiece where R_w = q_w/q. Typically, R_w may be as low as 5% in deep grinding or as high as 75% in conventional grinding.

- **The work-wheel fraction R_{ws}:** Some heat q_{ch} is carried away by the chips. The remaining heat $q - q_{ch}$ is shared between the wheel and the workpiece at the grain contacts. In short, $q - q_{ch} = q_s + q_{wg}$. Initially, heat q_{wg} goes into the workpiece but this heat is larger than the net heat into the workpiece q_w. This is because some heat immediately comes out from the workpiece again into the fluid. Therefore, the net heat flow into the workpiece q_w is less than q_{wg}. In other words: $q_{wg} = q_w + q_f$. The work-wheel fraction is $R_{ws} = q_{wg} / (q_s + q_{wg})$. The work-wheel fraction for conventional abrasives is of the order of 85% and for super-abrasives of the order of 50%.

- **Heat to the wheel:** Heat shared between the workpiece and the wheel yields the heat conducted into the wheel. Two different approaches have been employed: Wheel contact analysis and grain contact analysis.

2.6. Wheel contact analysis

An early technique was later abandoned for practical reasons. The early technique estimated the work-wheel fraction based on the thermal properties and speeds of the wheel and the workpiece using the expression $\sqrt{\beta_s v_s / \beta_w v_w}$ (Marinescu 2004). However, bulk thermal properties for the wheel are required and are not available from published data. This technique was later abandoned in favor of the grain contact analysis.

2.7. Grain contact analysis

A grain contact model allows the work-wheel fraction to be based on grain properties rather than bulk wheel properties. A good case can be made that grain properties are physically more relevant than bulk wheel properties since heat partition takes place at the grains. Initially, the conical grain model by Lavine (Lavine et al 1989) was incorporated within our heat partition approach. At the same time, a plane grain solution had been derived for comparison. (Rowe et al 1991) The plane grain model was found to be more accurate than the conical model (Rowe et al 1997). It was realized later that the steady-state version of our plane grain model only differed in minor detail from a very early steady-state assumption. In most cases, a steady-state model is sufficiently accurate. However, the plane grain model can be readily extended to a more accurate transient solution when required.

2.8. Heat input

Grinding power goes into the contact zone as heat. A negligible proportion accelerates the chips and a very small proportion is locked into the deformed material. Power per unit area is known as heat flux q. The heat is divided by the real contact length and the width of the grinding contact. $q = P/l_c.b_w$

- **Flash heating:** Heat enters the grinding contact in short bursts of intensive energy leading to flash temperatures. The flash temperatures occur in the extremely short time it takes for a grain to pass a point on the workpiece. A point on the workpiece has contact with an individual grain for approximately 1 micro-second. The heat enters the contact in a near-adiabatic process.
- **Grain heating:** A grain is heated at the grain-workpiece contact for much longer than a point on the workpiece. A grain typically moves across the whole contact length in 100 micro-seconds. The grain therefore experiences a heat pulse for a period approximately 100 times longer than a point on the workpiece. It can be shown that this allows the surface of the grain to reach quasi steady-state temperatures. The maximum grain temperature is close to the workpiece melting temperature.
- **Background heating:** Numerous flash contacts gradually heat up the whole workpiece contact area. It is usual therefore to make a distinction between flash temperatures at a grain contact and background temperatures over the whole contact area. The overall duration of energy pulses in the contact area that provides the background temperatures is of the order of 10,000 micro-seconds. This is the time it takes the wheel to move through the contact length. Many energy pulses lead to background temperature rise at depths up to and often exceeding1 mm.

3. Energy partition

An integral approximation solution of energy partition is studied in the grinding process. Flux into the chips was estimated from the limiting chip energy to raise the chip material close to melting.

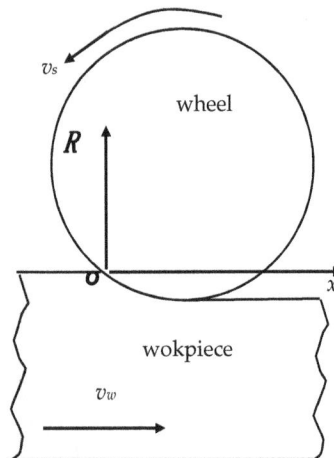

Figure 1. Coordinate system of grinding

The physical model and the coordinate system for a typical grinding wheel and workpiece are shown in Fig. 1. The total heat flux on the contact area of wheel and workpiece is qn(x). Therefore,

$$q_n(x) = q_{wh}(x) + q_{wb}(x) + q_f(x) + q_{ch}(x) \tag{1}$$

Where, $q_{wh}(x)$, $q_{wb}(x)$, $q_f(x)$ and $q_{ch}(x)$ are the heat fluxes into the grinding wheel, workpiece, fluid and chips, respectively as shown in Fig.2. Flux into the chips was estimated from the limiting chip energy to raise the chip material close to melting. Taking the mean value of specific heat from ambient temperature to melting point (about 1500°C), the partial limiting chip energy, e_{ch}, flows into workpiece material. Therefore

$$q_{ch} = e_{ch} a v_w / l_c \tag{2}$$

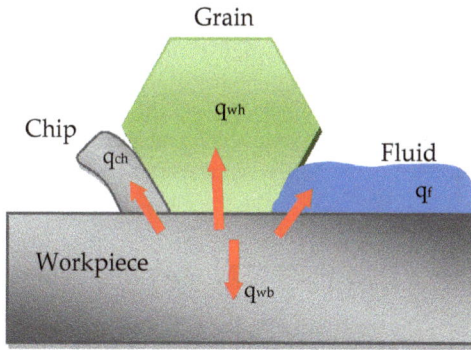

Figure 2. Heat flows to the workpiece, the grain, the chip and the fluid

Many grains of the wheel surface cut into the workpiece with a very high speed and a lot of heat is generated in the grain-workpiece interface. The heat generated at the grain-workpiece interface, ql(x), is assumed to be uniform in the present study, and will conduct into the workpiece and grain. Therefore

$$q_l(x) = q_g(x) + q_{wg}(x) \tag{3}$$

Where, $q_g(x)$ and $q_{wg}(x)$ are the heat fluxes into the grain and workpiece. The ratio between real contact area A_r of grain-workpiece and geometric contact area A_n of wheel-workpiece is A, then

$$q_l(x) = q_n(x) \frac{A_n}{A_r} = \frac{q_n(x)}{A} \tag{4}$$

The heat flux into the workpiece qwg(x) will be divided into three parts. First part remains in the workpiece, second part conducts into fluid, and third part is removed by chips, then

$$q_{wg}(x)A_r = q_{wb}(x)A_n + q_f(x)(A_n - A_r) + q_{ch}(x)A_n \tag{5}$$

For Ar<<An, from formula (3) and (5), then

$$q_l(x)A = q_g(x)A + q_{wb}(x) + q_f(x) + q_{ch}(x) \tag{6}$$

It is assumed that the grinding fluid fills the space around the abrasive grain to a depth that is larger than the thickness of the thermal boundary layer. Therefore, the grinding fluid flow over the workpiece surface can be considered as slug flow with grinding velocity vs. Finally, the grinding fluid can be modeled as a moving semi-infinite solid with a variable heat flux q_f (x) at its surface. The temperature, T_f, of the grinding fluid can be obtained by using an integral approximation solution,

$$T_f = \left[\frac{4}{3(K\rho C)_f (v_s \pm v_w) q_f(x)} \int_0^x q_f(x)dx \right]^{\frac{1}{2}} q_f(x) \tag{7}$$

Since the workpiece moves with table velocity, vw, the heat transfer into the workpiece background can be considered as a heat conduction problem in a moving semi-infinite solid. Therefore, the background temperature, Twb, of the workpiece can be expressed as

$$T_{wb} = \left[\frac{4}{3(K\rho C)_{wb} v_w q_{wb}(x)} \int_0^x q_{wb}(x)dx \right]^{\frac{1}{2}} q_{wb}(x) \tag{8}$$

The temperature distribution in the grain is assumed to be a cubic polynomial function. The grain temperature, Tg, at the grain-workpiece surface can be expressed as

$$T_g = \frac{r_0}{K_g} \frac{\eta}{\eta + 3} q_g(x) \tag{9}$$

An individual grain is modelled as a band heat source with width lg, causing a heat flux qwg(x) into workpiece surface. The average workpiece surface temperature, Twg, underneath a grain due to an individual heat source can be expressed as

$$T_{wg} = \frac{4}{3} \left[\frac{l_g}{\pi (K\rho C)_{wb} (v_s \pm v_w)} \right]^{\frac{1}{2}} q_{wg}(x) \tag{10}$$

In grinding arc area, the energy partition must satisfy the compatibility requirement that the temperature on the workpiece surface equals the temperature on the grain and fluid surface everywhere along the grinding zone. Therefore,

$$T_f = T_{wb} \tag{11}$$

$$T_g = T_{wb} + T_{wg} \tag{12}$$

From the formula (1) to (12), energy partition of chips, work piece, fluid and wheel can be expressed as

$$R_{ch} = \frac{e_{ch} a v_w b}{F_t v_s} \tag{13}$$

$$R_{wb} = \frac{N}{(1+U)(G+N)+WA}(1-R_{ch}) \tag{14}$$

$$R_f = \frac{q_f}{q_n} = \frac{NU}{(1+U)(G+N)+WA}(1-R_{ch}) \tag{15}$$

$$R_{wh} = \frac{q_{wh}}{q_n} = \frac{G(1+U)+WA}{(1+U)(G+N)+WA}(1-R_{ch}) \tag{16}$$

And

$$N = \frac{l_g}{\sqrt{\pi K_g}}\frac{\eta}{\eta+3} \tag{17}$$

$$W = \left[\frac{4}{3(K\rho C)_{wb} v_w q_{wb}(x)} \int_0^x q_{wb}(x)dx \right]^{\frac{1}{2}} \tag{18}$$

$$U = \left[\frac{(K\rho C)_f (v_s \pm v_w)}{(K\rho C)_{wb} v_w} \right]^{\frac{1}{2}} \tag{19}$$

$$G = \frac{2}{3}\left[\frac{4l_g}{\pi(K\rho C)_{wb}(v_s \pm v_w)} \right]^{\frac{1}{2}} \tag{20}$$

4. Numerical simulation

4.1. Finite element model

The finite element model proposed here is based on Jaeger's model. Grinding heat is entering the workpiece on the top surface of the work piece, in the form of heat flux q, input that moves along this surface. All the other sides of the work piece are considered to be adiabatic, and so no heat exchange takes place in these sides. The model has a length of

26mm and a height of 10mm, sufficient enough for the temperature fields to be fully deployed and observed in full length. A mesh method is applied on the proposed model, consisting of 2400 4-noded quadrilateral elements and 1501 nodes, as shown in Fig. 3.

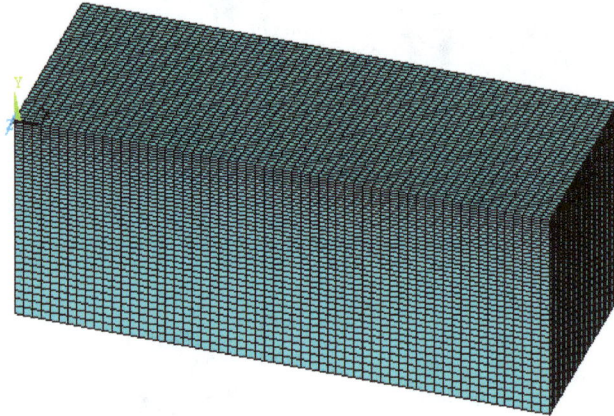

Figure 3. The meshed workpiece model

The mesh grid is denser towards the grinding surface, which is the thermally loaded surface, and, thus the most affected zone of the workpiece, allowing for greater accuracy to be obtained.

4.2. Grinding contact length

For the proposed model, except the work conditions and the material prosperities, contact length need to be calculated in order to determine the temperatures. The geometrical contact length between the workpiece and the grinding wheel is equal to

$$l_c^2 = l_f^2 + l_g^2 \tag{21}$$

Where l_f is the contact length between surfaces acted on by a normal force and l_g is the geometric contact length defined by Equation (21). The length l_f is evaluated from:

$$l_f = \sqrt{8F_n'(Ks + Kw)d_s} \tag{22}$$

Where F_n' is the specific normal force, and

$$Ks = \frac{1 - v_s^2}{\pi E_s} \tag{23}$$

$$Kw = \frac{1 - v_w^2}{\pi E_w} \quad (24)$$

Variables Ks and Kw are determined from the physical properties of materials in the contact. v_s and v_w are Poisson's ratios, E_s and E_w are the Young's Modulus. The real contact length can be expressed using a surface roughness approach or a contact area approach. The first yields more faithful results in comparison with the experimental results. Based on the roughness approach, the magnitude of the grinding contact length is represented as :

$$l_c = \sqrt{(l_{fr})^2 + l_g^2} = \sqrt{(R_r l_f)^2 + l_g^2} \quad (25)$$

Where l_{fr} is contact length for rough surfaces with normal force and R_r is roughness factor. The magnitudes of the roughness factor are acquired as experimental values from the tests. R_r is sensitive to the grinding conditions for some material combinations. For general analysis of the grinding conditions, where measured values of the roughness factor are not available, it is suggested that the value R_r is equal 8.

Combining equations (21), (22) and (25) yields the relationship

$$l_c = \sqrt{8R_r^2 F_n (Ks + Kw)d_s + ad_s} \quad (26)$$

Where a is the depth of cut and ds the diameter of the grinding wheel. Equation (26) determines the contact length between the wheel body and workpiece taking account not only elastic deflection and geometric effect but also roughness of both surfaces in the contact.

4.3. Heat flux profile

The heat flux q, that is distributed over the contact length can be calculated as

$$q = R_w \frac{F_t v_s}{b l_c} s(x) \quad (27)$$

where F_t is the tangential force, v_s is the wheel speed, b is the grinding width, and R_w is the heat partition of the energy entering the work piece. This kind of modelling is suitable for a grinding process with a very small depth of cut, since there is no modelling of the chip. The initial temperature of the workpiece is considered to be surrounding temperature.

As shown in Figure 4, the wheel speed is v_s and the table speed is v_w. When the grits begin to contact the workpiece, the vertical force and elastic deformation of the workpiece is small. The grits only rub the workpiece surface. As the table is going on, the vertical force becomes larger, and the grits are pressed into the workpiece surface. Because of material plastic deformation, the bits become and the grits plough the surface. When the depth of the grits into surface and the vertical force become much larger, the chips are cut from the surface. So there are three separate grain action: rubbing; ploughing; and cutting. At the stages of rubbing and ploughing, friction between grits and workpiece surface is the main action, and the rectangle heat flux profile is generated. At the stage of cutting, the cutting depth and the vertical force become lager and lager, which forms triangle heat flux profile. So the heat flux model should include rectangle and triangle heat flux profiles.

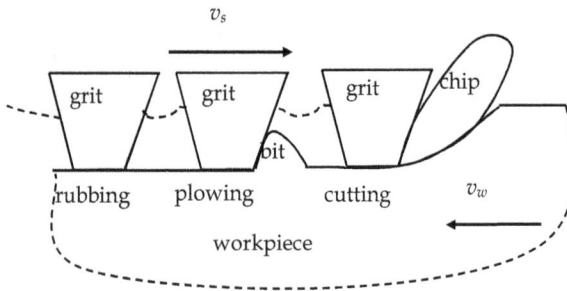

Figure 4. Grinding processes of grains

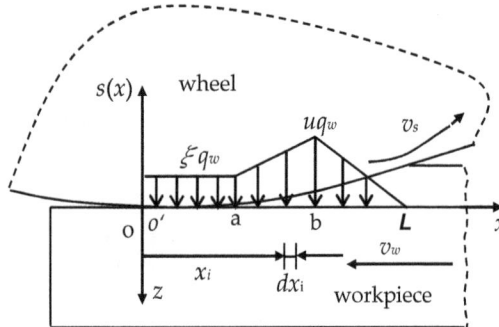

Figure 5. Heat flux profile

The coordinate system $xos(x)$ is fixed to the grinding zone with its origin o at the grinding entry point of workpiece, shown in Figure 5. The heat flux value into workpiece ξq_w is in rectangle heat flux profile and at the apex of triangle heat flux profile $u q_w$. The mean heat flux value q_w transfers into workpiece. The parameter ξ and u are non-dimensional parameters. The alphabet o$-$a is rectangle heat flux profile and the length is a. The alphabet a$-$L is triangle heat flux profile. The alphabet o$-$L is the grinding zone and the length is l_c. The alphabet b is the location of triangle heat flux apex. The length between o$-$b is b. The function $s(x)$ is the grinding heat flux distribution shape function in the grinding arc and can be expressed as

$$s(x) = \begin{cases} 0 & x \in (-\infty, 0) \\ \xi & x \in (0, a) \\ \dfrac{\xi(b-x) + u(x-a)}{b-a} & x \in (a, b) \\ \dfrac{u(x-L)}{b-L} & x \in (b, L) \\ 0 & x \in (L, +\infty) \end{cases} \tag{28}$$

The net heat flux per unit grinding width into the workpiece could be expressed as $q_w(x) = q_w$ $s(x)$. The alphabet a; b; l_c; ξ and u is five important parameters. The contact length l_c can be calculated by wheel dimension and cutting depth. If the workpiece is harder, cutting depth is larger, table speed is higher, wheel speed is lower, and grain is bigger and sharper, cutting in the grinding process is main stage and rubbing and ploughing can be neglected. Then $a=0$ and $\xi=0$, rectriangle heat flux model becomes triangle heat flux model. If the workpiece is softer, cutting depth is smaller, table speed is lower, wheel speed is higher, and grain is smaller and duller, for example polishing, rubbing and ploughing is main stage and cutting can be neglected. Then $a=l_c$ and $\xi=1$, rectriangle heat flux model becomes rectangle heat flux model. The grinding fluid can reduce rubbing stage. So the alphabet a and ξ are related to workpiece hardness; cutting depth; table speed; wheel speed; grain and grinding fluid. Besides the above factors, the parameter b is related to grinding operations, for example up grinding and down grinding.

4.4. Thermal transfer model

The first law of thermodynamics states that thermal energy is conserved. It will be assumed that all effects are in the global Cartesian system. Specializing this to a equation as

$$\rho c \left(\frac{\partial T}{\partial t} + v_x \frac{\partial T}{\partial x} + v_y \frac{\partial T}{\partial y} + v_z \frac{\partial T}{\partial z} \right) = \bar{q} +$$
$$\frac{\partial}{\partial x}\left(K_x \frac{\partial T}{\partial x} \right) + \frac{\partial}{\partial y}\left(K_x \frac{\partial T}{\partial y} \right) + \frac{\partial}{\partial z}\left(K_x \frac{\partial T}{\partial z} \right) \tag{29}$$

Where ρ-density; c-specific heat; T-temperature; t-time; K-conductivity in the element x, y, and z directions respectively; $\bar{q}$ -heat generation rate per unit volume.

Three types of boundary conditions are considered. Specified temperatures acting over surface, specified heat flows acting over surface, and specified convection surfaces acting over surface.

Dry grinding is needed in order to utilize the grinding heat in grind-hardening technology. The grinding is assumed to be quasi-stationary. The coordination system $xo'z$ is fixed to the workpiece and its origin o' is coincident with o. The width of one line heat source is defined as dx_i, and then the heat flux value of line heat source is $q_w\, s(x_i)dx_i$. Thus the equation of the induced temperature field in the workpiece can be written as

$$T = \frac{q_w}{\pi\kappa}\int_0^{l_c}\exp\left[-\frac{(x-x_i)v_w}{2\alpha}\right]K_0\left[\frac{v_w}{2\alpha}\sqrt{(x-x_i)^2+z^2}\right]s(x_i)dx_i \qquad (30)$$

where κ is thermal conductivity of the workpiece, α is thermal diffusivity of the workpiece, K_0 is the second order modified Bessel function.

5. Experiment set up and method

5.1. Grinding conditions

The experiment is performed on a M7120A surface grinder, using a vitrified aluminum oxide wheel (WA60L6V). The workpiece material is alloy steel 40Cr. Its chemical composition and thermal properties are listed in Table 1 and Table 2.

Material	Chemical composition of the material [%]					
	C	Si	Mn	P	S	Cr
40Cr	0.41	0.28	0.61	0.019	0.01	1.02

Table 1. Workmaterial and its chemical composition

Density	Thermal conductivity	Specific heat capacity
ρ [kg/m³]	k [W/m·ºC]	C [J/kg·ºC]
7850	Figure 6 (a)	Figure 6 (b)

Table 2. Thermal properties of workmaterial

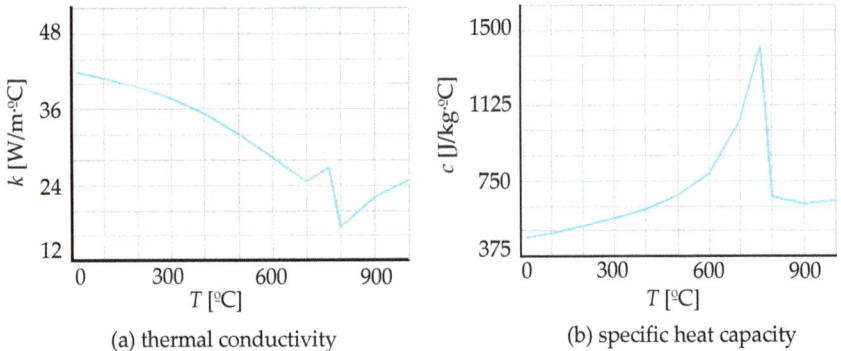

(a) thermal conductivity　　　(b) specific heat capacity

Figure 6. Thermal properties of 40Cr steel

Grinding conditions are listed in Table 3. The work piece is fixed in a clamp in advance and is then fixed on the dynamometer. The single-point diamond dresser is used to dress the grinding wheel.

Grinding conditions	Parameters
Wheel Diameter	250[mm]
Wheel speed	35[m/s]
Table speed	0.01, 0.03 and 0.05[m/s]
Depth of cut	0.3[mm]
Coolant	Dry
Grinding operation	Down-grinding

Table 3. Grinding conditions of surface grinding

5.2. Measurement methods

A three-component dynamometer of type YDXM-III97 is used to measure the grinding forces. The dynamometer is connected to a charge amplifier of type JY5002, and the electric potential signals are converted into digital signals by an analogue-to-digital transition board of type NI 6024E, then digital signals are transferred to computer, as shown in Figure 7.

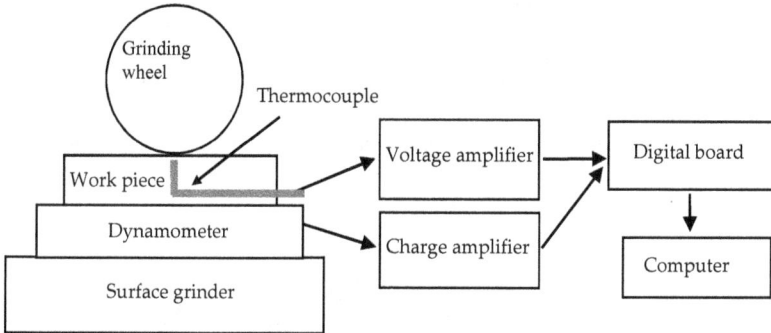

Figure 7. Schematic diagram of grind-hardening experiment

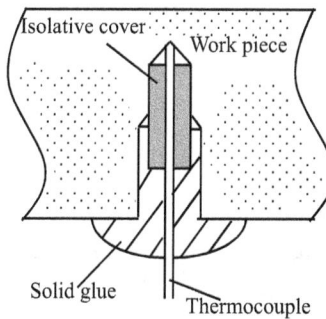

Figure 8. Schematic diagram of the thermocouple

A traditional static temperature calibration method is used in the experiment. The terminal-strand thermocouple comprising a constantan wire and an iron wire is used to measure the temperature. The standard thermocouple and the one waiting for calibration are put in a tube furnace. The thermo-electromotive force between the standard thermocouple and the other one is measured by a potentiometer. By means of the measured temperature of the standard thermocouple the temperature in the furnace is found. Using the measured thermo-electromotive force as a reference, the thermo-electricity curve of the thermo-couple that needs to be calibrated can be described.

The thermocouple comprises a constantan wire and a threadlike chip. The constantan wire is fixed to the isolative cover and the isolative cover on the thermocouple is fixed in the mouth of the blind hole in order to avoid movement of the constantan wire. The Schematic diagram of the thermocouple is shown in Figure 8.

A traditional static temperature calibration method is used before experiment. The standard thermocouple and the one waiting for calibration are put in a tube furnace. A potentiometer measures the thermo-electromotive force between the standard thermocouple and the calibrating one. By means of the measured temperature of the standard thermocouple, the temperature in the furnace is found. Using the measured thermo-electromotive force as a reference, the thermal-electricity curve of the thermocouple need to be calibrated is described. Figure 9 is the calibration method of the thermocouple. Two terminals on the thermocouple wires are connected to the amplifier of type NI SCXI-1102, and the electric potential signals are converted into digital signals by an analogue to digital transition board of type NI 6024E. A computer records the grinding force and temperature results.

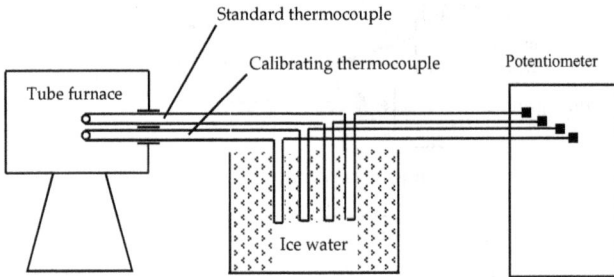

Figure 9. Calibration method of the thermocouple

6. Result and discussion

6.1. Grinding temperature field

The whole temperature field in the grinding process is shown in figure 10. The temperature gradient of work piece surface layer in grind-hardening is similar to the one in high-frequency quenching.

The comparisons between simulated temperature history obtained from the finite element model and experimental measurement reveal good agreement, as shown in Figure 11 (a), (b) and (c). The cooling rates derived from the simulated temperature history are higher than the critical cooling rate of the 40Cr steel phase transformation.

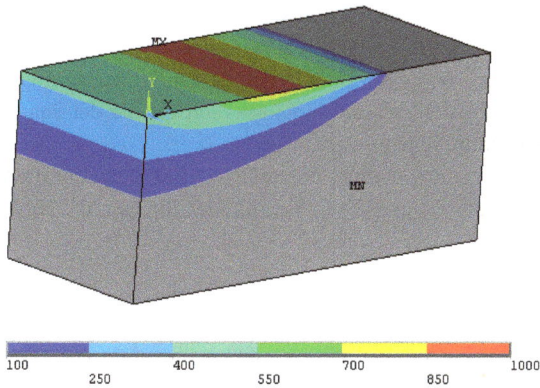

Figure 10. Grinding temperature field of workpiece

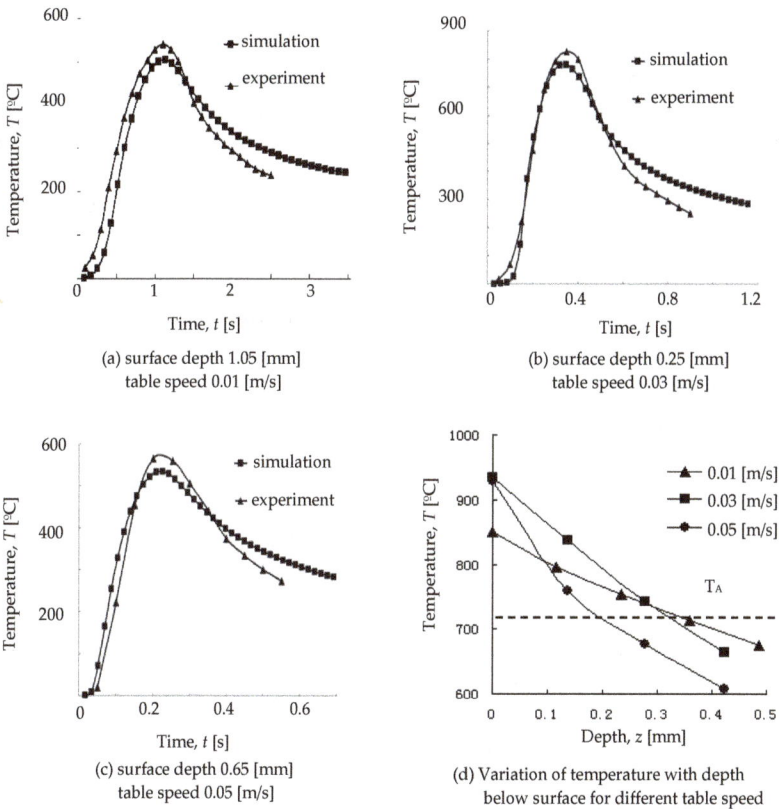

(a) surface depth 1.05 [mm]
table speed 0.01 [m/s]

(b) surface depth 0.25 [mm]
table speed 0.03 [m/s]

(c) surface depth 0.65 [mm]
table speed 0.05 [m/s]

(d) Variation of temperature with depth
below surface for different table speed

Figure 11. Comparison between simulation and experiment

The variation of the temperature in the work piece with the depth below the surface of the 40Cr steel is presented in Figure 11 (d) for table speed 0.01 m/s, 0.03 m/s and 0.05 m/s. In the figure, the critical temperature TA of austenitic transformation is indicated. Theoretically, the depth of austenitic transformation can be determined by this critical temperature. The maximum temperature of the work piece surface and the thermal gradient with the depth is down when the table speed is slow, but the hardness depth is deeper. When table speed changes from 0.05 m/s to 0.01 m/s, the hardness depth varies from 0.2 mm to 0.36 mm.

6.2. Energy partition

The temperature history underneath the ground surface of the workpiece was accurately measured by using the semi-natural thermocouple method. The theoretical values of temperature history calculated by using FEM and the experimental values of temperature history are compared in Fig. 12.

The experimental values are in qualitative agreement with the theoretical ones. An example is presented to illustrate energy partition variation along the grinding zone. The calculated energy partition distribution is shown in Fig. 13 under grinding conditions. It can be seen that the energy partition distribution deviates significantly from the traditional constant energy partition value. The energy partition within workpiece decreases from the leading edge to the trailing edge of the grinding zone, however energy partition within grinding wheel increases.

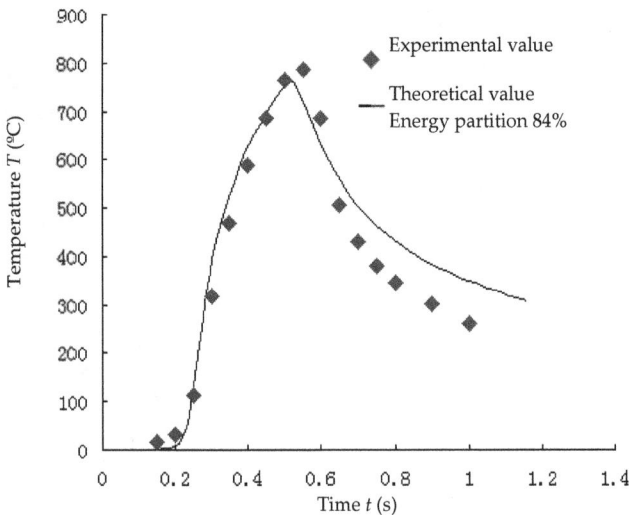

Figure 12. Compare theoretical value with experimental value of temperature history

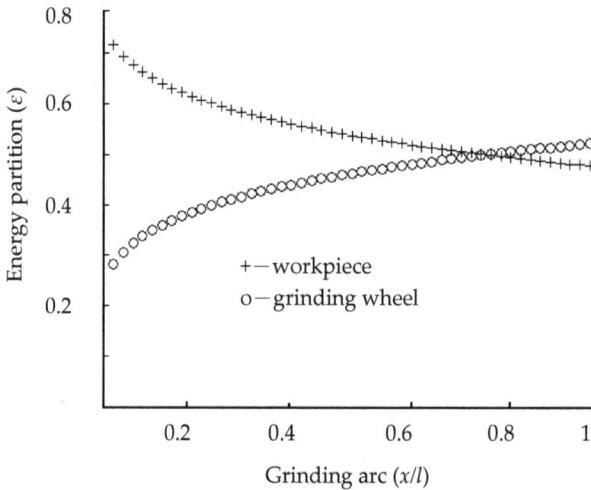

Figure 13. Energy partition within workpiece and wheel along the grinding arc zone

7. Conclusion

The finite element method was used to calculate the temperature field, maximal surface temperature and temperature history development in grind-hardening. The experiment and measurements were performed with aluminium oxide grinding wheel in different grinding conditions by using the semi-natural thermocouple method.

The austenitic and martensitic transformation layer depth of the workpiece can be predicted by the temperature history development calculated from the finite element model, the critical temperature of austenitic transformation and the critical cooling rate of martensitic transformation.

The maximum surface temperature and the thermal gradient values are lower when the table speed is slower, but the hardness depth is deeper. When table speed varies from 0.05 m/s to 0.01 m/s, the hardness depth varies from 0.2 mm to 0.36 mm.

Author details

Lei Zhang
School of Mechanical Engineering, Shandong University, P.R. China

Acknowledgement

The authors wish to thank the National Science Foundation of NSFC for financial support to the project.

8. References

Lei Zhang, Peiqi Ge and et al. 2011. Experiment and Simulation on Residual Stress of Surface
    Hardened Layer in Grind-Hardening. *Solid State Phenomena*. Vol. 175, pp. 166-170.
Lei Zhang and Jianlong Zhang. 2011. Temperature Measurement and Simulation in Surface
    Grinding. *IEEE 3rd International Conference on Computer and Network Technology*. Vols.2,
    pp. 589-592.
Zhang, L. and Ge, P.Q., 2004. Temperature field analysis and phase transformation research
    in grind-hardening. *Tool Engineering*, 38, pp. 15-19.
Brinksmeier, E. and Brockhoff, T., 1996. Utilization of grinding heat as a new heat treatment
    process. *Annals of the CIRP.*, 45, pp. 283-286.
Brockhoff, T., 1999. Grind-hardening: A comprehensive view. *Annals of the CIRP.*, 48, pp.
    255-260.
Zarudi, I. and Zhang, L.C., 2002. A revisit to some wheel-workpiece interaction problems in
    surface grinding. *J. Mach. Tools. Manu.*, 42, pp. 905-913
Zarudi, I. and Zhang, L.C., 2002. Mechanical property improvement of quenchable steel by
    grinding. *J. Mat. Sci.*, 37, pp. 3935-3943
Zarudi, I. and Zhang, L.C., 2002. Modeling the structure changes in quenchable steel
    subjected to grinding. *J. Mat. Sci.*, 37, pp. 4333-4341
Wang, G.C. and Liu, J.D., 2005. Study on forming mechanism of surface hardening in two-
    pass grinding 40Cr steel. *Key. Eng. Mat.*, 304-305,pp. 588-592
Liu, J.D. and Wang, G.C., 2004. Effect of original structure on the grind-hardened layer of
    40Cr steel. *Heat Treatment of Metals*, 38, pp. 61-65
Rowe, W.B. and Jin, T., 2001. Temperature in high efficiency deep grinding. *Annals of the
    CIRP.*, 50, pp. 205-208.
Ioan D. Marinescu, W. Brian Rowe, Boris Dimitrov and Ichiro Inasaki, 2004. Tribaology of
    Abrasive Machining Processes. *William Andrew publishing*.
Ramesh, M.V., Seetharamu, K.N. and Ganesan, N., 1999. Finite element modelling of heat
    transfer analysis in machining of isotropic materials. *Int. J. Heat. Mat. Trans.*, 42, pp.
    1569-1583.
Lavine, A.S., 2000. An exact solution for surface temperature in down grinding. *Int. J. Heat.
    Mat. Trans.*, 43, pp. 4447-4456.
Zhang, L. and Ge, P.Q., 2004. New heat flux model in surface grinding. *Mat. Sci. Forum.*, 471-
    472, pp. 298-301.
L. Zhang, P.Q. Ge, and et al. 2007. Experimental and simulation studies on temperature field
    of 40Cr steel surface layer in grind-hardening, *International Journal of Abrasive
    Technology*, Vol. 1, pp. 187-197
Lei Zhang, Wenbo Bi and et al. 2010. An Approximate Solution of Energy Partition in Grind-
    hardening Process. *Advanced Materials Research*. Vol. 135, pp. 298-302
Lei Zhang, and et al. 2010. Analysis of Grinding Parameters on Hardness Layer Depth.
    Applied Mechanics and Materials. Vols. 37-38, pp. 131-134

Developing 1-Dimensional Transient Heat Transfer Axi-Symmetric MM to Predict the Hardness, Determination LHP and to Study the Effect of Radius on E-LHP of Industrial Quenched Steel Bar

Abdlmanam S. A. Elmaryami and Badrul Omar

Additional information is available at the end of the chapter

1. Introduction

A study of the constitution and structure of all irons and steel must first start with the iron-carbon equilibrium diagram and the steel part of the phase diagram as shown in Figure 1.

Iron: is a general word used to describe metals that have pure iron as their main constituent.

Most iron wares around us are not made of chemically pure iron but are alloys, the most important of which is Carbon. Carbon is a major factor in understanding the difference between Iron, Steel and Cast iron. Adding some carbon to chemically pure iron makes steel. If the quantity of carbon is increased, cast iron will be obtained.

Steel: is the most commonly and widely used metallic material in today's society as shown in Figure 2. It is one of the most signification engineering material and can be classified plain carbon steel and alloy steel.

Steel is an alloy of iron and carbon. The amount of carbon dictates whether the steel is hard or tough. Adding Carbon makes the iron harder. Steels can be hardened by heat treatment.

Carbon in steel may be present up to 2.03 percent. Steels with carbon content from 0.025 percent to 0.8 percent are called hypo-eutectoid steel. Steel with a carbon content of 0.8 percent is known as eutectoid steel. Steels with carbon content greater than 0.8 percent are called hyper-eutectoid steel. There are three major categories of steel which are as follows:

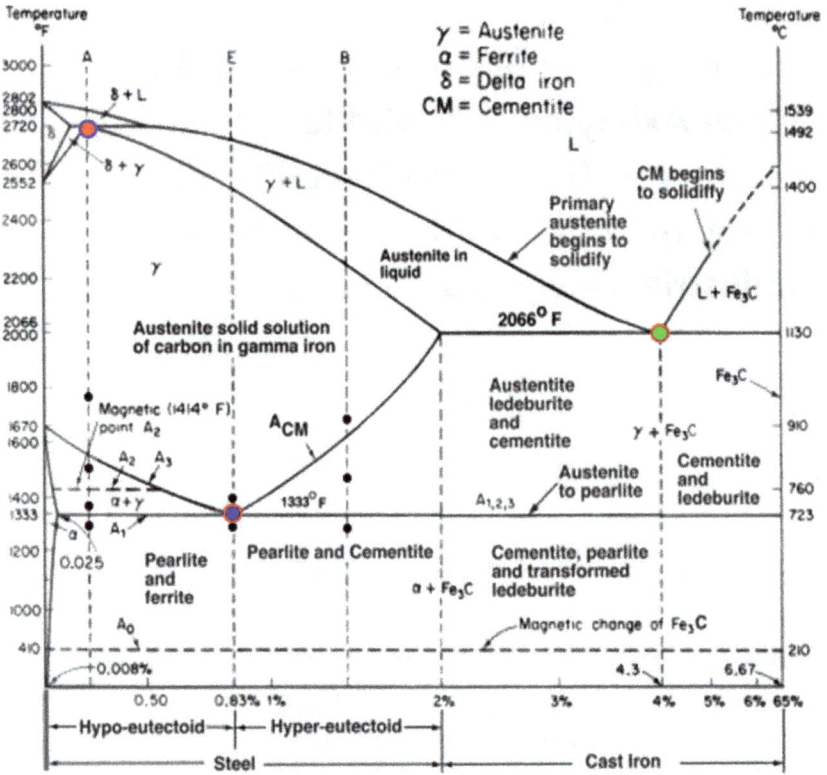

Figure 1. Fe-Fe₃C Phase Diagram, [23, 24].
Many of the basic features of this system influence the behaviour of even the most complex alloy steels. The iron-carbon diagram provides a valuable foundation on which to build knowledge of both plain carbon and alloy steels.
⬤ Peritectic point, ⬤ Eutectoid point, and ⬤ Eutectic point,
(i) Peritectic reaction equation may be written as:
Delta (δ) + Liquid (L) -->Austenite
(ii) Eutectoid reaction equation may be written as:
Solid (γ) --> Ferrite + Fe3C (Cementite) (Heating Eutectoid mixture (Pearlite))
(iii) Eutectic reaction equation may be written aS:
Liquid (L) --> Austenite + Cementite (Eutectic mixture (Ledeburite))

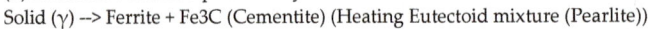

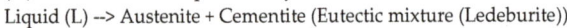

i.   Low carbon steels (carbon up to 0.3 percent)
ii. Medium carbon steels (carbon from 0.3 to 0.7 percent)
iii. High carbon steel (carbon more than 0.7 percent).

Alloy Steels: are basically carbon steels with certain chemical elements added to improve the properties of the metal for specific applications or end products. Alloying elements include carbon, copper, sulphur, manganese, phosphorus, nickel, molybdenum, boron and

Developing 1-Dimensional Transient Heat Transfer Axi-Symmetric MM to Predict the Hardness, Determination LHP and to Study the Effect of Radius on E-LHP of Industrial Quenched Steel Bar

155

chromium, and the resulting material is called an alloy steel as in the sample AISI-SAE 8650H, which will be studied in detail here. The 1-D mathematical model was developed to:

i.   Predict the hardness.
ii. Determine E-LHP.
iii. Study the effect of radius on E-LHP.

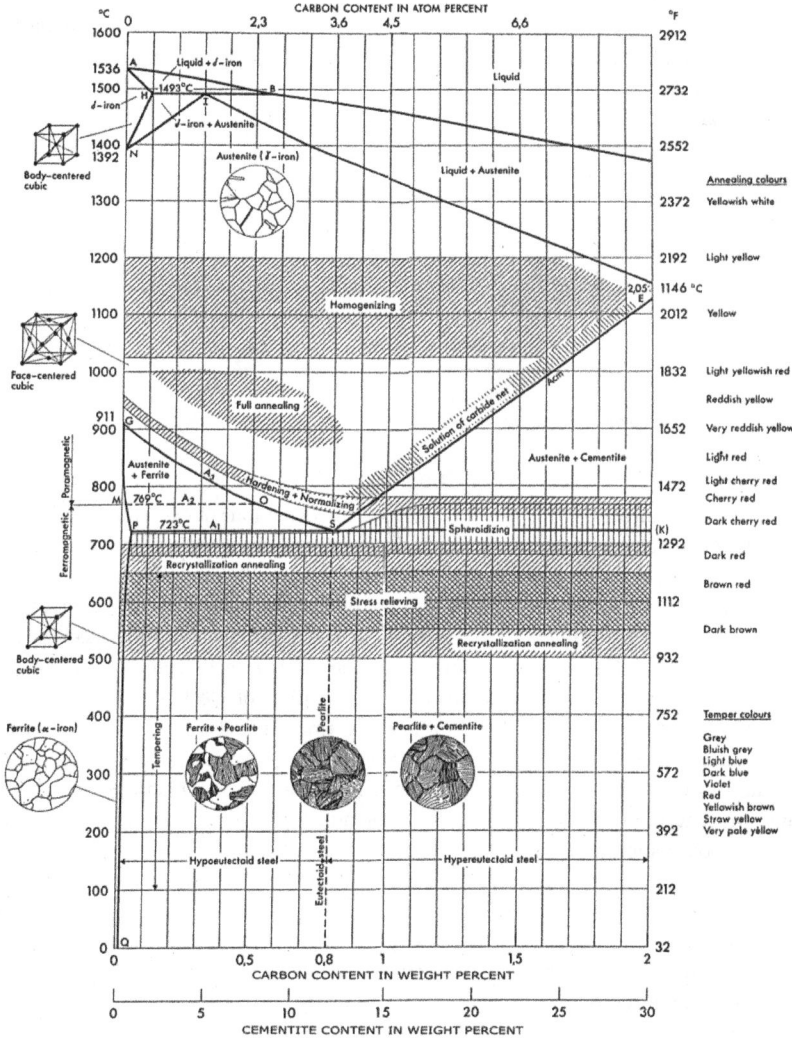

Figure 2. Fe-Fe3C Phase Diagram [Steel Part], [23, 25].

TTT Diagram: T (Time) T (Temperature) T (Transformation) diagram is a plot of temperature versus the logarithm of time for a steel alloy of definite composition as shown in Figure 3. It is used to determine when transformations begin and end for an isothermal (constant temperature) heat treatment of a previously austenitized alloy. TTT diagram indicates when a specific transformation starts and ends and it also shows what percentage of transformation of austenite at a particular temperature is achieved. In Figure 3 the cooling rates A and B indicate two rapid cooling processes. In this case curve A (water cooled) will cause a higher hardness than the cooling rate B (sea water cooled). The end product of both cooling rates will be martensite. Cooling rate B is also known as the Critical Cooling Rate, which is represented by a cooling curve that is tangent to the nose of the TTT diagram. Critical Cooling Rate is defined as the lowest cooling rate which produces 100% Martensite while minimizing the internal stresses and distortions.

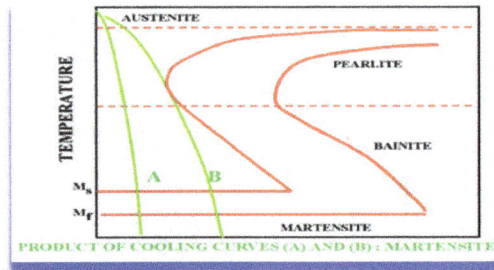

Figure 3. Rapid Quench, [23].

Many of the basic features of this system influence the behaviour of even the most complex alloy steels. For example, the phases found in the simple binary Fe-C system persist in complex steels, but it is necessary to examine the effects of alloying elements have on the formation and properties of these phases. The iron-carbon diagram provides a valuable foundation on which to build knowledge of both plain carbon and alloy steels in their immense variety [1]. Quenching is physically one of the most complex processes in engineering, and very difficult to understand. Simulation of steel quenching is thus a complex problem [2, 9-23]. Quenching is a heat treatment usually employed in industrial processes in order to control mechanical properties of steels such as toughness and hardness [9-23]. The process consists of raising the steel temperature above a certain critical value, holding it at that temperature for a specified time and then rapidly cooling it in a suitable medium to room temperature [9-23]. The resulting microstructures formed from quenching (ferrite, cementite, pearlite, upper bainite, lower bainite and martensite) depend on cooling rate and on chemical composition of the steel [3, 9-23]. Quenching of steels is a multi-physics process involving a complicated pattern of couplings among heat transfer, because of the complexity, coupled (thermal-mechanical-metallurgical) theory and non-linear nature of the problem, no analytical solution exists; however, numerical solution is possible by finite difference method, finite volume method, and the most popular one - finite element method (FEM) which will be used in this study [9-23]. A larger diameter rod quenched in a particular medium will obviously cool more slowly than a smaller diameter rod given with a similar heat treatment [4].

It will be more important to know the lowest hardness point E-LHP (node E) once the radius of the quenched steel bar increases, in other words the E-LHP will be lower than the hardness on the surface (node 4), thus the radius of the bar will be inversely proportional to the hardness at E-LHP. During the quenching process of the steel bar, the heat transfer is in an unsteady state as there is a variation of temperature with time [9-23]. The heat transfer analysis in this study will be carried out in 3-dimensions. The three dimensional analysis will be reduced into a 1-dimensional axisymmetric analysis to save cost and computing time [9-23]. This is achievable because in 1-dimensional axisymmetric conditions, there is no temperature variation in theta (Θ) and (z)-direction for 1-D as shown in Figure 4 (a)(b) and Figure 5, the temperature deviations is only in (r)-direction [9-23]. The Galerkin weighted residual technique will be used to derive the mathematical model to predict the hardness at any point (node) of the heat treated quenched steel bar. Therefore [E-LHP] can be calculated where it's exactly at half the length at the centre of the bar as shown in Figure 1. Experimentally, measurement of E-LHP is an almost impossible task using manual calculation techniques and furthermore the earlier methods only used hardness calculated at the surface (node 5) as shown in Figure 4 (a)(b) and Figure 5. This surface hardness is higher than E-LHP and this has negative consequences which can lead to the deformation and failure. In this chapter, 1-D line (radius) element will be used.

Quenching of steels in general has been and continues to be an important commercial manufacturing process for steel components. It is a commonly used heat treatment process employed to control the mechanical properties of steels. In this chapter hardness in specimen points was calculated by the conversion of calculated characteristic cooling time for phase transformation $t_{8/5}$ to hardness [5-8, 9-23]. Temperature histories must be performed to obtain more accurate transformation kinetics; an adequate tool has been produced for investigating the impact of process history on metallurgy and material properties. Mathematical modelling of axisymmetric industrial heat treated quenched steel bar on the finite element method has been developed to predict temperature history then the hardness at any point (node) even inside the bar can be determined and also the effect of the radius on the temperature history subsequently on LHP can be studied. The temperature history needs to be properly understood in order to efficiently produce high quality components.

It is clear that the first point (node) will be completely cooled after quenching (surface node) because it is located on the surface in contact with the cooling medium, then the other points (nodes) on the radial axis to the centre accordingly will be cooled and the last point will be completely cooled after quenching (centre node)[15, 34].

Thus the maximum hardness will be on the surface node subjected to fast cooling, then the hardness will decrease from the surface node on the radial axis to the centre node of the quenched steel bar. Consequently the lowest hardness point of the quenched steel bar will be detected at the centre node [15, 34].

The lowest hardness point (E-LHP) should be expected inside the heat treated quenched steel bar at the half of the length of the bar centre (centre node). To prove this experimentally is an almost impossible task using manual calculation techniques.

It will be more important to know the E-LHP (centre node) when the radius of the quenched steel bar increases because the lowest hardness point will be lower than the hardness on the surface (surface node). This means that increasing the radius of the bar is inversely proportional to LHP (centre node), while the hardness at the surface (surface node) will be the same [15, 34].

No published information are available till date on this aspect. This work represents a contribution towards understanding of steel behaviour at elevated temperature during quenching at the LHP (centre node) of the steel bar. The results of this study may prove useful to obtain the hardness of the lowest point of the steel bar in order to reach the maximum benefit against the deformation and failure of the component [15, 34].

2. Mathematical model

Three dimensional heat transfers can be analyzed using one dimensional axisymmetric element as shown in Figure 4 (a)(b) and Figure 5 [9-22].

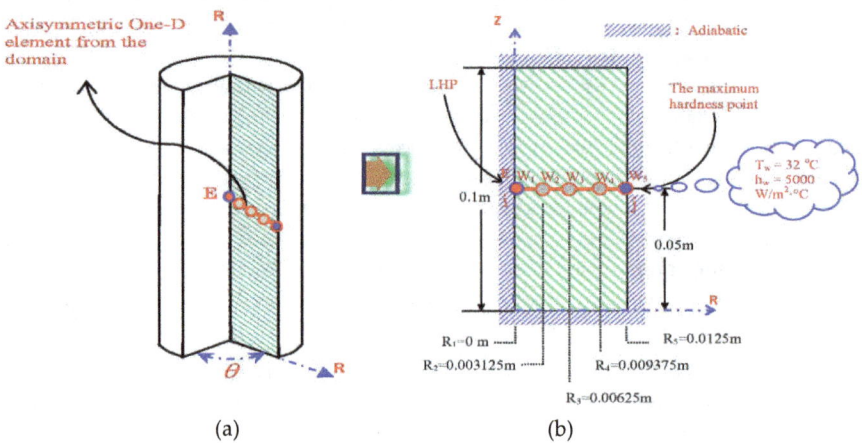

(a)   (b)

Figure 4. (a) The axisymmetric one dimensional line (radius) element from the domain, on the cylindrical steel bar which had been heated and then submerged in water. (b) The 1-dimensional element from the domain on the axisymmetric rectangular cross section when the radii equal 12.5mm, the selected 4 elements with 5 nodes and the boundary at node j [5] for an element 4.

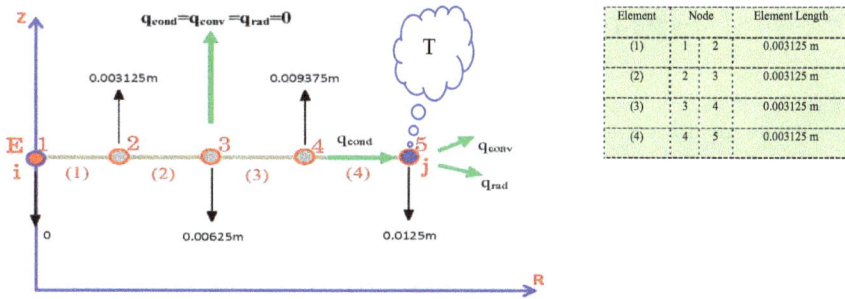

Element	Node		Element Length
(1)	1	2	0.003125 m
(2)	2	3	0.003125 m
(3)	3	4	0.003125 m
(4)	4	5	0.003125 m

Figure 5. The axisymmetric one dimensional line (radius) element from the domain, the selected 4 elements with 5 nodes and the boundary at node j [5] for an element 4.

2.1. Methodology of building the F. E. Model in details

The temperature distribution inside the cylindrical steel bar at thermal equilibrium will be calculated. These are special classes of three-dimensional heat transfer problem:

i.   Geometrically axisymmetric.
ii. Each thermal load is symmetrical about an axis.

This three-dimensional heat transfer problem may be analyzed using one-dimensional axisymmetric elements as shown in Figure 4 (a)(b) and Figure 5 [9-22].

The finite element method is applied to the one-dimensional cylindrical coordinates heat transfer problem.

The finite element formulation is developed with the Garlekin Weighted-residual method. The appropriate working expressions of the conductance matrix, capacitance matrix and thermal load matrix are derived in details.

The time dependent solution is obtained by applying the Backward Difference Scheme.

2.1.1. Meshing the engineering problem of the domain

Since the modelling work is on one-dimensional axisymmetric elements then line element has been selected in this study. Let us consider a cylindrical chromium steel bar as shown in Figure 4 (a) which had been heated and then submerged in water.

The linear temperature distribution for an element (radius) line, T is given by:

$$T^{(R)} = a_1 + a_2 R \qquad (1)$$

Where,
$T^{(R)}$ = nodal temperature as the function of R
a_1 and a_2 are constants.
R is any point on the (radius) line element.

2.1.2. Shape function of 1-D axisymmetric element

The shape functions were to represent the variation of the variable field over the element. The shape function of axisymmetric 1-Dimensional line (radius) is element expressed in terms of the r coordinate and its coordinate are shown in Figure 6;

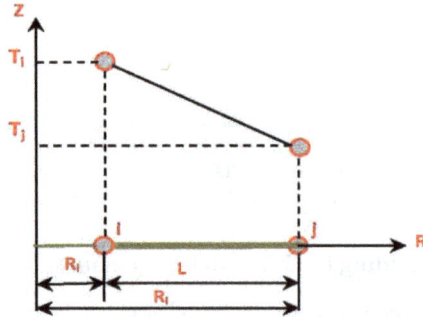

Figure 6. one-dimensional linear temperature distributions for an element (radius) line in global coordinate system.

They are derived to obtain the following shape functions [9-22];

$$S_i = \left(\frac{R_j - R}{R_j - R_i} \right) = \left(\frac{R_j - R}{L} \right) \qquad \text{(a)}$$

$$S_j = \left(\frac{R - R_i}{R_j - R_i} \right) = \left(\frac{R - R_i}{L} \right) \qquad \text{(b)}$$

(2)

Thus the temperature distribution of 1-D radius for an element in terms of the shape function can be written as:

$$T^{(R)} = S_i T_i + S_j T_j = S^{(r)}\{T\} \qquad (3)$$

Where $[S^{(r)}] = [S_i \quad S_j]$ is a row vector matrix and $\{T\} = \begin{Bmatrix} T_i \\ T_j \end{Bmatrix}$ is a column vector of nodal

temperature of the element.

Eq. (3) can also be expressed in matrix form as:

$$T^{(R)} = \begin{bmatrix} S_i & S_j \end{bmatrix} \qquad (4)$$

Thus for 1-dimensional element we can write in general:

$$\Psi^{(e)} = \begin{bmatrix} S_i & S_j \end{bmatrix} \begin{Bmatrix} \psi_i \\ \psi_j \end{Bmatrix} \qquad (5)$$

Where Ψ_i and Ψ_j represent the nodal values of the unknown variable which in our case is temperature. The unknown can also be deflection, or velocity etc.

2.1.3. Natural area coordinate

Using the natural length coordinates and their relationship with the shape function by simplification of the integral of Galerkin solution:

The two length natural coordinates L_1 and L_2 at any point p inside the element are shown in Figure 7 [9, 10, 19]. From which we can write:

$$L_1 = \frac{R_j - R}{R_j - R_i} = \frac{l_1}{L} \quad \text{(a)}$$

$$L_2 = \frac{R - R_i}{R_j - R_i} = \frac{l_2}{L} \quad \text{(b)}$$

(6)

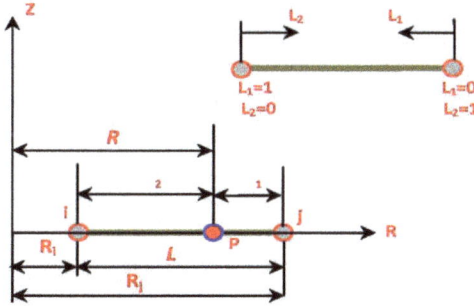

Figure 7. Two-node line element showing interior point p and the two naturals coordinates L_1 and L_2.

Since it is a one-dimensional element, there should be only one independent coordinate to define any point P. This is true even with natural coordinates as the two natural coordinates L_1 and L_2 are not independent, but are related as:

$$L_1 + L_2 = 1 \text{ or } L_1 + L_2 = \frac{l_1}{L} + \frac{l_2}{L} = 1 \quad (7)$$

The natural coordinates L_1 and L_2 are also the shape functions for the line element, thus:

$$S_i = \left(\frac{R_j - R}{R_j - R_i} \right) = \left(\frac{R_j - R}{L} \right) = L_1$$

$$S_j = \left(\frac{R - R_i}{R_j - R_i} \right) = \left(\frac{R - R_i}{L} \right) = L_2 \quad (8)$$

$$S_i = L_1, \; S_j = L_2,$$

$$R = R_j L_2 + R_i L_1 = R_i S_i + R_j S_j \tag{9}$$

$$\frac{\partial [S]_i}{\partial r} = \frac{\partial L_1}{\partial r} = \frac{-1}{R_j - R_i} = -\frac{1}{L} \tag{10}$$

$$\frac{\partial [S]_j}{\partial r} = \frac{\partial L_2}{\partial r} = \frac{1}{R_j - R_i} = \frac{1}{L} \tag{11}$$

2.2. Develop equation for all elements of the domain

Derivation of equation of heat transfer in axisymmetric 1-dimensional line (radius) elements [9-22]. By applying the conservation of energy to a differential volume cylindrical segment has been done. As shown in Figure 8;

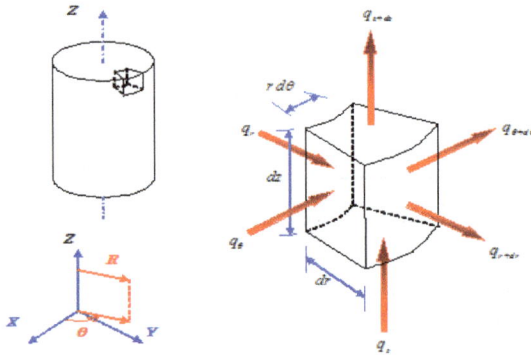

Figure 8. Axisymmetric element from an axisymmetric body.

$$E_{in} - E_{out} + E_{generated} = E_{stored} \tag{12}$$

The transient heat transfer within the component during quenching can mathematically be described by simplifying the differential volume term [4, 12]; the heat conduction equation is derived and given by;

$$\frac{1}{r}\frac{d}{dr}\left(K_r r \frac{dT}{dr}\right) + \frac{1}{r^2}\frac{d}{d\theta}\left(K_\theta \frac{dT}{d\theta}\right) + \frac{d}{dz}\left(K_z \frac{dT}{dz}\right) + q = \rho c \frac{dT}{dt} \tag{13}$$

Where;
k_r = heat conductivity coefficient in r-direction, W/m·°C.
k_θ = heat conductivity coefficient in θ-direction, W/m·°C.
k_z = heat conductivity coefficient in z-direction, W/m·°C.
T = temperature, °C.

q = heat generation, W/m³.
ρ = mass density, kg/m³.
c = specific heat of the medium, J/kg·K.
t = time, s.

2.3. The assumption made in this problem was

For axisymmetric situations one dimensional line (radius) element, there is no variation of temperature in the Z-direction as shown in Figure 4 (a)(b)and Figure 5. This is because we have already assumed that in steel quenching and cooling process of the steel bar is insulated from convection at the cross section of the front and back. It means that we have convection and radiation at one node only which is on the surface [node 5]. In our research we focus to calculate E-LHP which is at [node 1], where it is the last point that will be cooled. This gives the maximum advantage to make our assumption safer, because it is the last point that will be affected by convection and radiation from the front and back cross section of the steel bar.

Therefore we can write, $\left(\dfrac{\partial T}{\partial z} = 0 \right)$

For axisymmetric situations, there is no variation of temperature in the θ-direction, because it is clear from Figure 4 (a)(b) and Figure 5 that the temperature distribution along the radius will be the same if the radius moves with an angle θ, 360°.

Therefore, $\left(\dfrac{\partial T}{\partial \theta} = 0 \right)$.

The thermal energy generation rate $\dot{q}$ represents the rate of the conversion of energy from electrical, chemical, nuclear, or electromagnetic forms to thermal energy within the volume of the system. The conversion of the electric field will be studied with details in the 2nd part of our research, however in this manuscript no heat generation has been taken into accounted.

Therefore, $\dot{q} = 0$

After simplifying, the Eqn. (13) becomes;-

$$\frac{k}{r}\frac{\partial}{\partial r}\left(r\frac{\partial T}{\partial r} \right) = \rho \, c \frac{\partial T}{\partial t} \tag{14}$$

And also known as residual or partial differential equation

$$\{\Re\} = \frac{k}{r}\frac{\partial}{\partial r}\left(r\frac{\partial T}{\partial r} \right) - \rho \, c \frac{\partial T}{\partial t} = 0 \tag{15}$$

2.4. Galerkin weighted residual method formulation

From the derived heat conduction equation, the Galerkin residual for 1-dimensional line (radius) element in an unsteady state heat transfer can be obtained by integration the transpose of the shape functions times the residual which minimize the residual to zero becomes;

$$\int_V [S]^T \{\Re\}^{(e)} \, dv = 0 \tag{16}$$

Where, $[S]^T$ = the transpose of the shape function matrix

$\{\Re\}^{(e)}$ = the residual contributed by element (e) to the final system of equations.

$$\frac{k}{r}\overbrace{\int [S]^T \frac{\partial}{\partial r}\left(r\frac{\partial T}{\partial r}\right)dv}^{1} - \overbrace{\int [S]^T \rho c \frac{\partial T}{\partial t}dv}^{2} = 0 \tag{17}$$

2.5. Chain rule

The term 1 and 2 of Eqn. (17) can be re-arranged using the chain rule which states that;

$$\left(fg\right)^- = fg^- + gf^-$$

Therefore, $fg^- = \left(fg\right)^- - f^-g$ then $\dfrac{\partial}{\partial r}\left([S]^T r\dfrac{\partial T}{\partial r}\right) = [S]^T\left\{\dfrac{\partial}{\partial r}\left(r\dfrac{\partial T}{\partial r}\right)\right\} + r\dfrac{\partial T}{\partial r}\dfrac{\partial [S]^T}{\partial r}$

Term 1 of Eqn. 17 is rearranged thus;

$$[S]^T\left\{\frac{\partial}{\partial r}\left(r\frac{\partial T}{\partial r}\right)\right\} = \frac{\partial}{\partial r}\left([S]^T r\frac{\partial T}{\partial r}\right) - r\frac{\partial T}{\partial r}\frac{\partial [S]^T}{\partial r} \tag{18}$$

By substitute Eqn. (18) in to Eqn. (17), get

$$= \overbrace{\frac{k}{r}\int\left\{\frac{\partial}{\partial r}\left([S]^T r\frac{\partial T}{\partial r}\right)\right\}dv}^{A} - \overbrace{\frac{k}{r}\int\left\{\frac{\partial [S]^T}{\partial r}r\frac{\partial T}{\partial r}\right\}dv}^{B} - \underbrace{\int [S]^T\left\{\rho c\,\frac{\partial T}{\partial t}\right\}dv}_{C} \tag{19}$$

Term A is the heat convection terms and contributes to the conductance and thermal load matrix. Term B is the heat conduction terms and contributes to the conductance matrix. Term C is the transient equation and contributes to the capacitance matrix.

Where,

$$\overbrace{\frac{k}{r}\int \frac{\partial}{\partial r}\left(\left[S\right]^{T} r \frac{\partial T}{\partial r}\right)dV}^{A} =$$

$$\underbrace{-2\pi h r z \left[S\right]^{T}\left[S\right]\{T\}^{e}}_{(1)} + \underbrace{2\pi h r z \left[S\right]^{T} T_{f}}_{(2)} - \underbrace{2\pi r z \varepsilon_{s}\sigma\left[S\right]^{T}\left[S\right]\{T\}^{e}\left(\left[S\right]\{T\}^{e}\right)^{3}}_{(3)} + \underbrace{2\pi r z \varepsilon_{s}\sigma\left[S\right]^{T} T_{f}^{4}}_{(4)}$$

Note: term (1) and term (3) contributed to the conductance matrix since they contain the unknown temperature {T}. Term (2) and (4) contributed to thermal load matrix as T_f is the known fluid temperature. Term (3) and term (4) heat radiation very important if our heat treatment is Annealling [cooling in the furnace] or Normalizing [cooling in air or jet air], but it can be ignored if the cooling is quenching in water, sea water or oil as in this work.

From earlier explanations, derivation and after simplification we can formulate the conductance matrix in the r-direction for term B finally we get:

Term B (the conduction term) contributes to the conductance matrix

$$\overbrace{\frac{k}{L}\left(R_{j}+R_{i}\right)\begin{bmatrix}1 & -1\\-1 & 1\end{bmatrix}\begin{Bmatrix}Ti\\Tj\end{Bmatrix}}^{K_c} \tag{20}$$

Similarly, term C, the unsteady state (transient) which contributes to the Capacitance Matrix becomes:

Term C (heat stored) contributes to the Capacitance Matrix

$$\overbrace{\frac{L\rho c}{6}\begin{bmatrix}\left(3R_{i}+R_{j}\right) & \left(R_{i}+R_{j}\right)\\\left(R_{i}+R_{j}\right) & \left(R_{i}+3R_{j}\right)\end{bmatrix}\begin{Bmatrix}\dot{T}_{i}\\\dot{T}_{j}\end{Bmatrix}}^{C} \tag{21}$$

Term A (Heat Convection)

Term A₁ -Contributes to Conductance Matrix

Term A₁ (the convection term) contributes to the conductance matrix

$$\overbrace{-2hR_{j}\begin{bmatrix}0 & 0\\0 & 1\end{bmatrix}\begin{Bmatrix}Ti\\Tj\end{Bmatrix}}^{K_h} \tag{22}$$

Term A₂ -Contributes to thermal load Matrix

Term A₂ (the convection term) contributes to thermal load matrix

$$2hR_jT_w \overbrace{\begin{bmatrix} 0 \\ 1 \end{bmatrix}}^{f_h} \tag{23}$$

2.6. Construct the element Matrices to the Global Matrix

The global, conductance, capacitance and thermal load matrices and the global of the unknown temperature matrix for all the elements in the domain are assembled i.e. the element's conductance; capacitance and thermal load matrices have been derived. Assembling these elements are necessary in all finite element analysis [9-22].

Constructing these elements will result into the following finite element equation:

$$[K]^{(G)}\{T\}^{(G)} + [C]^{(G)}\{\dot{T}\}^{(G)} = \{F\}^{(G)} \tag{24}$$

Where:

$[K] = [K_c] + [K_h]$: is conductance matrix due to Conduction (Elements 1 to 4) and heat loss through convection at the element's boundary (element 4 node 5) as shown in Figure 1, Figure 2 and Figure 3.

$\{T\}$: is temperature value at each node, °C.

$[C]$: is capacitance matrix, due to transient equation (heat stored)

$\{\dot{T}\}$: is temperature rate for each node, °C/s.

$\{F\} = \{F_h\} + \{F_{\dot{q}}\}$: is heat load due to heat loss through convection at the element's boundary (element 4 node 5) and internal heat generation (element 4 node 5).

2.7. Euler's method

Two points recurrence formulas will allow us to compute the nodal temperatures as a function of time. In this paper, Euler's method which is known as the backward difference scheme (FDS) will be used to determine the rate of change in temperature, the temperature history at any point (node) of the steel bar [27, 28, 32].

If the derivative of T with respect to time t is written in the backward direction and if the time step is not equal to zero ($\Delta t \neq 0$), we have;

$$\{[K]^{(G)}\}\{T(t)\}^{(G)} + [C]^{(G)}\left\{\frac{T(t) - T(t - \Delta t)}{\Delta t}\right\}^{(G)} = \{F(t)\}^{(G)} \tag{25}$$

With;

$\dot{T}$ = temperature rate (°C/s); T (t) = temperature at t s (°C); T (t-Δt) = temperature at (t-Δt) s, (°C)
Δt = selected time step (s) and t = time (s) (at starting time, t = 0)

By substituting the value of $\{\dot{T}\}$ into the finite element global equation, we have that;

$$\left[K\right]^{(G)}\left\{T(t)\right\}^{(G)} + \left[C\right]^{(G)}\left\{\frac{T(t) - T(t - \Delta t)}{\Delta t}\right\}^{(G)} = \left\{F(t)\right\}^{(G)} \tag{26}$$

Finally, the matrices become;

$$\left[\left[K\right]^{(G)}\Delta t + \left[C\right]^{(G)}\right]\left\{T\right\}^{(G)}_{i+1} = \left[C\right]^{(G)}\left\{T\right\}^{(G)}_{i} + \left\{F\right\}^{(G)}_{i+1}\Delta t \tag{27}$$

From Eqn. (27) all the right hand side is completely known at time t, including t = 0 for which the initial condition apply.

Therefore, the nodal temperature can be obtained for a subsequent time given the temperature for the preceding time.

Once the temperature history is known the important mechanical properties of the low carbon steel bar can be obtained such as hardness and strength.

3. Application

3.1. Calculation of the temperature history

The present developed mathematical model is programmed using MATLAB to simulate the results of the temperature distribution with respect to time in transient state heat transfer of the industrial quenched chromium steel AISI 8650H. The cylindrical chromium steel bar has been heated to 850°C. Then being quenched in water with $T_{sea\ water}$ = 32°C and the convection heat transfer coefficient, $h_{sea\ water}$ = 1250 W/m2·°C.

The temperature history for the selected nodes of the cylindrical chromium steel AISI 8650H after quenching is being identified in Figure 9 and Figure 10.

The cylindrical bar was made from chromium steel AISI 8650H, with properties as seen below [26].

Thermal capacity, ρc (J/m³·°C)

$0 \leq T \leq 650°C$, $\rho c = (0.004T + 3.3) \times 10^6$, $650 < T \leq 725°C$, $\rho c = (0.068T - 38.3) \times 10^6$
$725 < T \leq 800°C$, $\rho c = (-0.086T + 73.55) \times 10^6$, $T > 800°C$, $\rho c = 7.55 \times 10^6$

Thermal conductivity, k (W/m·°C)

$0 \leq T \leq 900°C$, $k = -0.022T + 48$, $T > 900°C$, $k = 28.2$

Where in our case the global conductance matrix $[K]^{(G)}$, the global capacitance matrix $[C]^{(G)}$ and the global thermal load matrix $\{F\}^{(G)}$ can be computed easily as follow:

$$[K]^{(G)} = [K_c]^{(1)} + [K_c]^{(2)} + [K_c]^{(3)} + [K_c]^{(4)} + [K_h]^{(4)} \qquad (28)$$

Where $[K_c]^{(1)}$, $[K_c]^{(2)}$, $[K_c]^{(3)}$, $[K_c]^{(4)}$ are the conductance matrices due to conduction in 1-D element for the 1st element, the 2nd, the 3rd, and the 4th element respectively, while $[K_h]^{(4)}$ because we note that there is convection in element 4 at node j(5) only as shown clearly in Figure 5 and Figure 9.

$$[C]^{(G)} = [C]^{(1)} + [C]^{(2)} + [C]^{(3)} + [C]^{(4)}$$

Where $[C]^{(1)}$, $[C]^{(2)}$, $[C]^{(3)}$, $[C]^{(4)}$ are the capacitance matrices due to transient [unsteady state] in 1-dimensional line (radius) element.

$$\{F\}^{(G)} = \{F_h\}^{(4)}$$

We have convection in element 4 at node $j^{(5)}$ only as shown clearly in Figure 5 and Figure 9.

With the input data and boundary conditions provided, a sensitivity analysis is carried out with the developed program to obtain the temperature distribution at any point (node) of the quenched steel bar. As an example the transient state temperature distribution results of the selected five nodes from the center [W_1] to the surface [W_5] of the quenched chromium steel AISI 8650H which were computed as shown in Figure 10 with dimension as in Figure 9. [26].

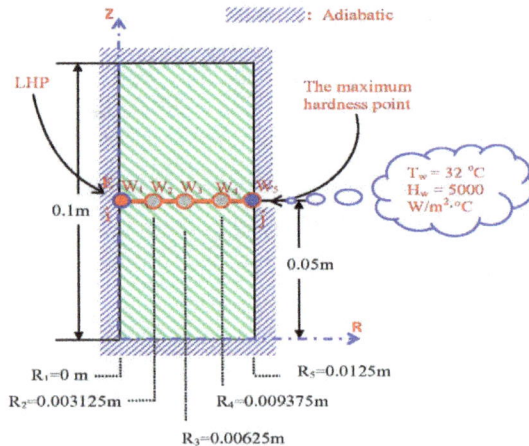

Figure 9. The axisymmetric one dimensional line (radius) element from the domain when the radius equal 12.5 mm, the selected 4 elements with 5 nodes and the boundary at node j [5] for an element 4 of chromium steel AISI 8650H.

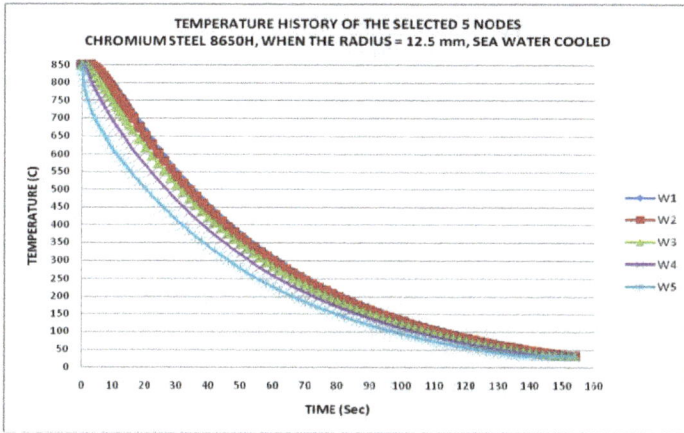

Figure 10. Graph of temperature history along WW cross-section when the radius = 12.5 mm from MATLAB program.

3.2. LHP Calculation

3.2.1. Calculating the cooling time required

In this study, we choose to calculate the cooling time between 800°C and 500°C [5-8, 27-31]. Where, the characteristic cooling time, relevant for phase transformation in most structural steels is the time of cooling from 800 to 500°C (time $t_{8/5}$) [5-8, 27-31].

$$t_c = t_{800} - t_{500}$$

Interpolation Method:

From Figure 10 we can *determine* the time taken for node W_5 to reach 800°C,

By interpolation method as the following:

Node W_5	:	$t = 0.6$ s when T = 800.151°C
Nolde W_5	:	$t = t_{800}°c$ when T = 800°C
Node W_5	:	$t = 1$ s when T = 781.6454°C

t_1	T_1
t_{800}	800
t_3	T_3

Solving for t_{800}

$$\frac{t_{800} - t_1}{t_3 - t_1} = \frac{T_{800} - T_1}{T_3 - T_1}$$

Thus, $t_{800} = \dfrac{(T_{800} - T_1)}{(T_3 - T_1)}(t_3 - t_1) + t_1 = \dfrac{(800 - 800.151)}{(781.6454 - 800.151)}(1 - 0.6) + 0.6 = 0.603\,\text{sec}$

$t_{800°C} = 0.603\,\text{sec}$

The time taken for node W5 to reach 500°C

$= \dfrac{(500 - 502.9792)}{(483.933 - 502.9792)}(22 - 20) + 20 = 20.313\,\text{sec thus } t_{500°C} = 20.313\,\text{sec}$

So the Cooling time t_c for node W5;

$t_c = t_{500\,°C} - t_{800\,°C} = 20.313 - 0.603 = 19.71\,\text{sec}$

For nodes W1 to W4, the cooling time t_c calculated by the same way, the final results shown in Table 1.

3.2.2. Calculating the Jominy distance from Standard Jominy distance versus cooling time

Cooling time, t_c obtained will now be substituted into the Jominy distance versus cooling time curve to get the correspondent Jominy distance. Jominy distance can also be calculated by using polynomial expressions via polynomial regression.

In this chapter the standard Table [Cooling rate at each Jominy distance (Chandler, H., 1998)] will be used. Then Jominy distance of nodes W1 to W5 will be calculated by using the data from [Cooling rate at each Jominy distance (Chandler, H., 1998)] [33]. The final results shown in Table 1, where the Rate of Cooling, ROC, was defined as;

$$\text{ROC} = \frac{800°C\text{-}500°C}{t_c} = \frac{800°C\text{-}500°C}{t_{500°C} - t_{800°C}}\left(°C\!\Big/\!_{Sec}\right)$$

3.2.3. Predict the hardness of the quenched steel bar

The HRC of chromium steel AISI 8650H can be calculated by using the relation between the J-Distance and the HRC from the Practical date Handbook, the Timken Company 1835 Duebex Avenue SW Canton, Ohio 44706-2798 1-800-223, the final results shown in Figure 11 & Table 1:

Node	tc (s)	ROC (°C /s)	Jominy-distance (mm)	Hardness (HRC)
W₁	25.744	11.653200	16.209	51.688
W₂	25.451	11.787356	16.028	51.859
W₃	25.091	11.956478	15.837	52.054
W₄	23.551	12.738312	15.277	52.758
W₅	19.170	15.220700	13.888	54.506

Table 1. Cooling time, Cooling rate, Jominy distance and HRC for the nodes W1 to W5, sea water cooled by 1-D mathematical model when the radius = 12.5 mm.

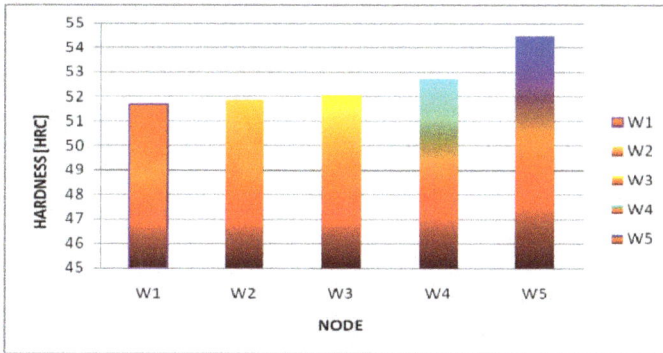

Figure 11. Hardness distribution along WW cross section for the nodes W₁ to W₅ from the centre to the surface respectively at half the length at the centre of the quenched steel bar sea water cooled by 1-D mathematical model, when the radius = 12.5 mm.

4. Mathematical model verification

The same data input for the steel properties and boundary condition used in the mathematical model is applied to the ANSYS software to verify the temperature simulation results. The temperature distribution from the ANSYS analysis is depicted figuratively as shown in Figure 12(a) and Figure 12(b); Figure 12(a) shows the temperature distribution just before the steel bar becomes completely cooled and Figure 12(b) shows the temperature distribution at the moment that the entire steel bar becomes completely cooled after 1175s.

The temperature time graph from the ANSYS analysis is depicted as shown in Figure 13;

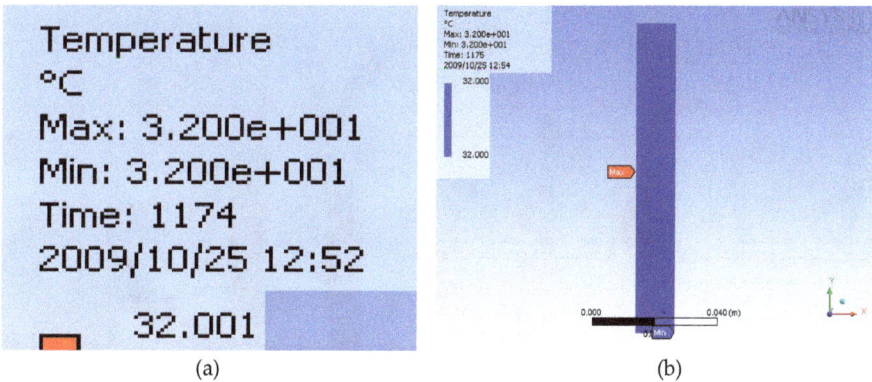

Figure 12. (a)   (b)

From the graphs shown in Figure 10 by mathematical model and Figure 13 by ANSYS, it can be clearly seen that the temperature history of the quenched steel bar has the same pattern. The heat transfer across the steel bar is uniform. From Figure 13 the cooling time, Jominy-distance and consequently the hardness of the quenched Chromium steel 8650H at any point (node), even the lowest hardness point (E-LHP) is determined by ANSYS too, the final results shown in Table 2 and Figure 14.

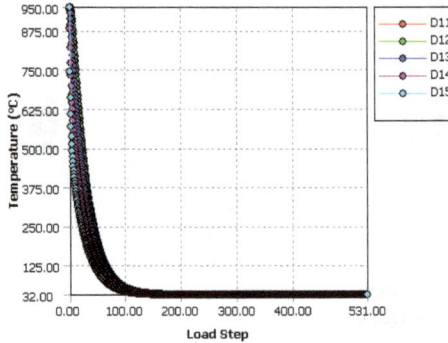

Figure 13. Temperature-time graph from ANSYS.

Node	Cooling time,	Cooling rate	J-distance (mm)	HRC
D_{11}	28.262996	10.61459	17.609	50.359
D_{22}	28.046194	10.69664	17.542	50.423
D_{33}	27.224717	11.01940	17.120	50.823
D_{44}	23.998865	12.50059	15.440	52.553
D_{55}	20.855801	14.38449	14.29	54

Table 2. Cooling time, Cooling rate, Jominy distance and HRC for the nodes D_{11} to D_{55}, sea water cooled by ANSYS.

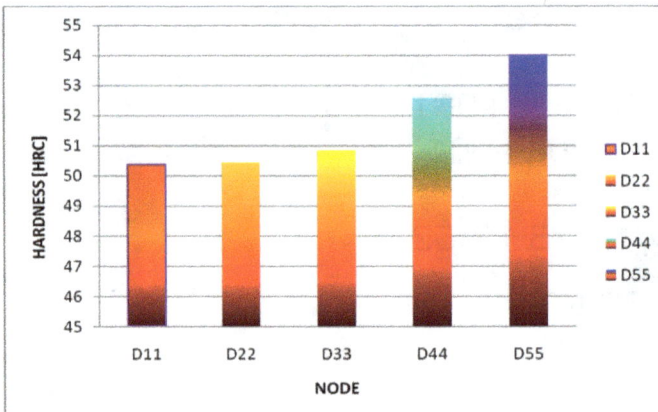

Figure 14. Hardness distribution by ANSYS along DD cross section for the nodes D_{11} to D_{55} from the centre to the surface respectively at the half length at the centre of the quenched chromium steel bar where sea water is cooled, when the radius = 12.5 mm.

From the above results it was found that in the mathematical model for the 1st node with W_1 in the center, we found that HRC = 51.688. While by ANSYS for the same node A_1, it was found that HRC = 50.359. And for the nodes on the surfaces W_5 and A_5, it was found that HRC = 54.506 and 54 for the mathematical model and ANSYS respectively. From the above, it can be seen that there is a strong agreement between both results. The difference between all the results of the mathematical model and the Ansys simulations can be accounted due to the fact that the ANSYS software is commercial purpose, and thereby has some automated input data. But the developed mathematical model is precisely for a circular steel bar axisymmetric cross section. However, there is strong agreement between both results and thereby the result is validated where, the comparison indicated reliability of the proposed model. Also the results showed that the node on the surface will be the 1st which completely cooled after quenching because it is in the contact with the cooling medium then the other nodes on the radial axis to the centre respectively and the last point will be completely cooled after quenching will be at half the length at the centre. Hence E-LHP will be at half the length at the centre of the quenched industrial Chromium steel bar.

5. Effect of the radius on the temperature history

Where there is strong agreement between the mathematical model and the Ansys simulation results and thereby the result is validated where, the comparison indicated reliability of the proposed model thereby; we will apply the mathematical model to study the effect of radius on the temperature history and on E-LHP. By the same way the temperature history of the 5 selected nodes has been obtained when the radii 18.75, 25, 37.5 and 50 mm is determined. The final results are shown in Figure 15, Figure 16, Figure 17, and Figure 18 respectively.

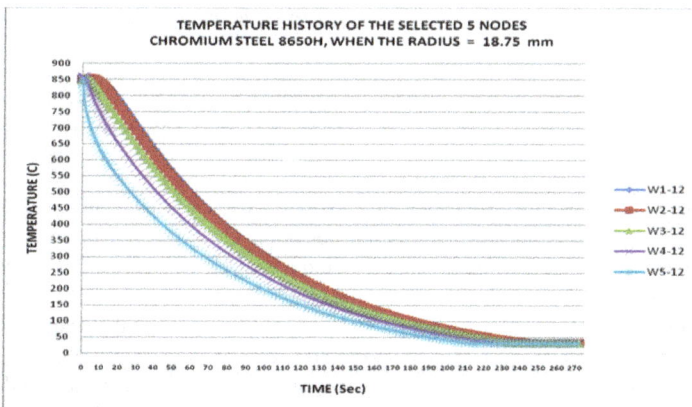

Figure 15. Graph of temperature history along WW cross-section when the radius = 25 mm from MATLAB program.

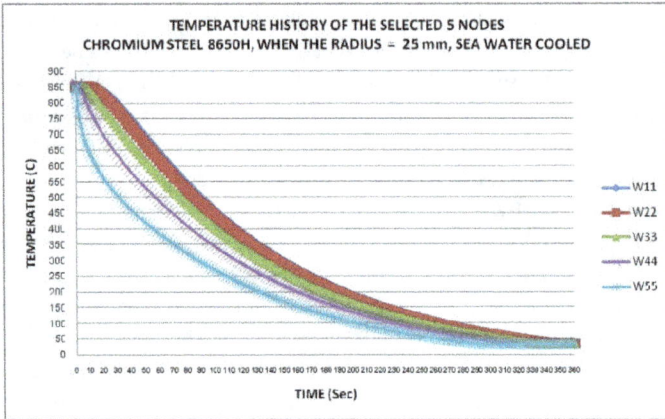

Figure 16. Graph of temperature history along WW cross-section when the radius = 25 mm from MATLAB program.

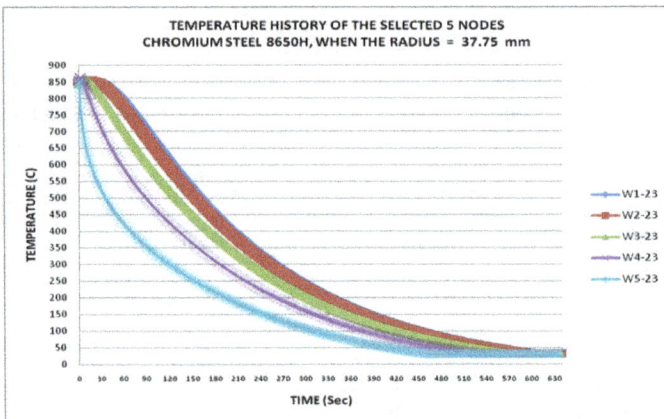

Figure 17. Graph of temperature history along WW cross-section when the radius = 37.5 mm from MATLAB program.

It is clear from the results that the nodes on the surfaces cooled faster than the nodes at half the lengths at the centres in other words the nodes on the surfaces will be the 1st to be completely cooled after quenching because it is in the contact with the cooling medium then the other points (nodes) on the radial axis to the centre respectively while the last point that will be completely cooled after quenching will be at half the length at the centre because the cooling time tc of nodes W_5, W_{5-12}, W_{55}, W_{5-23} and W_{555} less than tc of nodes W_1, W_{1-12}, W_{11}, W_{1-23} and W_{111}, respectively. A larger diameter rod quenched in a particular medium will obviously cool more slowly than a smaller diameter rod. where Figure 10, Figure 15, Figure 16, Figure 17, and Figure 18 showed that when the radii equals 12.5, 18.75, 25, 37.5 and 50 mm it required 154, 246, 354, 606 and 924 sec respectively to be cooled decreasingly from the austenitizing temperature [850°C] to the fluid temperature [32°C]. Based on the above results we expect that if the radius is 100 mm, 3000 sec is required to reach the fluid temperature.

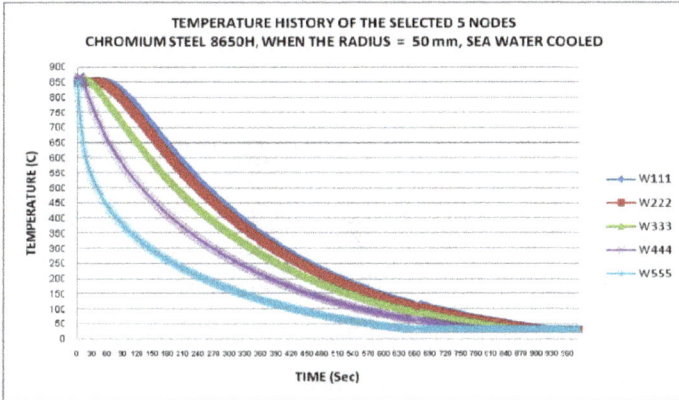

Figure 18. Graph of temperature history along WW cross-section when the radius = 50 mm from MATLAB program.

6. Effect of the radius on E-LHP

As explained above and from Figure 15, Figure 16, Figure 17, and Figure 18 of the temperature history the hardness when the radii equals 18.75, 25, 37.5 and 50 mm is determined, the final results shown in Figures 19, Table 3, Figure 20, Table 4, Figures 21, Table 5, Figure 22 and Table 6.

Node	tc (s)	ROC (°C /s)	Jominy-distance (mm)	Hardness (HRC)
$W_{1\text{-}12}$	41.386	7.248828	21.567	47.125
$W_{2\text{-}12}$	41.272	7.268850	21.531	47.159
$W_{3\text{-}12}$	40.287	7.446570	21.218	47.455
$W_{4\text{-}12}$	36.027	8.327088	19.935	48.443
$W_{5\text{-}12}$	25.930	11.56961	16.323	51.579

Table 3. Cooling time, Cooling rate, Jominy distance and HRC for the nodes $W_{1\text{-}12}$ to $W_{5\text{-}12}$, sea water cooled by 1-D mathematical model, when the radius = 18.75 mm.

Node	t_c (s)	ROC (°C/s)	Jominy-distance (mm)	Hardness (HRC)
W_{11}	59.584	5.03490	27.035	43.614
W_{22}	59.424	5.04846	26.992	43.624
W_{33}	57.339	5.23202	26.425	43.758
W_{44}	48.847	6.14162	23.718	45.087
W_{55}	30.004	9.99866	18.142	49.857

Table 4. Cooling time, Cooling rate, Jominy distance and HRC for the nodes W_{11} to W_{55}, sea water cooled by 1-D mathematical model, when the radius = 25 mm.

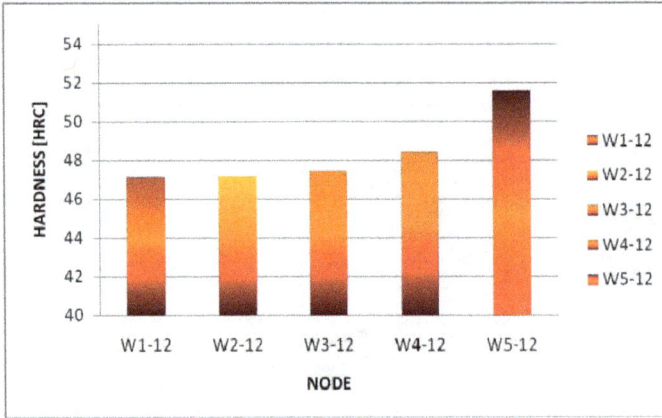

Figure 19. Hardness distribution along WW cross section for the nodes W_{1-12} to W_{5-12} from the centre to the surface respectively at half the length at the centre of the quenched steel bar sea water cooled by 1-D mathematical model, when the radius = 18.75 mm.

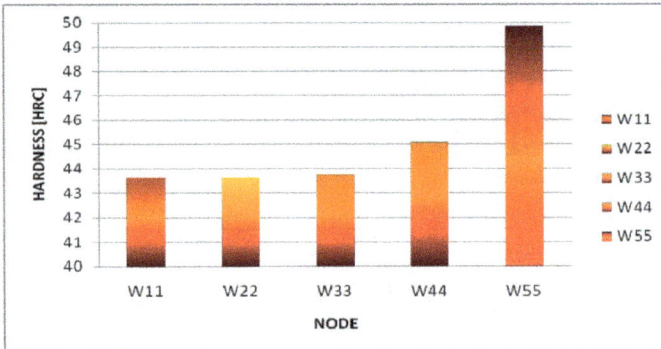

Figure 20. Hardness distribution along WW cross section for the nodes W_{11} to W_{55} from the centre to the surface respectively at half the length at the centre of the quenched steel bar sea water cooled by 1-D mathematical model, when the radius = 25 mm.

Node	tc (s)	ROC (°C /s)	Jominy-distance (mm)	Hardness (HRC)
W_{1-23}	102.680	2.921698	37.162	41.222
W_{2-23}	102.429	2.928858	37.109	41.234
W_{3-23}	96.943	3.094601	35.957	41.506
W_{4-23}	75.4300	3.977197	31.344	42.596
W_{5-23}	34.6140	8.667013	19.527	48.700

Table 5. Cooling time, Cooling rate, Jominy distance and HRC for the nodes W_{1-23} to W_{5-23}, sea water cooled by 1-D mathematical model, when the radius = 37.5 mm.

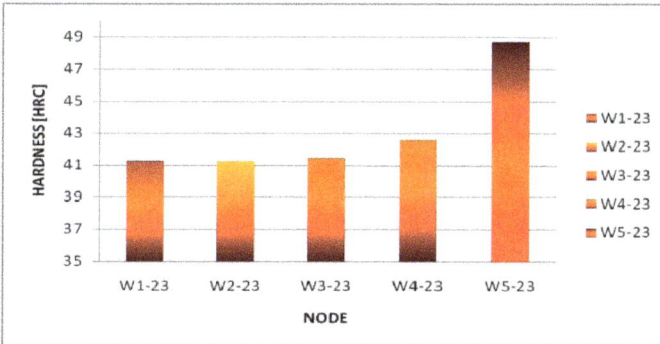

Figure 21. Hardness distribution along WW cross section for the nodes W_{1-23} to W_{5-23} from the centre to the surface respectively at half the length at the centre of the quenched steel bar sea water cooled by 1-D mathematical model, when the radius = 37.5 mm.

Node	tc (s)	ROC (°C /s)	Jominy-distance (mm)	Hardness (HRC)
W_{111}	154.772	1.93833	49.291	38.931
W_{222}	154.420	1.94275	49.213	38.942
W_{333}	143.679	2.08798	46.835	39.275
W_{444}	103.513	2.89818	37.337	41.180
W_{555}	37.1610	8.07297	20.264	48.236

Table 6. Cooling time, Cooling rate, Jominy distance and HRC for the nodes W_{111} to W_{555}, sea water cooled by 1-D mathematical model, when the radius = 50 mm.

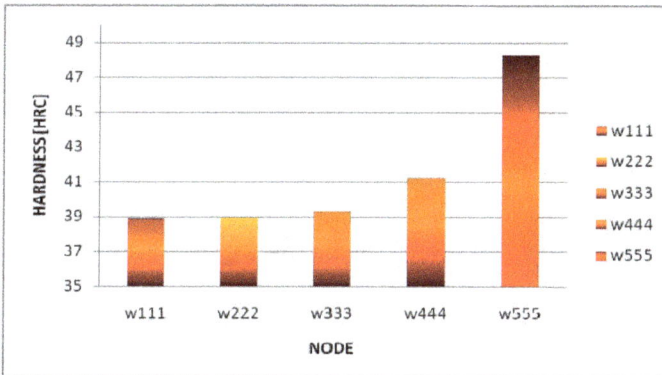

Figure 22. Hardness distribution along WW cross section for the nodes W_{111} to W_{555} from the centre to the surface respectively at half the length at the centre of the quenched steel bar sea water cooled by 1-D mathematical model, when the radius = 50 mm.

From the above results it was found that the hardness on the surfaces nodes will be higher than the hardness on the centres nodes as shown in Figures 11, 19, 20, 21 and 22 where the hardness on the surfaces at the nodes W_5, W_{5-12}, W_{55}, W_{5-23} and W_{555} equals 54.506, 51.579.

49.857, 48.700 and 48.236 respectively while the hardness at half the length at the centres W_1, W_{1-12}, W_{11}, W_{1-23} and W_{111} equals 51.688, 47.125, 43.614, 41.222 and 38.931 respectively. Hence E-LHP will be at half the length at the centre of the quenched industrial steel bar. Hence lowest hardness point [E-LHP] will be at centre of half length of the quenched industrial chromium steel bar.

It is clear from the results as shown in Figure 23 and Table 7 that increasing the radius of the bar inversely proportional with E-LHP, then it's more important to know E-LHP once the radius of the quenched steel bar is large because this has negative consequence which can result of the deformation and failure of the component.

Node	Radius (mm)	t_c (s)	ROC($^\circ$C/s)	J-distance	E-LHP [HRC]
W_1	12.5	25.744	11.6532	16.209	51.688
W_{1-12}	18.75	41.386	7.24882	21.567	47.125
W_{11}	25	59.584	5.03490	27.035	43.614
W_{1-23}	37.5	102.68	2.92169	37.162	41.222
W_{111}	50	154.772	1.93833	49.291	38.931

Table 7. Effect of radius on E-LHP of transient heat transfer axi-symmetric industrial quenched chromium steel AISI 8650H, sea water cooled by 1-D mathematical model.

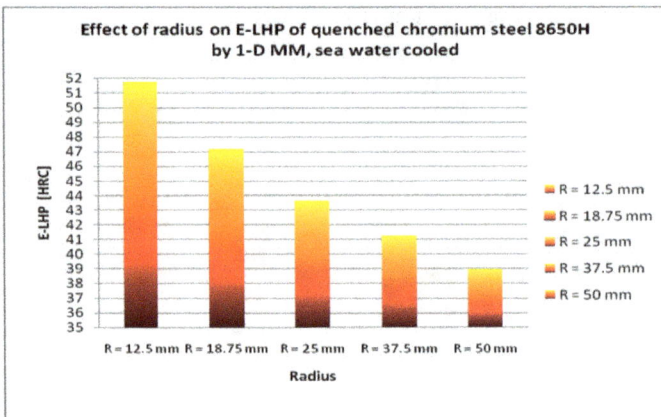

Figure 23. Effect of the radius on E-LHP

From the above explanation we can say that the developed 1-Dimensional mathematical modelling in this chapter solved design problems in transient heat transfer axi-symmetric industrial quenched chromium steel 8650H bar, where experimental calculation of E-LHP is an almost impossible task using manual calculation techniques. Also the earlier methods only used hardness calculated at the surface, which is higher than E-LHP.

7. Conclusion

A mathematical model of steel quenching has been developed to compute E-LHP of the quenched Chromium steel 8650H at any point (node) in a specimen with cylindrical geometry and to study the effect of radius on E-LHP. The model is based on the finite element Galerkin residual method. The numerical simulation of quenching consisted of numerical simulation of temperature transient field of cooling process. This mathematical model was verified and validated by comparing the hardness results with ANSYS software simulations. From the mathematical model and ANSYS results, it is clear that the nodes on the surface [W_5, $W_{5\text{-}12}$, W_{55}, $W_{5\text{-}23}$, W_{555} and D_{55}] respectively cools faster than the nodes on the center [W_1, $W_{1\text{-}12}$, W_{11}, $W_{1\text{-}23}$, W_{111} and D_{11}] because t_{CW5}, $t_{CW5\text{-}12}$, t_{CW55}, $t_{CW1\text{-}23}$, t_{CW555} and t_{CD55} less than t_{CW1}, $t_{CW1\text{-}12}$, t_{CW11}, $t_{CW1\text{-}23}$, t_{CW111} and t_{CD11}, this means that the mechanical properties will be different such as hardness where the hardness on the surfaces nodes [W_5, $W_{5\text{-}12}$, W_{55}, $W_{5\text{-}23}$, W_{555} and D_{55}] will be higher than the hardness on the center nodes [W_1, $W_{1\text{-}12}$, W_{11}, $W_{1\text{-}23}$, W_{111} and D_{11}] respectively.

Author details

Abdlmanam S. A. Elmaryami* and Badrul Omar
University Tun Hussein Onn Malaysia, Mechanical Engineering Department, Batu Pahat, Johor, Malaysia

Acknowledgement

The authors would like to thank {Ministry of Science, Technology and Innovation, Malaysia} for supporting this research under the Science Fund Grant with grant number {03-01-13-SF0071}, The corresponding author is grateful to the Postgraduate Centre of UTHM for supporting this research under the university PhD scholarship scheme.

8. References

[1] Ref. Key to Metals AG. Doldertal 328032 Zürich, Switzerland All Rights Reserved, TechSupport@keytometals.com.

[2] K. Funtani and g. Totten, (1997). Present Accomplishments and Future Challenges of Quenching Technology, The 6th International Seminar of IFHT, Kyongju.

[3] Robert K. N. (2001). "Quenching and Tempering of Welded Steel Tubular." July 29, 2001. The FABRICATOR articles.

[4] Budinski, K. G. (1992). "Engineering Material: Properties and Selection." 4th ed. Prentice-Hall International. Inc. 285-309.

[5] A. Rose et al, Atlas zur Wärmebe- handlung der Stähle I, Verlag Stahleisen, Düsseldorf, 1958.

---

* Corresponding Author

[6] B. Smoljan, Mathematical modelling of austenite decomposition during the quenching, 13th International Science Conference, The Polish Academy of Science 2004.

[7] B. Smojan (2006). Prediction of mechanical properties and microstructure distribution of quenched and tempered steel shaft, journal of materials processing technology, volume 175, Issu1-3, pp. (393-397).

[8] B. Smoljan, D. Iljkić, S. Smokvina Hanza, Computer simulation of working stress of heat treated steel specimen, Journal of Achievements in Materials and Manufacturing Engineering 34/2 (2009) 152-156.

[9] Abdlmanam S. A. Elmaryami and Badrul Omar, (2011). "Modeling the lowest hardness point in steel bar during quenching". Journal of ASTM International, Vol. 9, 2011, No. 5, ID JAI104386. [Impact Factor 0.279].
     http://www.astm.org/DIGITAL_LIBRARY/JOURNALS/MPC/PAGES/MPC104386.htm

[10] Abdlmanam S. A. Elmaryami and Badrul Omar, (2011). "Developing 1-D MM of Axi-symmetric Transient Quenched Cr-Steel to Determine LHP, water cooled". Journal of Metallurgy Volume 2012 (2012), Article ID 539823, 9 pages doi:10.1155/2012/539823
     http://www.hindawi.com/journals/jm/2012/539823/

[11] Abdlmanam S. A. Elmaryami and Badrul Omar, (2011). "Developing 1-D MM of Axisymmetric Transient Quenched Boron Steel to Determine LHP". European Journal of Scientific Research. [Impact Factor: 0.43, 2010].

[12] Abdlmanam S. A. Elmaryami and Badrul Omar, (2011). "Effect of Radius on Temperature History of Transient Industrial Quenched Cr-Steel 8650H by Developing 1-D MM". Journal of Applied Mathematical Sciences, [Impact Factor 0.275].

[13] Abdlmanam S. A. Elmaryami and Badrul Omar, (2011) "Developing of Unsteady State Axi-symmetric FEMM to Predict the Temperature of Industrial Quenched Steel", Journal of Metals Science and Heat Treatment, [Impact Factor 0.34].

[14] Abdlmanam S. A. Elmaryami and Badrul Omar, (2011). "Developing 1-D MM of Axi-symmetric Transient Quenched Mo-STEEL AISI-SAE 4037H to Determine LHP". Journal of Metallurgy and Materials Science, Vol. 53, No. 3, PP. 289-303.
     http://www.indianjournals.com/ijor.aspx?target=ijor:jmms&volume=53&issue=3&article=008

[15] Abdlmanam S. A. Elmaryami and Badrul Omar, (2011). "Unsteady State Computer Simulation of 2 Chromium Steel at 925°C as Austenitizing Temperature to Determine LHP". Metalurgia-Journal of Metallurgy, Metallurgical & Materials Engineerign, Vol 18 (2) 2012 pp. 79-91, [Impact Factor: 0.3].
     http://metalurgija.org.rs/mjom/vol18/No2/1_Elmaryami_MME_1802.pdf
     http://metalurgija.org.rs/mjom/vol18.html

[16] Abdlmanam. S. A. Elmaryami and Badrul Omar, (2011). Determination LHP of axisymmetric transient industrial quenched chromium steel 8650H by developing 1-D MM, sea water cooled. International Journal of Applied Engineering Research, [impact factor: 2].

[17] Abdlmanam S. A. Elmaryami and Badrul Omar, (2011). "Computer Simulation to Predict the Hardness of Transient Axi-Symmetric Industrial Quenched Steel Bar at

Different Radial Axises". International Journal of Emerging Technology in Science and Engineering.

[18] Abdlmanam S. A. Elmaryami and Badrul Omar, (2011) "Transient Computer Simulation of Industrial Quenched Steel Bar to Determine LHP of Molybdenum and Boron Steel at 850°C as Austenitizing Temperature Quenched in Different Medium" International Journal of Material Science.

[19] Abdlmanam S. A. Elmaryami and Badrul Omar, (2011). "LHP Calculation by Developing MM of Axisymmetric Transient Quenched Boron Steel, Sea Water Cooled". International Journal of Engineering Science and Technology (IJEST™). [Impact Factor 1.85 in 2009, Index Copernicus (IC Value)-3.14, 2011-12]

[20] Abdlmanam S. A. Elmaryami and Badrul Omar, (2011). The lowest hardness point calculation by transient computer simulation of industrial steel bar quenched in oil at different austenitizing temperatures. "International Conference on Management and Service Science, MASS", Wuhan, China, Article number 5999335, Indexed by Ei Compendex, Sponsors: IEEE. SCOPUS indexed.

[21] Badrul Omar, Mohamed Elshayeb and Abdlmanam. S.A. Elmaryami, (2009). "Unsteady state thermal behavior of industrial quenched steel bar", 18th World IMACS Congress and MODSIM09 International Congress on Modelling and Simulation: Interfacing Modelling and Simulation with Mathematical and Computational Sciences, MODSIM09; Cairns, QLD; 13 July 2009 through 17 July 2009; ISBN: 978-097584007-8, Proceedings, pp. 1699–1705, 2009, Code 86475, SCOPUS indexed.

[22] Abdlmanam S. A. Elmaryami, Sulaiman Bin Haji Hasan, Badrul Omar and Mohamed Elshayeb, (2009). "Unsteady state hardness prediction of industrial quenched steel bar [one and two dimensional]". Materials Science and Technology Conference and Exhibition, (MS & T'09), October 25-29, 2009, David L. Lawrence Convention Centre, Pittsburgh, Pennsylvania, USA, ISBN: 978-161567636-1, vol. 3, pp. 1514–1520, Code 79396, SCOPUS indexed.

[23] Abdlmanam S. A. Elmaryami, (2010). "Heat treatment of steel by developing finite element mathematical model and by simulation". Master's thesis. University Tun Hussein Onn, Malaysia.

[24] Herman W. Pollack, (1988). Materials Science and Metallurgy, 4th ed, Prentice-Hall, Englewood Cliffs, N.J.

[25] MDME, Manufacturing Design, Mechanical Engineering, Cambridge University. Available: http://www.ejsong.com/mdme/memmods/MEM30007A/steel/steel.html,

[26] Abdlmanam S. A. Elmaryami, (2011). "Determination LHP of axisymmetric transient industrial quenched chromium steel 8650H by developing 1-D MM, sea water cooled". International Journal of Engineering Research.

[27] Elshayeb Mohamed & Yea Kim Bing (2000). Application of finite difference and finite element methods. University Malaysia Sabah Kota Kinabalu.

[28] Saeed Moaveni (2008). Finite element analysis; theory and application with ANSYS, Pearson education international.

[29] B. Donnay, J.C Herman and V.Leroy (CRM, Belgium) U. Lotter, R. Grossterlinden and H. Pircher (Thyssen Stahl AG, Germany), Microstructure Evolution of C-Mn Steels in the Hot Deformation Process : The Stripcam Model.

[30] Bozidar Liscic (2010) System for process Analysis and hardness prediction when quenching Axially-Symmetrical workpieces of any shape in liquid quenchants, Journal of materials science form (vol. 638-642).

[31] Hsieh, Rong-Iuan; Liou, Horng-Yih; Pan, Yeong-Tsuen (2001). Effect of cooling time and alloying elements on the microstructure of the gleeble-simulated heat-affected zone of 22% Cr duplex stainless steels, journal of materials engineering and performance, volume 10, ssue 5, pp. 526-536.

[32] Saeed Moaveni, 1999, 2003. Finite Element Analysis. A Brief History of the Finite Element Method and ANSYS. 6-8.Pearson Education, Inc.

[33] Chandler, H., 1999. Hardness Testing Applications. Hardness Testing Second Edition: 111-133. United States of America: ASM International.

[34] Abdlmanam S. A. Elmaryami and Badrul Omar, (2011). "Determination LHP of Axi-symmetric Transient Molybdenum Steel-4037H Quenched in Sea Water By Developing 1-D Mathematical Model". Metalurgia-Journal of Metallurgy, Metallurgical & Materials Engineering, Vol 18 (3) 2012 p. 203-221, [Impact Factor: 0.3].
http://metalurgija.org.rs/mjom/vol18/No3/5_Elmaryami_MME_1803.pdf
http://metalurgija.org.rs/mjom/vol18.html

Chapter 7

Computer Simulation of Thermal Processing for Food

Rudi Radrigán Ewoldt

Additional information is available at the end of the chapter

1. Introduction

This chapter presents some mathematical tools that will allow a computer to make decisions in various thermal processes such as thermal conduction and diffusion applied to the preparation of processed foods and their study in the particular case of Canning.

The purpose of this chapter is to introduce the reader to the use and determination of heat transfer properties allowing a real approximation to the phenomenon under study.

Foods are complex systems that exhibit anisotropic behavior, which hinders real modeling of the phenomena that occurs within them or their interaction with the environment.

After several years of study of thermal phenomena in foods, we have identified some routes and mathematical algorithms, which are modeled on personal computers with Intel architecture. We have also corroborated the results of these models with real data of canning, pasteurizing, cooking and freezing experiments, showing less than 3% deviation between theoretical and real value.

Example Case:

A computer-aided engineering model is described. This model is capable of simulating the thermal sterilization of canned foods. The use of the model to find optimum processing conditions, physical properties and container geometry is reported.

This example describes the use of thermal properties in the development of computer models that simulate conduction heat transfer in canned foods. These models can be used to mix physical and thermal properties of the material under study for the prediction of the product temperature and processing time, considering different geometries from the metallic containers for foods and conditions of operation in pressure and temperature of the steam. The results obtained in the simulations are compared with the real processes of canning in the pilot plant. A second advantage is that the retort's temperature need not to be held constant, but can vary in any prescribed manner throughout the process and the model will predict the correct product temperature history at the can's center. The use of these models has become invaluable for simulating the process conditions in sterilizer system. Another important application of these models is the rapid evaluation of an unscheduled process deviation, such as when an unexpected drop in retort temperature occurs during the process. The model can quickly predict the product center temperature history in response to such a deviation, and calculate the delivered sterilizing value (Fo) comparison with the target value specified for the product. Specific objectives of this chapter are to briefly describe how the model was developed and to use it in process optimization and on-line computer control applications.

2. Principles of thermal processing

Generally, thermal processing is not meant to destroy all microoganisms in a packaged product. Such a process would result in low product quality due to the long heating required. Instead, the pathogenic microorganisms in a hermetically sealed container are destroyed an enviroment is created inside the package which does not support the growth of spoilage type microooganisms. In order to determine the extent of heat treatment, several factors must be known [1], (1) type ab heat resistance of the target microorganisms or enzime present in the food; (2) pH of the food; (3)heating conditions; (4) thermo-physical properties of the food and the container shape and size; and (5) storage conditions following the process.

Foods have different microooganisms and/or enzymes that the thermal process is designed to destroy. In orden to determine the type of microooganims on which the process should be based several factors must be considered. With reference to thermal processing, the most important distinction in pH classication is the dividing line between acid and low acid food. Most laboratories deading with thermal processing devote special attention to *Clostridium botulinum* which is a highly heat resistance, rod-shaped, spore-forming, anaerobic pathogen that produce botulism toxin. It has been generally accepted than C.botulinum does not grow and produce toxins below a pH of 4.6. Hence, pH as 4.5 is taken the dividing line between the low acid and acid foods. There are other microorganisms, for example *Bacillus thermoacidurans*, *B. stearotermophilus*, and *C.thermosaccolyticum*, which are more heat resistance than *C.butulinum*. These are generally thermophilic in nature (50-55ºC), and hence are not of much concern if the processed cans are stored at temperatures below 30ºC.

The phrase "minimal thermal process" was introduced by tha US Food and Drugs Administrations in 1977 and defined as "the application of heat to food, either before or after sealing in a hermetically sealed container, for a period of time and at temperature acientifically determined to be adequate to ensure the destruction of microorganims of public health concern"[2].

The C.butulinum is a microorganims of public health low-acid foods and due to this high-heat resistance, temperatures of 115-125ºC are commonly employed for processing these foods.With reference to the acid and medium-acid foods, the process is usuallly based on the heat-resistant spoilage-type vegetative bacteria or enzyme which are easily destroyed even at temperatures below 100ºC. The thermal processes for such foods are therfore normally carried out in boiling water.

3. Thermal resistance of microoganimsms

The thermal resistance of microorganisms (vegetative cells or spores) is dependent upon a number of factors: 1) the growth characteristics of the microorganisms, 2) the nature of the food in which the microorganisms are heated, and 3) the kind of food in which the heated microorganisms are allowed to grow. Because of the variability of any biological entity, thermobacteriology is a highly complex science, and variations in any of these factors can affect the heat resistance of microorganisms.

3.1. Thermal death time (TDT) tests

The amount of heat required to destroy microorganisms in a product can be determined through thermal death time (TDT) tests. TDT tests are conducted by thermobacteriologists in a laboratory. Very few food processing establishments have the facilities to conduct TDT tests on-site.

The instruments and equipment used for TDT tests include TDT retorts, tubes, and/or cans; three-neck flask, oil baths, sealed plastic pouches, and/or capillary tubes. The equipment and instrumentation used will depend on the type of product being tested – whether it is low-acid, acidified, thick puree, solid or a liquid. TDT tests involve heating a known amount of microorganisms in a buffer solution or food at several temperatures and for several time intervals at each temperature. The results from the TDT tests are used to calculate D- and z-values. These values are used to define the heat resistance of the microorganisms of concern.

3.2. Determination of D- and z-values

In conducting TDT tests, the thermal characteristics (D- and z-values) of the microorganisms will be determined. The D-value is defined as the time at a particular temperature required to reduce a known number of microorganisms by 90% or to result in a 1-log reduction. This is also termed the decimal reduction time because exposure for this length of time decreases the population by 90%, thus shifting the decimal point in the number of microorganisms remaining one place to the left. For example, if you had 100,000 spores and if exposing them to a temperature of 240°F for 3 minutes reduced the count to 10,000 spores, the $D_{240°F}$ would be 3 minutes.

The D-value decreases as the temperature increases, since it takes less time to destroy the microorganisms at the higher temperature. By determining the D- values at various temperatures, a z-value can be determined from the slope of the line that results from plotting the log of D-values versus temperature (Figure 1a). The z-value, indicative of the change in the death rate based on temperature, is the number of degrees between a 10-fold changes (1 log cycle) in an organism's resistance (Figure 1b). As an example, suppose that z = 18°F and D232°F = 3 minutes. The D250°F would be 0.3 minutes. (Because 232°F + 18°F = 250°F and 3 minutes / 10 = 0.3 minutes.) Both D- and z-values are indirectly used to establish processing conditions.

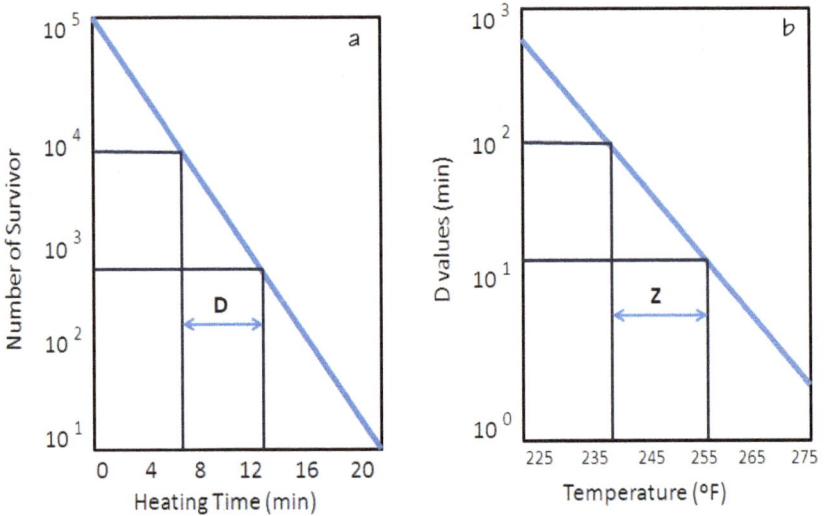

Figure 1. (A) Typical survivor curve. (B) A typical thermal resistance curve.

In other words the D value represents a heating time that causes 90% destruction of the existing microbial population. Graphically, this represents the time between which the survival curve passes through one logarithmic cycle (figure 1). Mathematically

$$D = (t_2 - t_1)/[log(a) - log(b)] \tag{1}$$

Where a and b represent the survivor counts following heating for t_1 and t_2 min, respectively.

Using regression techniques, z value can be obtained as the negative reciprocal slope of the thermal resistance curve (regression of log D values vs. temperature). Mathematically

$$Z = (T_2 - T_1)/[log(D_1) - log(D_2)] \tag{2}$$

Where D_1 and D_2 are D values at T_1 and T_2 respectively. The D values at any give temperature can be obtained from a modified formulation of the above equation using a reference D value (D_0 at a reference temperature, T_0 usually 250°F for thermal sterilization).

$$D = D_0 10^{(T_0-T)/z} \tag{3}$$

Equations 3 also can be written with reference to TDT values and z values can be obtained from:

$$Z = (T_2 - T_1)/[log(TDT_1) - log(TDT_2)] \tag{4}$$

Where TDT_1 and TDT_2 are TDT values at T_1 and T_2 respectively.

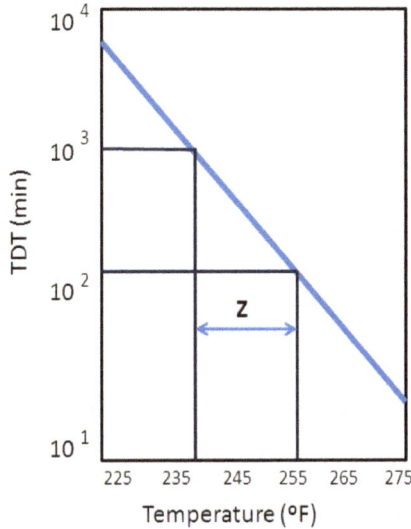

Figure 2. A typical TDT curve

Graphically, as with the D value approach, the z values can be obtained as the negative reciprocal slope of log TDT vs. temperature curve (Figure 2). When using this approach, it is advisable to plot the longest survivor time and shortest destruction time (on logarithmic scale) vs. temperature (linear scale). The regression line could be necessary to make sure that the TDT curve is above all survivor data point. The TDT curve should be parallel to the general trend of the survival and destruction points.

3.3. Lethality concept

Lethality (F value) is a measure of the treatment or sterilization processes. In order to compare the relative sterilization capacities of heat processes, a unit of lethality needs to be established. For convenience, this is defined as an equivalent heating of 1 min at a reference temperature, which is usually 250ºF (121.1ºC) for sterilization processes. Thus, the F value would represent a certain multiple or fraction of D values depending on the type of microorganisms; therefore, a relationship like Equation 4 also holds with reference to F value.

$$F_o = F10^{(T-T_o)/z} \tag{5}$$

The F_o in this case will be the F value at the reference temperature (T_o).A reference TDT curve is defines as curve parallel to the real TDT or thermal resistance curve.

4. pH of the food

Almost every food, with the exception of white eggs and soda crackers, has a pH value of less than 7. Foods can be broadly categorized on the basis of their pH as high acid, acid, medium acid or low acid. Examples of each category include:

high acid (3.7) : apples, lemons, raspberries
acid (3.7 to 4.6) : oranges, olives, tomatoes (some)
medium acid (4.6 to 5.3) : bread, cheese, carrots
low acid (over 5.3) : meat, fish, most vegetables

Most micro-organisms grow best in pH range of 6.5 to 7.5. Yeasts and moulds are capable of growing over a much broader pH range than bacteria. Few pathogens will grow below pH 4.0, such as this valuable information helps in determination of food stability with respect to microbial spoilage.[3]

5. Heating conditions

The heat transfer rate of a solid to a fluid can be expressed by Newton's Law of cooling:

$$Q = hA_s\Delta T \qquad (6)$$

Where Q is the heat flow rate (J/s), A_s is the surface area (m^2), ΔT is the temperature gradient ($^\circ$C) and the proportional constant h is the heat transfer coefficient or surface heat conductance (W/m^{2}K). The surface heat conductance depends on the thermophysical properties of fluid and solid (density, specific heat, thermal conductivity), characteristics of the solid (shape, dimensions, surface temperature, surface roughness, outgoing fluxes), and the characteristics of fluid flow (velocity, turbulence intensity) and the systems (heat transfer equipment) [4]. Although heat transfer coefficient is not a property of food materials, but it is an important parameters for designing and controlling food processing equipment where fluids (air, nitrogen, steam, water, or oil), are used as heating, cooling, frying, freezing or cooking media.

The following data are normally obtained from the heat penetration curves and heating condition for calculation purpose (Figure 3).

Autoclaves or retorts do not reach the specified operating temperature immediately after the steam is turned on, but require a measurable heating time until they reach operating temperature. The time measure from steam -on until the unit reaches the specified operating temperature is called the "come-up period"; the objective of the heat penetration test is to obtain data for the product-container system that can be used to design a sterilization process.

In processes where water is used as the heating medium, if come-up time [CUT], is long and the size of the container is small, meaningful fh and j values for the product-container unit cannot always be contained. To have the results of a heat penetration test yield meaningful fh and jch values, the CUT should preferably be less than 0.5fh.

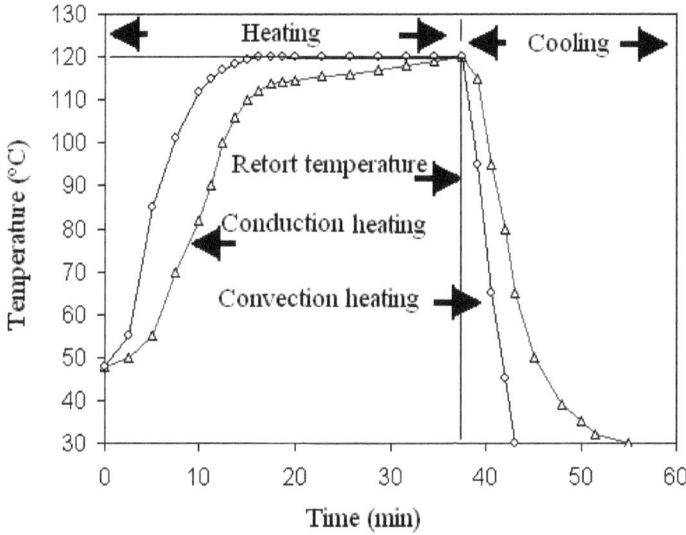

Figure 3. Heat penetration profiles of conduction and convection heating foods.

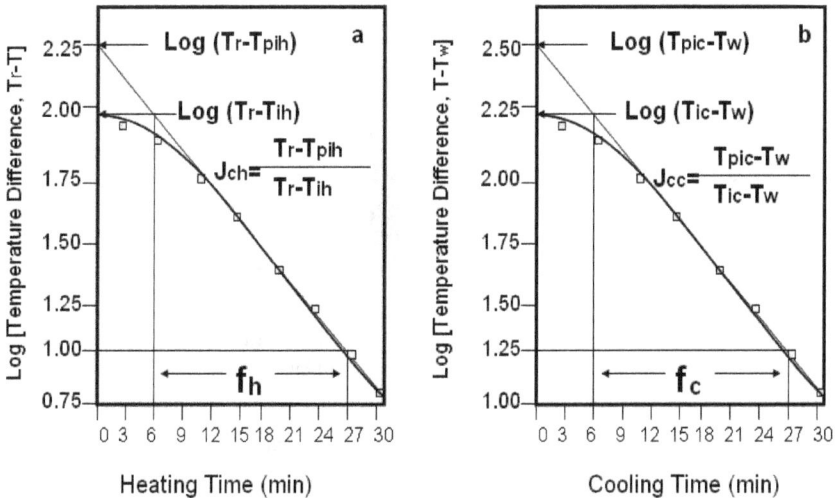

Figure 4. Heating curve and heating parameters (a), cooling curve and cooling parameters (b).

The true j value of a product container unit for ideal condition at time zero, the retort is turned on and is immediately at the operating temperature. For example the autoclave reaches the operating temperature of 121ºC after 5 min and remains at this temperature throughout the remainder of the process. The CUT correction indicates that 2 min of the 5-minutes CUT can be considered time as heating-medium temperature. The net result is the replacement of the first 5 min, the CUT in this example, with 0.42 x tc, which means neglecting the first 0.58 x tc. Therefore, in this example, the corrected zero is 2 min before the time when the retort reached the operating temperature and at 3 min after turning on the stream.

6. Thermo-physical properties of the food

Thermophysical properties, a well-known group of thermal and related properties, aere necessary for the design and prediction of heat transfer operation during handling,processing, canning, and distribution of foods. In this chapter, the most important properties associated with the transfer of heat in foods are defined. Measurement techniques, available empirical equations, and mathematical models used for prediction of density, porosity, specific heat, thermal conductivity, and thermal diffusivity are presented and condensed in tables, figures, and graphs.

6.1. Microestructure

The micro structure in foods are essential for heat flux, how figure 5 show, when seeing the different microstructures we can see some centers that absorb and soon they generate heat, changing the heat flow to inside the food

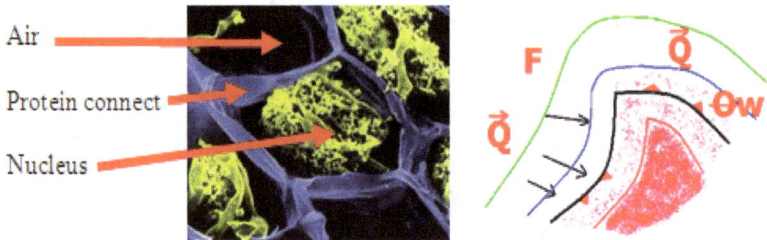

Figure 5. Vortex of Koch Function

The velocity of heat flux depends on the chaotic distribution of the center, with or without complicity of flow and thermal properties which complicity or not the flow and thermal properties affecting directly inside the food. [5]

Real products are rarely of a regular geometry, have thermal properties which vary with temperature and have different heat resistances along the boundary. For example, in retorts, when condensing steam is used as the heating method, condensation may adversely affect the uniformity of heat transfer to the product surface; heat transfer to a dry surface will be very high, but the presence of a film of liquid will reduce the heat transfer rate [6]

Non-isotropic aspects of conductive cooking have been addressed, for example, by Pan *et al.* [7] in the modeling of the cooking of frozen hamburgers. Their approach, which involved unequal cooking to both the major external surfaces of the patty, considered the enthalpy changes associated with the melting of ice and fat as well as resulting mass transfer effects.

In the numerical data analysis to the heat equation incorporated the function of Koch, for the model heat transfer with equation:

$$Cm\frac{T_m^{i+1} - T_m^i}{\Delta t} = \sum_n \frac{T_n^i - T_m^i}{R_{mn}^{cond}} + \sum_n \frac{T_n^i - T_m^i}{R_{mn}^{Rad}} + Q_v''' \Delta V_m \tag{7}$$

When:

$$R_{mn}^{Rad} = \frac{1}{A_n \Gamma_{mn} \sigma \left(T_n^2 + T_m^2\right)\left(T_n T_m\right)} \tag{8}$$

Iterative equation:

$$T_m^{i+1} = \left(1 - \frac{\Delta t}{Cm} \sum_n \frac{1}{R_{mn}}\right) T_m^i + \frac{\Delta t}{Cm} \sum_n \frac{T_n^i}{R_{mn}} \tag{9}$$

Valid for all coordinates as:

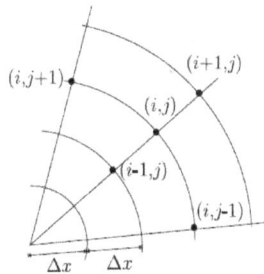

Figure 6. Generated, 160 x 51 iterative spaces nodes in two dimensions (X,Y)

The result of the iterative generates thermograms (8211 nodes per layer), this represents the distribution of heat inside of the cans. [8]

7. Example case

7.1. Introduction

This case describes the use of thermal properties in the development of computer models that simulate conduction heat transfer in canned foods. These models can be used to mixture physical and thermal properties of the material to predict the product temperature and processes time, considering different geometries from the metallic containers for foods and conditions of operation in pressure and temperature of the steam; the results obtained in the simulations compare with the real processes of the canning in the pilot plant. A second advantage is that the retort temperature need not to be held constant, but can vary in any prescribed manner throughout the process and the model will predict the correct product temperature history at the can center. Use of these models has become invaluable for simulating the process conditions experience in sterilizer system, in which cans pass from the can wall. Another important application of these models is the rapid evaluation of an unscheduled process deviation, such as when an unexpected drop in retort temperature occurs during the process. The model can quickly predict the product center temperature history in response to such a deviation, and calculate the delivered sterilizing value (Fo) comparison with the target value specified for the product. Specific objectives of this chapter are to briefly describe how the model was developed and use in process optimization and on-line computer control applications.

7.2. Method of model development

An attempt was made to define all the physical aspects of the mathematical models developed by Ball [9], using numerical methods, the trapezoidal rule of Patashnik [10]. The disadvantage that raises these traditional methods combined is the absence of the physical properties of foods, all the preceding models to considered a heat coefficient global and deals with it like a solid or block metal, the disadvantage appears when the selected system of heat transference is by conduction or convection.

The foods do not have a linear or logical behavior but an anisotropic behavior and it is a big obstacle, therefore it is little probable to design a system that models the real phenomena of heat transference in no stationary system. The simulation model considers the following aspects:

a. Generation of data composed of format of tins, temperatures of operation and steam, physical properties of the product and liquid of cover.
b. Calculation of physical values of the canning, and verifies the conditions heat transference to the interior of the package.
c. It generates point to point the increase of heat in the cold point and the time necessary to arrive at the temperature of operation
d. It determines the increase of time in optimum conditions of sterilization.

The model consists of four main programs and an information administrator, which work sequentially according to the directives of the user, to include better the process describes next to the sequences and postulates [11].

7.3. Transitory thermal response

The supposition is that it is hoped that the temperature gradients within the system are insignificant when the internal resistance to the heat transfer is small compared with the external resistance, that is to say the heat conduction by its length divided by the thermal conductivity, this relation gives origin to the adimensional Number (Biot);this number represents the relation between the form (plate and infinite cylinder) and the transitory answer, a value of Biot<0.1, it guarantees that the temperature in center does not differ more from a 5%.

For the model the number of Biot equal to 1000 guarantees that the thermal center temperature is different to the surface temperature of the product, and it should be find two functions, one for a plate and other for a cylinder, when uniting these infinite bodies generates a body finite. as shows figure to it 7. The resulting function is the sum of figures 8 and 9.

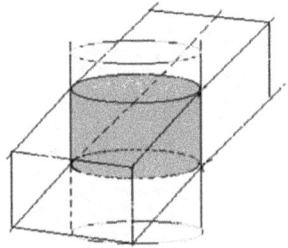

Figure 7. Finite cylinder

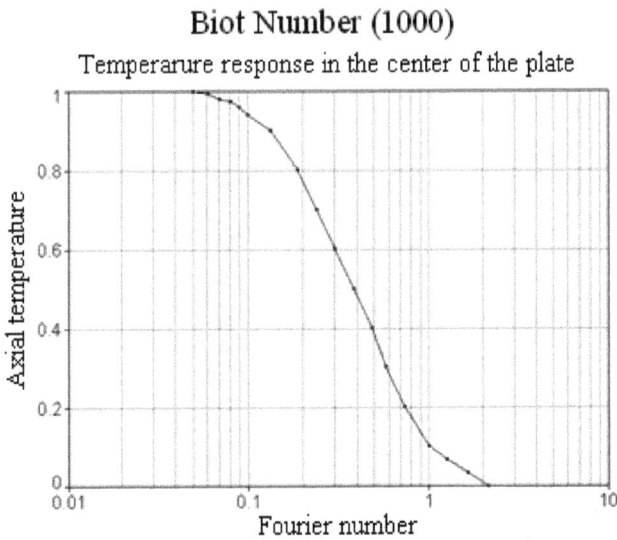

Figure 8. Biot number for infinite plate

Of these figures two dependent equations of Fourier are generated:

Plate:

$$Y = a + b(lnx) + c(lnx)^2 + d(lnx)^3 + e(lnx)^4 +$$

$$+f(lnx)^5 + g(lnx)^6 + h(lnx)^7 + i(lnx)^8 + j(lnx)^9 \tag{10}$$

Biot Number (1000)

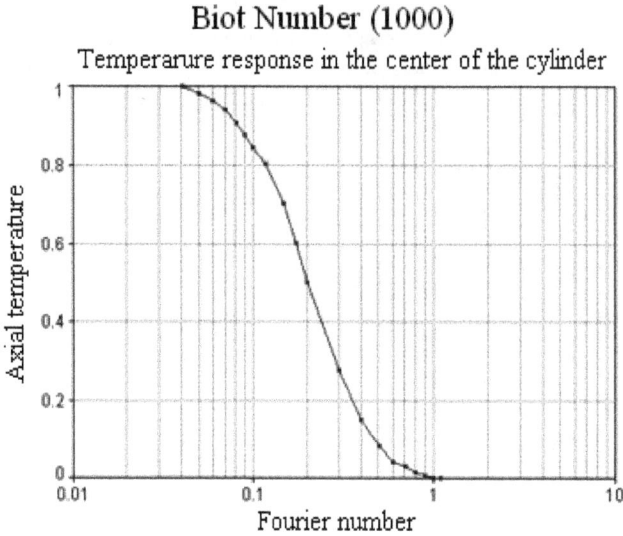

Figure 9. Biot number for Infinite Cylinder

Where:

a= 0.10168949 **b**= -0.21732379 **c**= 0.3032664 **d**= -0.10976533 **e**= -0.36311129
f= -0.052222376 **g**= 0.21755337 **h**= 0.16261937 **i**= 0.04545908 **j**= 0.0045689637

With r²= 0.999859613

Cylinder:

$$y^{-1} = a + bx + cx^2 + dx^3 + ex^4 + fx^5 \tag{11}$$

Where:

a= 0.89545265 **b**= 3.193429 **c**= -26.00579 **d**= 272.66439 **e**= -553.74231 **f**=540.91268

With r²= 0.9998774042

7.3.1. Solution of heat transference in multidimensional no stationary state.

The temperature distribution T (x, and, t) of an infinite body that submerges in a solution where St is defines as:

$$\vartheta = \frac{T - Te}{To - Te} \tag{12}$$

Adimensional temperature with boundary conditions $-L_1 \le x \le L_1$, y, $-L_2 \le x \le L_2$

$$\frac{\partial \vartheta}{\partial t} = \alpha \left(\frac{\partial^2 \vartheta}{\partial x^2} + \frac{\partial^2 \vartheta}{\partial y^2} \right) \quad t = 0: \qquad \vartheta = 1: \ x = 0 \ : \frac{\partial \vartheta}{\partial x} = 0 \ : \ y = 0 \ : \frac{\partial \vartheta}{\partial y} = 0 :$$

$$x = L_1 : -k \frac{\partial \vartheta}{\partial x} = h1\vartheta : \ y = L_2 : -k \frac{\partial \vartheta}{\partial y} = h2\vartheta \tag{13}$$

Using mathematical method

$$\vartheta(t,x,y) = T(t)X(x)Y(y),$$

$$\vartheta 1 = \frac{T1 - Te}{To - Te} \qquad \vartheta 2 = \frac{T2 - Te}{To - Te}$$

$$\frac{\partial \vartheta 1}{\partial t} = \alpha \left(\frac{\partial^2 \vartheta}{\partial x^2} \right) \qquad \frac{\partial \vartheta 2}{\partial t} = \alpha \left(\frac{\partial^2 \vartheta}{\partial y^2} \right) \tag{14}$$

$$t = 0: \quad J1 = 1 \qquad\qquad t = 0: \quad J2 = 1$$

$$x = 0 \ : \frac{\partial \vartheta 1}{\partial x} = 0 \qquad y = 0 \ : \frac{\partial \vartheta 2}{\partial y} = 0$$

$$x = L_1 : -k \frac{\partial \vartheta 1}{\partial x} = h1\vartheta 1 \qquad y = L_2 : -k \frac{\partial \vartheta 2}{\partial y} = h2\vartheta 2$$

The product of the solutions satisfies the original problem:

$$\vartheta(t,x,y) = \vartheta 1(t,x)\vartheta 2(t,y)$$

As it is a finite cylinder and it is the result of the union of an infinite cylinder and an infinite slab, the previous equation stays expressed as:

$$\vartheta = P(t,x)C(r,t)$$

$$x = L1 : -k \frac{\partial \vartheta}{\partial x} = \vartheta 2 \left(-k \frac{\partial \vartheta 1}{\partial x} \right) = \vartheta 2 \left(hc1\vartheta 1 \right) = hc1\vartheta \tag{15}$$

$$y = L2 : -k \frac{\partial \vartheta}{\partial y} = \vartheta 1 \left(-k \frac{\partial \vartheta 2}{\partial x} \right) = \vartheta 1 \left(hc2\vartheta 2 \right) = hc2\vartheta$$

Where:

h_c = convective coefficient

y = radio

As we can see the law of Fourier is implicit in the previous equations and for that reason we will only denote the use of the physical parameters in the use of these:

$$F_o = \frac{\alpha t}{L_2} \ o \ F_o = \frac{t}{t_c} \tag{16}$$

And

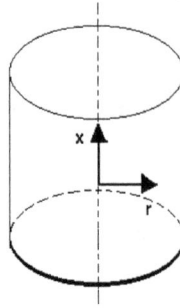

Figure 10. Finite Cylinder

$$\alpha = \frac{K}{(Cp*\rho)} \tag{17}$$

Where $t_c = L^2/\alpha$

α = thermal diffusivity
K= thermal conductivity
Cp= Specific heat
ρ = density appears.

7.4. Conduction or convection

One of the controversial subjects in the thermal transmission to the interior of the tin, since this allows us to know the coldest point the interior of the tin, and is there where the microorganisms proliferate and contaminates to the product, in practice it is said that if the product is solid the transference is by conduction and if he is liquid is convection, but What happens to food when liquids materials and solids materials are mixed ?, the usual thing to do is work them like solids, but this is not correct. A model for a porous and semisolid material considers the total factor of porosity of the package like:

$$PF = \frac{(\rho_{solid} - \rho_{average})}{\rho_{solid}} \tag{18}$$

In this way it can be compared the critical volume (V_c) of the product with the corrected volume and if the relation of the absolute value (V_t), when $V_c < 3/4V_T$ then transference is convective and when $V_c \geq 3/4V_T$ then the transference is conductive, this empirical relation allows us to increase the time necessary to assure a suitable commercial sterilization in the cold point.

7.5. Results

The output data of the simulation are graphical and shows the behavior through a temperature curve, which shows the temperature of operation of the retort and the thermal center. Once finalized it calculates the time necessary to equal the temperatures and to incorporate the time of commercial sterilization, as it shows the following figure 12 and table 1:

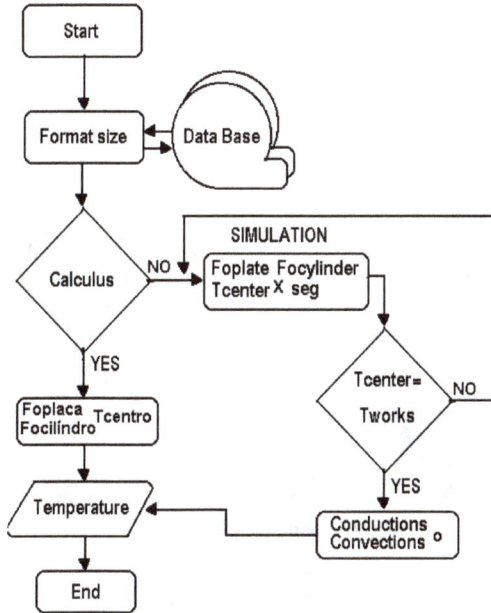

Figure 11. Flow Chart

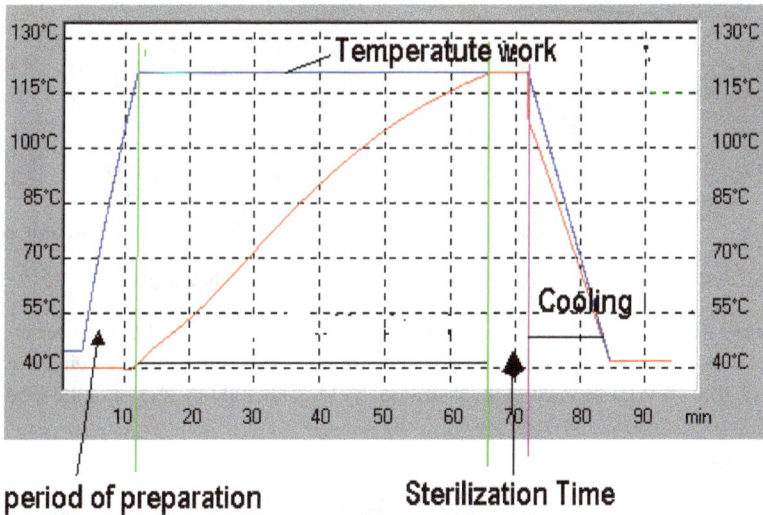

Figure 12. Characteristic line of a process of commercial sterilization

Product	Real Time	Simulation Time	St.Deviation
Pears	25min	24.99min	-0.01
Peach trees	35min	35min	0.00
Seafood's	55 min	55.1min	0.10
Peas with Bacon	56min	56min	0.00

Table 1. Process Simulated by TDT and Real date

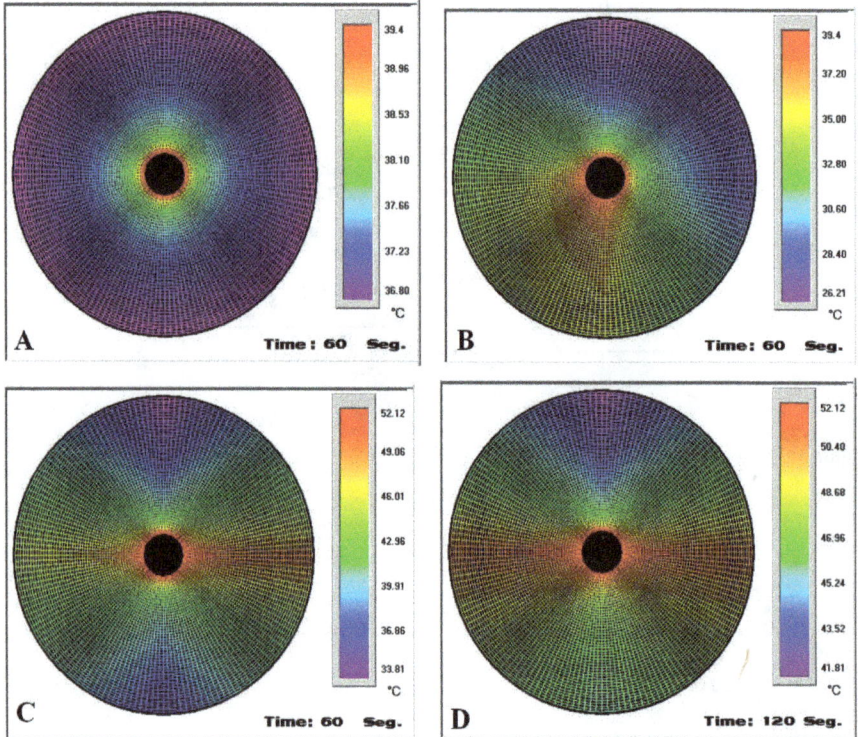

Figure 13. A Theory model, B Real flux heat in apple, C and D Real flux heat in pear

The theory analysis generates heat flux concentric like figure 13 without concerning the material, but when incorporating the Koch equation thermogram changes their form with the fig 13, B, C and D, this must to the distribution of the fiber to the interior of the food

8. Conclusion

As the results showed, the model proposed to the reader is a better approximation than the models of common usage. In practice, usually you overestimate the sterilization time for guarantee a better cleaning of pathogenic organisms and as consequence the loss of nutritional quality of the product. With the new model this is not a problem because of the better approximations in time, usage of physical properties and for the transference heat mechanisms.

In comparison, the data obtained from the simulation tests with real data obtained from experiments in the pilot plant were very close to the real data.

In computational terms, the new model showed a considerable improvement of the simulation average execution time (10 seconds.) This was compared with a normal simulation process that is about 25 minutes to 60 minutes.

An advantage of the new simulation is that from the thermograms shows second by second what is happening with the product during the process. This is not possible with the methods used normally.

As said before the average time of the simulations were low, making it a good choice for decision making in terms of industrial processes as for commercial decisions.

The numerical methods well formulated are a powerful tool for the decision making. In this case, finite differences and the finite volumes were used for the development of the work and the study that is shown in this chapter.

Nomenclature

$0.58l$ Effective beginning of the process; the retort come-up period varies from one process to the other and from one retort; to the other; in process evaluation procedures, about 42% of this come-up period generally considered as time at retort temperature because the product temperature increases even during this period.

α Thermal diffusivity

A_n Opacity factor

A_s Surface area (m^2)

B   Thermal process time; Ball-corrected for come-up period.

Cm   Control node

cp   Specific heat

CUT   Come-up time

D   Represents a heating time that 90% destruction of the existing microbial population

Γ_{mn} Emissivity angular and radial

ϱ Density appears

$\varrho_{average}$ Density appears of average

ϱ_{solid} Density appears of solid

F   Lethality value

f_c Cooling rate index; the time required for the straight line portion of the cooling curve (Figure 4 b) to pass through one log cycle; also the negative reciprocal slope of the cooling rate curve.

f_h	Heating rate index; the time required for the straight line portion of the heating curve,(Figure 4 a) to pass through one log cycle; also the negative reciprocal slope of the heating rate curve.
F_o	Initial lethality value
Fo	Fourier number
ϑ	Temperature distribution
g	Difference between the retort temperature and food temperature at time t.
g_c	The value of g at the end of heating or beginning of cooling.
h	Proportional constant of heat transfer coefficient
hc	Convective coefficient
I_c	Difference between the cooling water temperature and food temperature at the start of the cooling process.
I_h	Difference between the retort temperatures at the start of the heating process
j_{cc}	Cooling rate lag factor; a factor which when multiplied by Ic, locate the intersection of the extension of the straight-line portion of semilog cooling curve and the vertical line representing start of the cooling process.
j_{ch}	Heating rate lag factor, a factor which, when multiplied by I_h locates the intersection of the extension of the straight-line portion of the semilog heating curve and the vertical line representing the effective beginning of the process.
k	Thermal conductivity
l	Come-up period; in batch processing operation, the retort requires some time for reaching the operating condition; the time from stream to when the retort reached T_r is called the come-up period
PF	Porosity factor
pH	Hydrogen potential
P_t	Operator's process time
Q	Heat low rate
R_{mn}^{cond}	Conduction resistance radial and angular
R_{mn}^{Rad}	Radiation resistance radial and angular
t	Time (min)
T	Temperature ($^{\circ}$C)
TDT	Thermal death time
Te	Centre temperature
T_{ic}	Food temperature when cooling started
T_{ih}	Initial food temperature when heating is started.
To	Initial temperature
T_{pic}	Pseudo-initial temperature during cooling; temperature indicated by the intersection of the extension of the cooling curve and the vertical line representing the start of cooling
T_{pih}	Pseudo-initial temperature during heating; temperature indicated by the intersection of the extension of the heating curve and the vertical line representing the effective beginning of the process (0.58l).
T_r	Retort temperature

T_w	Cooling water temperature
T_m^i	Iteratively radial temperature
T_n^i	Iteratively angular temperature
ΔT	Temperature gradient
ΔV_m	Volume radial gradient
V_c	Critical volume
V_t	Absolute volume
x	Length (m)
y	Radio (m)
Z	Is the change in death rate based on temperature

Author details

Rudi Radrigán Ewoldt
Faculty of Agricultural Engineering, Agroindustries Department,
Development of Agroindustries Technology Center, University of Concepción, Chile

9. References

[1] Fellows P. Food Processing Technology: Principles and Practices, Ellis Horwood, Chichester England;1988.

[2] Lopez A. A complete Course in Canning and Related Processes, 12th ed, Baltimore, 1987.

[3] Ministry of Agriculture, Food and rural Affairs, Ontario Canada. http://www.omafra.gov.on.ca/english/food/industry/food_proc_guide_html/chapter_5.htm (accessed 8 February 2012)

[4] Arce J.Sweat V. Survey of published heat transfer coefficients encountered in food refrigeration processes, Ashrae Trans 1980;86,228

[5] NASA,Bach to Chaos:Chaotic variations on a Classical Theme, Sciene News, 1994; p428

[6] VERBOVEN P, NICOLAΪ¨ B, SCHEERLINCK N, DE BAERDEMAEKER J . The local surface heat transfer coefficient in thermal food process calculations: a CFD approach. *J. Food Eng.*1997.., 33, 15–35.

[7] PAN Z, SINGH P, RUMSEY R Predictive modeling of contact-heating process for cooking a hamburger patty. *J. Food Eng.*2000., 46, 9–19

[8] Radrigán R. Prototipo para la determinación de propiedades termo físicas en alimentos semisólidos PhD, ETSI-UNED España 2007

[9] Ball Co. Thermal process time for canned food. Bulletin of the National Resources Council 1923, 7.1

[10] Graham, Ronald L, Knuth, Donald E, and Patashnik, Oren. Concrete Mathematics: A Foundation for Computer Science, 2nd ed. Reading, MA: Addison-Wesley, 1994

[11] Radrigán R, Cañumir J. Computer simulation of thermal processing for canned food, 7th World Congress on Computers in Agriculture Conference Proceedings, 22-24 June 2009, Reno, Nevada

Analytical Solution of Heat Transfer

Analytical Solutions to 3-D Bioheat Transfer Problems with or without Phase Change

Zhong-Shan Deng and Jing Liu

Additional information is available at the end of the chapter

1. Introduction

Theoretical analysis on the bioheat transfer process has been an extremely important issue in a wide variety of bioengineering situations such as cancer hyperthermia, burn injury evaluation, brain hypothermia, disease diagnostics, thermal comfort analysis, cryosurgery and cryopreservation etc. In this chapter, the theoretical strategies towards exactly solving the three-dimensional (3-D) bioheat transfer problems for both cases with and without phase change were systematically illustrated based on the authors' previous works. Typical closed form analytical solutions to the hyperthermia bioheat transfer problems with space or transient heating on skin surface or inside biological bodies were summarized. In addition, exact solutions to the 3-D temperature transients of tissues under various phase change processes such as cryopreservation of biomaterials or cryosurgery of living tissues subject to freezing by a single or multiple cryoprobes were also outlined. Such solution is comprehensive enough by taking full account of many different factors such as generalized initial and boundary conditions, blood perfusion heat transfer, volumetric heating of hyperthermia apparatus or heat sink of cryoprobes etc. For illustrating the applications of the present methods, part of the solutions were adopted to analyze the selected bioheat transfer problems. The versatility of these theoretical approaches to tackle more complex issues was also discussed. The obtained solutions are expected to serve as the basic foundation for theoretically analyzing bioheat transfer problems.

2. Motivations of analytical solutions to bioheat transfer problem

Analytical solutions to bioheat transfer problems are very important in a wide variety of biomedical applications [1]. Especially, understanding the heat transfer in biological tissues involving either raising or lowering of temperature is a necessity for many clinical practices such as tumor hyperthermia [2], burn injury evaluation [3, 4], brain hypothermia resuscitation [5], disease thermal diagnostics [6], thermal comfort analysis [7], cryosurgery planning [8, 9], and cryopreservation programming [10]. The bioheat transfer problems involved in the above applications can generally be divided into two categories: with and without phase change. In this chapter, the phase change especially denotes the solid-liquid phase transition of biological hydrated tissues. The cases without phase change usually include tumor hyperthermia, burn injury evaluation, brain hypothermia resuscitation, disease diagnostics, and thermal comfort analysis, while the cases with phase change include cryosurgery and cryopreservation.

To guarantee optimal clinical outputs for such applications, it is essential to predict in advance the transient temperature distribution of the target tissues. For example, in a tumor hyperthermia process, the primary objective is to raise the temperature of the diseased tissue to a therapeutic value, typically above 43°C, and then thermally destroy it [11]. Temperature prediction would be used to find an optimum way either to induce or prevent such thermal damage to the target tissues. In contrast to the principle of hyperthermia, cryosurgery realizes its clinical purpose of controlled tissue destruction through deep freezing and thawing [12]. Applications of this treatment are quite wide in clinics owning to its outstanding virtues such as quick, clean, relatively painless, good homeostasis, and minimal scaring. An accurate understanding of the extent of the irregular shape of the frozen region, the direction of ice growth, and the temperature distribution within the ice balls during the freezing process is a basic requirement for the successful operation of a cryosurgery. Therefore, solving the bioheat transfer problems involved is very important for both hyperthermia and cryosurgery. Moreover, in thermal diagnostics, thermal comfort analysis, brain hypothermia resuscitation, and burn injury evaluation, similar bioheat transfer problems are also often encountered [13].

It is commonly accepted that mathematical model is the basis for solving many practical problems. Because modeling bioheat transfer is of the utmost importance in many biomedical applications such as proper device or heating/cooling protocol design, a number of bioheat transfer equations for living tissue have been proposed since the landmark work by Pennes published in 1948 [14], in which the perfusion heat source/sink was introduced. Until now, the classical Pennes equation is also commonly accepted as the best practical approach for modeling bioheat transfer in view of its simplicity and excellent validity [15]. This is because most of the other models either still lack sound experimental grounding or just appear too complex for mathematical solution. Although the real anatomical geometry of a biological body can be incorporated, the Pennes equation remains the most useful model for characterizing the heat transport process in most biomedical applications. For brevity, here only cases for space-dependent thermal properties will be mainly discussed. Then a generalized form of the Pennes equation for this purpose can be written as:

$$\rho c \frac{\partial T(\mathbf{X},t)}{\partial t} = \nabla \cdot k(\mathbf{X}) \nabla \left[T(\mathbf{X},t) \right] + \rho_b c_b \omega_b(\mathbf{X}) \left[T_a - T(\mathbf{X},t) \right] \\ + Q_m(\mathbf{X},t) + Q_r(\mathbf{X},t), \quad \mathbf{X} \in \Omega \tag{1}$$

where, ρ, c are the density and the specific heat of tissue, respectively; ρ_b and c_b denote the density and the specific heat of blood, respectively; $\mathbf{X}$ contains the Cartesian coordinates x, y and z; Ω denotes the analyzed spatial domain; $k(\mathbf{X})$ is the space dependent thermal conductivity; and $\omega_b(\mathbf{X})$ is the space dependent blood perfusion. The value of blood perfusion represents the blood flow rate per unit tissue volume and is mainly from microcirculation including the capillary network plus small arterioles and venules. T_a is the blood temperature in the arteries supplying the tissue and is often treated as a constant at 37°C; $T(\mathbf{X},t)$ is the tissue temperature; $Q_m(\mathbf{X},t)$ is the metabolic heat generation; and $Q_r(\mathbf{X},t)$ the distributed volumetric heat source due to externally applied spatial heating.

From the historical viewpoint, we can find that the development of the bioheat transfer's art and science can be termed as one to modify and improve the Pennes model [15]. Among the many efforts, the blood perfusion term in the Pennes equation has been substantially studied which led to several conceptually innovative bioheat transfer models such as Wulff's continuum model [16], Chen-Holmes model addressing both the flow and perfusion properties of blood [17], and the Weinbaum-Jiji three-layer model to characterize the heat transfer in the peripheral tissues [18]. The bioheat transfer equation and its extended forms can be directly used to characterize the thermal process of the biological bodies subject to various external or interior factors such as convective interaction with a heated or cooled fluid, radiation by fire or laser, contact with a heating or freezing apparatus, electromagnetic effect, or a combination among them. Such issues can be treated using different boundary conditions as well as spatial heating or freezing patterns. Generally, the geometric shape, dimensions, thermal properties and physiological characteristics for tissues, as well as the arterial blood temperature, can be used as the input to the Pennes equation for a parametric study. According to a specific need in clinics, the bioheat transfer model can even be modified by taking more factors into concern [19]. Traditionally, for solving bioheat transfer problems, people relied too heavy on numerical approaches such as finite difference method, finite element method, and boundary element method etc. Numerical simulation is necessary when the analytical solutions are not available. But if both analytical and numerical solutions can be obtained for the same issue, the analytical one is often preferred. Except for its simplicity being used to compile computer codes, the analytical solution is very attractive since its efficiency depends weakly on the dimensions of the problem, in contrast to the numerical methods. For analytical method, solution at a desired point can be performed independently from that of the other points within the domain, which can be an asset when temperatures are needed at only some isolated sites or times. But for most of the conventional numerical methods (except Monte Carlo simulation), the temperatures at all mesh points must be simultaneously computed even when only the temperatures at a single point are needed [20]. In this sense, the analytical solution will save computational time greatly, which is valuable in clinical practices.

Based on the above considerations, we aimed in this chapter to present several typical closed form analytical solutions to bioheat transfer problems with or without phase change, in which relatively complex boundary or heating/cooling conditions, and existence of discrete large blood vessel were included. Derivation of the solutions was mainly based on the Green's function method, which is beneficial for dealing with the non-homogeneous problems with spatial or transient heating source and initial temperature distribution, as well as complex cooling or boundary conditions. For generalized and practical purpose, complex bioheat transfer problems encountered in several typical clinical applications as well as basic studies such as tumor hyperthermia, cryosurgery, cryopreservation, and interpretation of physiological phenomena etc. will be especially addressed.

3. Bioheat transfer problems without phase change

3.1. Generalized analytical solutions to 3-D bioheat transfer problems

Derivation of the solutions was based on the Green's function method, since the Green's function obtained for the differential equation is independent of the source term. Therefore it can be flexibly used to calculate the temperature distribution for various spatial or temporal source profiles. Furthermore, the Green's function method is capable of dealing with the transient or space-dependent boundary conditions. Up to now, quite a few studies have applied the Green's function method to solve the bioheat transfer problems [21-25]. However, in most of the existing analytical studies, the available solutions to the bioheat transfer problem are for the cases with one dimensional geometry, steady state, infinite domain, constant heating, or heat conduction equations not considering blood perfusion, which may not be practical for some real bio-thermal situations. In this section, the generalized analytical solutions, which have incorporated relatively complex situations such as the 3-D tissue domain, the transient or space-dependent boundary conditions, and volumetric heating, were especially addressed. Such solutions are expected to be very useful in a variety of bio-thermal practices. The 3-D computational domain with widths $s_1 = s_2 = 0.08$m and height $L = 0.03$m was depicted as the shadowed region in Fig. 1, where s_1 and s_2 were widths of the tissue domain to be analyzed in y and z directions, respectively; the skin surface was defined at $x=0$ while the body core at $x=L$.

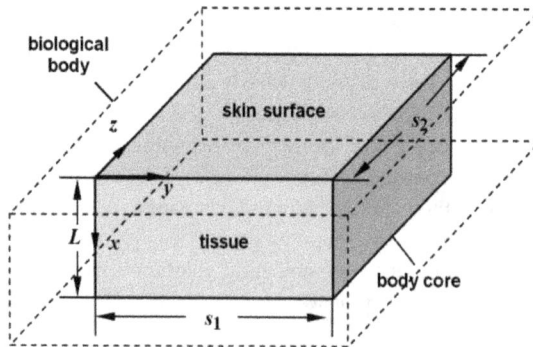

Figure 1. Calculation geometry for 3-D case [13]

For brief, only 3-D case with constant thermal parameters will be particularly studied, which is a good approximation when no phase change occurred in tissue. The corresponding 3-D Pennes equation can be derived from Equation (1) as:

$$\rho c \frac{\partial T}{\partial t} = k \frac{\partial^2 T}{\partial x^2} + k \frac{\partial^2 T}{\partial y^2} + k \frac{\partial^2 T}{\partial z^2} + \omega_b \rho_b c_b (T_a - T) + Q_m + Q_r(x,y,z;t) \tag{2}$$

The generalized boundary conditions (BCs) often encountered in a practical clinical situation can be written as:

$$-k \frac{\partial T}{\partial x} = f_1(y,z;t), \ x=0 \tag{3}$$

or

$$-k \frac{\partial T}{\partial x} = h_f \left[f_2(y,z;t) - T \right], \ x=0 \tag{4}$$

where, $f_1(y,z;t)$ is the time-dependent surface heat flux, $f_2(y,z;t)$ is the time-dependent temperature of the cooling medium, and h_f is the heat convection coefficient between the medium and the skin surface. In this chapter, Equation (3) was named the second BC and Equation (4) the third BC.

The body core temperature was regarded as a constant (T_c) on considering that the biological body tends to keep its core temperature to be stable, i.e.

$$T = T_c, \ x=L \tag{5}$$

The BCs at y and z directions can be expressed as

$$-k \frac{\partial T}{\partial y} = 0, \ y=0 \tag{6}$$

$$-k \frac{\partial T}{\partial y} = 0, \ y=s_1 \tag{7}$$

$$-k \frac{\partial T}{\partial z} = 0, \ z=0 \tag{8}$$

$$-k \frac{\partial T}{\partial z} = 0, \ z=s_2 \tag{9}$$

The reason for adopting the adiabatic conditions in the two ends of the y and z directions is from the consideration that at the positions far from the beam center of the heat deposition apparatus, the temperatures there were almost not affected by the external heating, which generally has a strong decay in y and z directions. However, it should be mentioned that more generalized boundary conditions in y and z directions can also be dealt with by the present approach, but they were not listed here for brevity.

The initial temperature is

$$T(x,y,z;0) = T_0(x,y,z), \quad t = 0 \tag{10}$$

where, $T_0(x,y,z)$ can be approximated by the 1-D solution, representing the initial temperature field for the basal state of biological bodies, which can be obtained through solving the following equation sets:

$$
\begin{cases}
k\dfrac{d^2 T_0(x)}{dx^2} + \omega_b \rho_b c_b \left[T_a - T_0(x) \right] + Q_m = 0 \\
T_0(x) = T_c, \qquad x = L \\
-k\dfrac{dT_0(x)}{dx} = h_0 \left[T_f - T_0(x) \right], \qquad x = 0
\end{cases}
\tag{11}
$$

where, h_0 is the apparent heat convection coefficient between the skin surface and the surrounding air under physiologically basal state and is an overall contribution from natural convection and radiation, and T_f the surrounding air temperature.

The solution to Equation (11) is:

$$
T_0(x) = T_a + \frac{Q_m}{\omega_b \rho_b c_b} + \frac{\left(T_c - T_a - \dfrac{Q_m}{\omega_b \rho_b c_b} \right) \cdot \left[\sqrt{A}\,ch\left(\sqrt{A}x\right) + \dfrac{h_0}{k} sh\left(\sqrt{A}x\right) \right]}{\sqrt{A}\,ch\left(\sqrt{A}L\right) + \dfrac{h_0}{k} sh\left(\sqrt{A}L\right)}
$$

$$
+ \frac{\dfrac{h_0}{k}\left(T_f - T_a - \dfrac{Q_m}{\omega_b \rho_b c_b} \right) \cdot sh\left[\sqrt{A}(L-x) \right]}{\sqrt{A}\,ch\left(\sqrt{A}L\right) + \dfrac{h_0}{k} sh\left(\sqrt{A}L\right)}
\tag{12}
$$

where, $A = \omega_b \rho_b c_b / k$. Through using the following transformation [13]

$$T(x,y,z;t) = T_0(x,y,z) + W(x,y,z;t) \cdot \exp\left(-\frac{\omega_b \rho_b c_b}{\rho c} t \right) \tag{13}$$

Equation (2) was transformed to the following form:

$$\frac{\partial W}{\partial t} = \alpha \frac{\partial^2 W}{\partial x^2} + \alpha \frac{\partial^2 W}{\partial y^2} + \alpha \frac{\partial^2 W}{\partial z^2} + \frac{Q_r(x,y,z;t)}{\rho c} \cdot \exp\left(\frac{\omega_b \rho_b c_b}{\rho c} t \right) \tag{14}$$

where, $\alpha = k/\rho c$ is the thermal diffusivity of tissue. The corresponding boundary and initial conditions are:

$$-k\frac{\partial W}{\partial x} = g_1(y,z;t), \quad x = 0 \tag{15}$$

$$-k\frac{\partial W}{\partial x} = h_f\left[g_2(y,z;t) - W\right], \quad x = 0 \tag{16}$$

$$W = 0, \quad x = L \tag{17}$$

$$-k\frac{\partial W}{\partial y} = 0, \quad y = 0 \tag{18}$$

$$-k\frac{\partial W}{\partial y} = 0, \quad y = s_1 \tag{19}$$

$$-k\frac{\partial W}{\partial z} = 0, \quad z = 0 \tag{20}$$

$$-k\frac{\partial W}{\partial z} = 0, \quad z = s_2 \tag{21}$$

$$W(x,y,z;0) = 0, \quad t = 0 \tag{22}$$

where,

$$g_1(y,z;t) = \left[k\frac{\partial T_0}{\partial x}\bigg|_{x=0} + f_1(y,z;t)\right] \cdot \exp\left(\frac{\omega_b \rho_b c_b}{\rho c}t\right) \tag{23}$$

$$g_2(y,z;t) = \left[\frac{k}{h_f}\frac{\partial T_0}{\partial x}\bigg|_{x=0} - T_0\big|_{x=0} + f_2(y,z;t)\right] \cdot \exp\left(\frac{\omega_b \rho_b c_b}{\rho c}t\right) \tag{24}$$

Using Green function method, $W(x,y,z;t)$ can be solved from the combined Equations (14-24). The Green's functions G_1 and G_2 to the second and third BCs can finally be obtained:

$$G_1(x,y,z,t;\xi,\vartheta,\varsigma,\tau) =$$
$$\frac{2}{Ls_1 s_2}\sum_{l=1}^{\infty}\sum_{m=0}^{\infty}\sum_{n=0}^{\infty} R_1 R_2 e^{-\alpha(\beta_l^2+\gamma_m^2+\psi_n^2)(t-\tau)} \cos\beta_l x \cos\beta_l\xi \cos\gamma_m y \cos\gamma_m\vartheta\cos\psi_n z \cos\psi_n\varsigma \tag{25}$$

$$G_2(x,y,z,t;\xi,\vartheta,\varsigma,\tau) =$$
$$\sum_{p=1}^{\infty}\sum_{m=0}^{\infty}\sum_{n=0}^{\infty} c_{pmn} e^{-\alpha(\beta_p^2+\gamma_m^2+\psi_n^2)(t-\tau)} \sin\beta_p(L-x)\sin\beta_p(L-\xi)\cos\gamma_m y \cos\gamma_m\vartheta\cos\psi_n z \cos\psi_n\varsigma \tag{26}$$

where,

$$\beta_l = \frac{2l-1}{2L}\pi, l = 1,2,3,... \tag{27}$$

$$\gamma_m = \frac{m\pi}{s_1}, m = 0,1,2,... \tag{28}$$

$$\psi_n = \frac{n\pi}{s_2}, n = 0,1,2,... \tag{29}$$

$$R_1 = \begin{cases} 1, & m=0 \\ 2, & m=1,2,3,... \end{cases} \tag{30}$$

$$R_2 = \begin{cases} 1, & n=0 \\ 2, & n=1,2,3,... \end{cases} \tag{31}$$

$$c_{pmn} = \frac{2\left[\beta_p^2 + \left(h_f/k\right)^2\right]R_1 R_2}{\left\{L\left[\beta_p^2 + \left(h_f/k\right)^2\right] + h_f/k\right\}s_1 s_2} \tag{32}$$

The Eigen-values β_p are the positive roots of the following equation

$$\beta_p \cdot \cot(\beta_p L) = -h_f/k \tag{33}$$

Then, the solution of Equation (14) can be easily obtained. For the second BC at the skin surface, one has

$$W(x,y,z;t) = \frac{\alpha}{k}\int_0^t d\tau \int_0^{s_1}\int_0^{s_2} G_1\left(x,y,z,t;\xi,\vartheta,\varsigma,\tau\right)\Big|_{\xi=0} g_1\left(\vartheta,\varsigma;\tau\right)d\vartheta d\varsigma +$$
$$\int_0^t d\tau \int_0^L \int_0^{s_1}\int_0^{s_2} G_1\left(x,y,z,t;\xi,\vartheta,\varsigma,\tau\right)\cdot\frac{Q_r\left(\xi,\vartheta,\varsigma;\tau\right)}{\rho c}\exp\left(\frac{\omega_b \rho_b c_b}{\rho c}\tau\right)d\xi d\vartheta d\varsigma \tag{34}$$

For the third BC, the solution is

$$W(x,y,z;t) = \frac{\alpha}{k}\int_0^t d\tau \int_0^{s_1}\int_0^{s_2} G_2\left(x,y,z,t;\xi,\vartheta,\varsigma,\tau\right)\Big|_{\xi=0} h_f \cdot g_2\left(\vartheta,\varsigma;\tau\right)d\vartheta d\varsigma +$$
$$\int_0^t d\tau \int_0^L \int_0^{s_1}\int_0^{s_2} G_2\left(x,y,z,t;\xi,\vartheta,\varsigma,\tau\right)\cdot\frac{Q_r\left(\xi,\vartheta,\varsigma;\tau\right)}{\rho c}\exp\left(\frac{\omega_b \rho_b c_b}{\rho c}\tau\right)d\xi d\vartheta d\varsigma \tag{35}$$

Then the tissue temperature field can be constructed as:

$$T(x,y,z;t) = T_0(x,y,z) + W(x,y,z;t) \cdot \exp\left(-\frac{\omega_b \rho_b c_b}{\rho c} t\right) \quad (36)$$

Clearly, the above method can also be extended to solve some other three-dimensional problems such as in spherical and cylindrical coordinates. But they will not be listed here for brevity. To illustrate the application of the above analytical solutions, a selective 3-D hyperthermia problem with point heating sources was particularly studied as an example. Accordingly, the temperature distribution of tissue subject to the point heating in volume was analytically solved. Practical examples for the point heating can be found in clinics where heat was deposited though inserting a conducting heating probe in the deep tumor site. Previously, such problems received relatively few attentions in compared with other heating patterns. Here, the point-heating source to be studied can be expressed as:

$$Q_r(x,y,z,t) = P_1(t)\delta(x-x_0)\delta(y-y_0)\delta(z-z_0) \quad (37)$$

where, $P_1(t)$ is the point-heating power, δ is the Dirac function, (x_0,y_0,z_0) is the position of the point-heating source.

The results were given in Fig. 2, which represent the temperature distribution in biological bodies heated by one and two-point sources, respectively. In calculations, the typical tissue properties were applied as given in Table 1. In Fig. 2(a), the single heating source was fixed at position (0.021m, 0.04m, 0.04m); in Fig. 2(b), the two point-heating sources were at (0.021m, 0.032m, 0.04m) and (0.021m, 0.048m, 0.04m), respectively. It makes clear that the maximum temperatures of the tissues occur at the positions of the point-heating sources. Further, one can still observe that the temperature for the tissues surrounding the point-heating sources can fairly be kept at a lower temperature on the whole. This is very beneficial for the hyperthermia operation since one can then selectively control the temperature level at the diseased tissue sites while the healthy tissues at the surrounding area will just stay below the safe threshold. This may be one of the most attractive features why the invasive heating probes are frequently used to thermally kill the tumor in the deep tissue, although they may cause mechanical injury. The above solutions are expected to be valuable for such hyperthermia treatment planning.

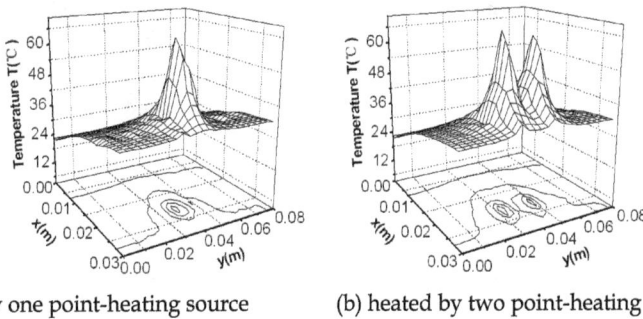

(a) heated by one point-heating source   (b) heated by two point-heating sources

Figure 2. Temperature distribution at cross-section $z = 0.04m$ after 1200s' heating

	Unit	Value
Air temperature (T_f)	°C	25
Artery blood temperature (T_a)	°C	37
Blood perfusion of tissue (ω_b)	ml/s/ml	0.0005
Body core temperature (T_c)	°C	37
Density of tissue (ρ)	Kg/m³	1000
Density of blood (ρ_b)	Kg/m³	1000
Heat convection coefficient (h_0)	W/m²·°C	10
Heat convection coefficient (h_f)	W/m²·°C	100
Metabolic heat generation of tissue (Q_m)	W/m³	33800
Specific heat of tissue (c)	J/Kg·°C	4200
Specific heat of blood (c_b)	J/Kg·°C	4200
Temperature of cooling medium (f_2)	°C	15
Thermal conductivity of tissue (k)	W/m·°C	0.5

Table 1. Typical thermophysical properties of soft biological tissues [13].

3.2. Analytical solutions to 3-D bioheat transfer involved in hyperthermia for prostate

Localized transurethral thermal therapy has been widely used as a non-surgical modality for treatment of benign prostatic hyperplasia [26]. One of the critical issues in clinical application is to effectively heat and cause coagulation necrosis in target tissue while simultaneously preserving the surrounding healthy tissue, especially the prostatic urethra and rectum. This requires administration of an optimal thermal dose which can induce the desired three dimensional tissue temperature distributions in the prostate during the therapy. In this section, the analytical approach to solving the transient 3-D temperature field was illustrated, which can be used to predict point-by-point tissue temperature mapping during the heating.

The transurethral microwave catheter (T3 catheter) was used as the heating apparatus in this section. Geometric presentation of the prostate with the inserted T3 catheter was shown in Fig. 3. It was modeled as a cylinder of 3.4cm in diameter and 3cm in length with constant temperature T_∞ at the surface which is near the body core temperature. The catheter was represented by the inner cylinder. The induced volumetric heating in tissue is [26]:

$$Q_r(r,\theta,z) = C_t Q \frac{\left[2\varepsilon(r - s\cos\theta) + (N-2)\right] e^{-2\varepsilon(r-s\cos\theta)}}{(r-s\cos\theta)^N} e^{-(z-L/2)^2/z_0^2} \tag{38}$$

where Q is the applied microwave power, C_t is a proportional constant, ε is the microwave attenuation constant in tissue, and z_0 is the critical axial decay length along the catheter while L is the total length of the prostate.

Practically, the microwave antenna is located with an offsets from the geometrical center to produce an asymmetric microwave field, which can prevent overheating the rectum. The chilled water at a given temperature flows between the antenna and the inner catheter wall.

The Pennes equation for the 3-D temperature field in the prostate can then be applied as:

$$k\frac{1}{r}\frac{\partial}{\partial r}\left(r\frac{\partial T}{\partial r}\right)+k\frac{1}{r^2}\frac{\partial^2 T}{\partial\theta^2}+k\frac{\partial^2 T}{\partial z^2}+\omega_b\rho_b c_b\left(T_a-T\right)+Q_m+Q_r=\rho c\frac{\partial T}{\partial t} \tag{39}$$

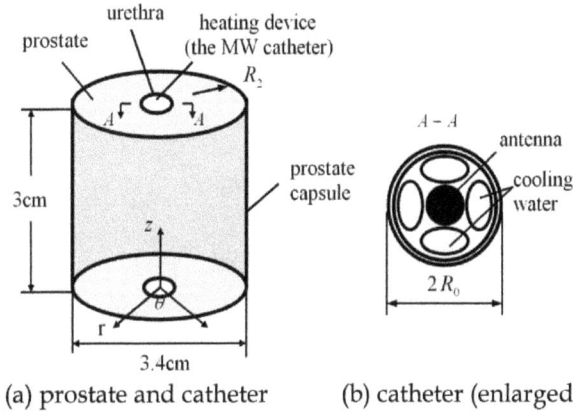

(a) prostate and catheter   (b) catheter (enlarged)

Figure 3. 3-D configuration of the prostate under microwave heating [26]

To obtain an analytical solution, all these parameters were assumed to be uniform throughout the prostate and remained constant except for ω_b which varied with respect to the heat power. The cooling effect from the chilled water running inside the catheter was modeled by an overall convection coefficient h. The external boundary at the capsule was prescribed as the body core temperature T_∞. Therefore, one has:

$$k\frac{\partial T}{\partial r}=h\left(T-T_f\right),\quad r=R_0 \tag{40}$$

$$T=T_\infty,\quad r=R_2 \tag{41}$$

where, R_0 and R_2 are the urethra and prostate radius, respectively; T_f is the coolant temperature. Considering the Gaussian distribution of microwave power deposition along the z direction, adiabatic conditions can be used at two ends of the prostate:

$$k\frac{\partial T}{\partial z}=0,\ z=0, L \tag{42}$$

The initial temperature is

$$T(r,\theta,z,0) = T_0(r,\theta,z), \quad t = 0 \tag{43}$$

Using transformation:

$$T = \Delta + T_\infty + (T_\infty - T_f) \frac{\ln(r/R_2)}{\ln(R_2/R_0) + k/(hR_0)} \tag{44}$$

One can rewrite the above equations (Equations (39-43) as:

$$\frac{1}{r}\frac{\partial}{\partial r}\left(r\frac{\partial \Delta}{\partial r}\right) + \frac{1}{r^2}\frac{\partial^2 \Delta}{\partial \theta^2} + \frac{\partial^2 \Delta}{\partial z^2} - \frac{\omega_b \rho_b C_b \Delta}{k} + \frac{Q^*(r,\theta,z)}{k} = \frac{1}{\alpha}\frac{\partial \Delta}{\partial t} \tag{45}$$

$$k\frac{\partial \Delta}{\partial r} = h\Delta, \quad r = R_0 \tag{46}$$

$$\Delta = 0, \quad r = R_2 \tag{47}$$

$$k\frac{\partial \Delta}{\partial z} = 0, \quad z = 0, L \tag{48}$$

$$\Delta(r,\theta,z,0) = F(r,\theta,z), \quad t = 0 \tag{49}$$

where,

$$Q^*(r,\theta,z) = Q_r(r,\theta,z) + Q_m + \omega_b \rho_b C_b\left[T_a - T_\infty - (T_\infty - T_f)\frac{\ln(r/R_2)}{\ln(R_2/R_0) + k/(hR_0)}\right] \tag{50}$$

$$F(r,\theta,z) = T_0(r,\theta,z) - T_\infty - (T_\infty - T_f)\frac{\ln(r/R_2)}{\ln(R_2/R_0) + k/(hR_0)} \tag{51}$$

The Green's function for the above equation sets can be obtained as:

$$G(r,\theta,z,t;\xi,\eta,\lambda,\tau) =$$
$$-\sum_{m=0}^{\infty}\sum_{n=0}^{\infty}\sum_{l=0}^{\infty}\left[J_m\left(\mu_n^* r\right) + A_n N_m\left(\mu_n^* r\right)\right]\cos\left[m(\theta-\eta)\right]\cos\left(\gamma_l z\right)e^{-\alpha\beta_i^2(t-\tau)}\frac{\kappa \alpha F_{mn}(\xi)\cos(\gamma_l\lambda)H(t-\tau)}{L} \tag{52}$$

where

$$\kappa = \begin{cases} 1, & l = 0 \\ 2, & l = 1,2,3... \end{cases} \tag{53}$$

$$A_n = -J_m\left(\mu_n^* R_2\right)\Big/N_m\left(\mu_n^* R_2\right) \tag{54}$$

$$\gamma_l = \frac{l\pi}{L}, \quad l = 0,1,2,3... \tag{55}$$

β_l can be found by

$$\beta_l^2 = \gamma_l^2 + \mu_n^2, l = 0,1,2,3\ldots \tag{56}$$

μ_n can be calculated from

$$\mu_n = \sqrt{\mu_n^{*2} + \frac{\omega_b \rho_b C_b}{k}} \tag{57}$$

μ_n^* are the positive roots of the following equation:

$$k\left[J_m{}'\left(\mu_n^* R_0\right) + A_n N_m{}'\left(\mu_n^* R_0\right) \right] = h\left[J_m\left(\mu_n^* R_0\right) + A_n N_m\left(\mu_n^* R_0\right) \right] \tag{58}$$

Here J_m and N_m are the Bessel functions of the first kind and the second kind, respectively.

$$F_{mn} = \begin{cases} \dfrac{\xi\left[J_0\left(\mu_n^* \xi\right) + A_n N_0\left(\mu_n^* \xi\right) \right]}{2\pi} \Big/ \displaystyle\int_{R_0}^{R_2} r\left[J_0\left(\mu_n^* r\right) + A_n N_0\left(\mu_n^* r\right) \right]^2 dr, \quad m = 0 \\[2ex] \dfrac{\xi\left[J_m\left(\mu_n^* \xi\right) + A_n N_m\left(\mu_n^* \xi\right) \right]}{\pi} \Big/ \displaystyle\int_{R_0}^{R_2} r\left[J_m\left(\mu_n^* r\right) + A_n N_m\left(\mu_n^* r\right) \right]^2 dr, \quad m = 1,2,3,\ldots,\infty \end{cases} \tag{59}$$

$H(t-\tau)$ is a heavy-side unit step function which has the following properties:

$$\frac{dH(t)}{dt} = \delta(t) \tag{60}$$

$$H(t) = \begin{cases} 1 & for \quad t > 0 \\ 0 & for \quad t \leq 0 \end{cases} \tag{61}$$

Finally, the temperature field was constructed as:

$$\begin{aligned} \Delta(r,\theta,z,t) &= \int_0^t d\tau \int_{R_0}^{R_2} \int_{-\pi}^{\pi} \int_0^L \frac{-Q*(\xi,\eta,\lambda)}{k} \cdot G(r,\theta,z,t;\xi,\eta,\lambda,\tau) d\xi d\eta d\lambda \\ &+ \int_{R_0}^{R_2} \int_{-\pi}^{\pi} \int_0^L \frac{-G(r,\theta,z,t;\xi,\eta,\lambda,0) \cdot F(\xi,\eta,\lambda)}{\alpha} d\xi d\eta d\lambda \end{aligned} \tag{62}$$

where, $Q*(r,\theta,z)$ and $F(r,\theta,z)$ were given in Equations (50-51), respectively.

This analytical solution has been applied to perform parametric studies on the bioheat transfer problems involved in prostate hyperthermia [26].

3.3 Analytical solutions to bioheat transfer with temperature fluctuation

Contributed from microcirculation including the capillary network plus small arterioles and venules of less than $100\mu m$ in diameter, blood perfusion plays an important role in the transport of oxygen, nutrients, pharmaceuticals, and heat through the body [27]. Although generally treated as a constant, blood perfusion is in fact a transient value even under physiological basal state. This is due to external perturbation and the self-regulation of biological body. The pulsative blood flow behavior also makes blood perfusion a fluctuating quantity. It is well accepted that blood perfusion is a fluctuating quantity around a mean value. Corresponding to the pulsative blood flow and very irregular distribution of blood vessels with various sizes, temperatures in intravital living tissues also appear fluctuating. To address this issue, we have obtained before an analytical model to characterize the temperature fluctuation in living tissues based on the Pennes equation [27]. It provides a theoretical foundation to better understanding the temperature fluctuation phenomena in living tissues. The Pennes equation used for the analysis can be rewritten as

$$\rho C \frac{\partial T}{\partial t} = K\nabla^2 T + W_b C_b \left(T_a - T\right) + Q_m \tag{63}$$

Considering that small perturbations of arterial blood temperature, blood perfusion, and metabolic heat generation will result in tissue temperature fluctuation, each of these parameters can be expressed as the sum of a mean and a fluctuation value, i.e.

$$T = \overline{T} + T' \tag{64}$$

$$T_a = \overline{T_a} + T'_a \tag{65}$$

$$W_b = \overline{W_b} + W'_b \tag{66}$$

$$Q_m = \overline{Q_m} + Q'_m \tag{67}$$

where, symbol " − " represents the mean value, and " ′ " the fluctuation one. Here, temporal averaging is adopted and defined as:

$$\overline{A(x,y,z;t)} = \frac{1}{\Gamma}\int_{t-\Gamma/2}^{t+\Gamma/2} A(x,y,z;\tau)d\tau \tag{68}$$

where, $A(x,y,z;t)$ denotes the transient physical quantity at the vicinity of point (x,y,z), and Γ the temporal averaging period.

Compared with the mean value, the fluctuation value is generally a small quantity. Then one has the following statistical relation:

$$\overline{T'} = 0, \quad \overline{T'_a} = 0, \quad \overline{W'_b} = 0, \quad \overline{Q'_m} = 0 \tag{69}$$

Substituting Equations (64-67) into Equation (63) leads to:

$$\rho C \frac{\partial(\overline{T}+T')}{\partial t} = K\nabla^2(\overline{T}+T') + (\overline{W_b}+W'_b)C_b\left[(\overline{T_a}+T'_a)-(\overline{T}+T')\right] + \overline{Q_m}+Q'_m \tag{70}$$

Further,

$$\overline{\rho C \frac{\partial(\overline{T}+T')}{\partial t}} = \overline{K\nabla^2(\overline{T}+T')} + \overline{(\overline{W_b}+W_b')C_b\left[(\overline{T_a}+T_a')-(\overline{T}+T')\right]} + \overline{Q_m}+Q_m' \tag{71}$$

Using Equation (69), Equation (71) was simplified as:

$$\rho C\frac{\partial \overline{T}}{\partial t} = K\nabla^2\overline{T} + \overline{W_b}C_b(\overline{T_a}-\overline{T}) + C_b\overline{W_b'T_a'} - C_b\overline{W_b'T'} + \overline{Q_m} \tag{72}$$

Subtracting Equation (72) from Equation (70) leads to:

$$\rho C\frac{\partial T'}{\partial t} = K\nabla^2 T' + W_b'C_b(\overline{T_a}-\overline{T}) - C_b\overline{W_b'T_a'} + C_b\overline{W_b'T'} \\ - W_b'C_bT' - \overline{W_b}C_bT' + \overline{W_b}C_bT_a' + W_b'C_bT_a' + Q_m' \tag{73}$$

Equations (72) and (73) consist of the theoretical models for characterizing the temperature fluctuation in living tissues. Derivation of the perturbation Equation (73) is similar to that of the well known Reynolds equation in fluid mechanics. Compared with the Pennes equation, there are two additional terms appearing in Equation (72) both of which have explicit physical meaning: $C_b\overline{W_b'T_a'}$ and $-C_b\overline{W_b'T'}$ represent the mean transferred energy due to perfusion perturbations and temperature fluctuations in tissue and arterial blood, respectively. It is from Equation (73) that the temperature fluctuation (T') and the pulsative blood perfusion, arterial blood temperature and metabolic heat generation (W', T_a', and Q_m') were correlated. Clearly, since the mean tissue temperature $\overline{T}$ is a space dependent value, T' is expected to be different at various tissue positions. To solve for Equation (72) and (73), dimension analysis was performed to simplify the model. One can express the orders of magnitude of the mean and pulsative physical quantities as:

$$\begin{array}{cccc} \overline{T}\sim O(1) & \overline{T_a}\sim O(1) & \overline{W_b}\sim O(1) & \overline{Q_m}\sim O(1) \\ T'\sim O(\delta) & T_a'\sim O(\delta) & W_b'\sim O(\delta) & Q_m'\sim O(\delta) \end{array} \tag{74}$$

where $O(\delta)$ stands for the value far less than $O(1)$, since the pulsative physical quantities investigated in this study is a small value.

Omitting those terms less than $O(1)$ in Equation (72) and those less than $O(\delta)$ in Equation (73), the Equations (72) and (73) can be respectively simplified as:

$$\rho C\frac{\partial \overline{T}}{\partial t} = K\nabla^2\overline{T} + \overline{W_b}C_b(\overline{T_a}-\overline{T}) + \overline{Q_m} \tag{75}$$

$$\rho C\frac{\partial T'}{\partial t} = K\nabla^2 T' + W_b'C_b(\overline{T_a}-\overline{T}) - \overline{W_b}C_bT' + \overline{W_b}C_bT_a' + Q_m' \tag{76}$$

For the interpretation of temperature fluctuation in living tissues, it is reasonable to apply the 1-D degenerated forms of Equations (75) and (76). The boundary condition at the skin surface can be chosen as convective case which is often encountered in reality, i.e.

$$-K\frac{\partial T}{\partial x} = h_f(T_f - T), \quad x = 0 \tag{77}$$

At the body core, a symmetrical or adiabatic condition can be used, namely

$$-K\frac{\partial T}{\partial x} = 0, \quad x = L \tag{78}$$

Then using the relations in Equations (64-67), the boundary and initial conditions of Equations (75) and (76) can be respectively obtained as:

$$-K\frac{\partial \overline{T}}{\partial x} = h_f(T_f - \overline{T}), \quad x = 0 \tag{79}$$

$$-K\frac{\partial \overline{T}}{\partial x} = 0, \quad x = L \tag{80}$$

$$\overline{T} = T_0(x), \quad t = 0 \tag{81}$$

$$K\frac{\partial T'}{\partial x} = h_f T', \quad x = 0 \tag{82}$$

$$-K\frac{\partial T'}{\partial x} = 0, \quad x = L \tag{83}$$

$$T' = 0, \quad t = 0 \tag{84}$$

where, $x = 0$ is defined as the skin surface while $x = L$ the body core; h_f is the apparent heat convection coefficient between the skin and the environment which is the contribution from the natural convection and radiation; T_f the environment temperature; and $T_0(x)$ the initial temperature distribution. The following transformation was introduced [27]

$$\overline{T(x,t)} = \overline{R(x,t)} \cdot \exp\left(-\frac{\overline{W_b C_b}}{\rho C}t\right) \tag{85}$$

Then Equations (75) and (79-81) were respectively rewritten as:

$$\frac{\partial \overline{R}}{\partial t} = \alpha\frac{\partial^2 \overline{R}}{\partial x^2} + \frac{\overline{W_b C_b}\overline{T_a} + \overline{Q_m}}{\rho C} \cdot \exp\left(\frac{\overline{W_b C_b}}{\rho C}t\right) \tag{86}$$

$$-K\frac{\partial \overline{R}}{\partial x} = h_f\left[T_f \cdot \exp\left(\frac{\overline{W_b C_b}}{\rho C}t\right) \cdot H(t) - \overline{R}\right], \quad x = 0 \tag{87}$$

$$-K\frac{\partial \overline{R}}{\partial x}=0, \quad x=L \tag{88}$$

$$\overline{R}=T_0(x), \quad t=0 \tag{89}$$

where, $\alpha = K/\rho C$ is the diffusivity of tissue, and $H(t)=\begin{cases}0, & t<0 \\ 1, & t>0\end{cases}$ the Heaviside function.

If the Green's function for the above equation system is obtained, its transient solution can thus be constructed [13]. Through introducing an auxiliary problem corresponding to Equations (86-89), the Green's function G can finally be obtained as:

$$G_{\overline{R}}\left(x,t;\xi,\tau\right)=\sum_{n=1}^{\infty}e^{-\alpha\beta_n^2(t-\tau)}\cdot\frac{2\left[\beta_n^2+\left(h_f/K\right)^2\right]H\left(t-\tau\right)}{L\left[\beta_n^2+\left(h_f/K\right)^2\right]+h_f/K}\cdot\cos\beta_n(L-x)\cdot\cos\beta_n(L-\xi) \tag{90}$$

where, the Eigen-values β_n are the positive roots of the following equation

$$\beta_n\cdot\tan(\beta_n L)=h_f/K \tag{91}$$

Then, the solution to Equation (86) can be obtained as

$$\overline{R(x,t)}=\frac{\alpha}{k}\int_0^t G_{\overline{R}}\left(x,t;\xi,\tau\right)\Big|_{\xi=0}\cdot h_f T_f \exp\left(\frac{\overline{W_bC_b}}{\rho C}\tau\right)\cdot H(\tau)d\tau+$$
$$\int_0^t d\tau\int_0^L G_{\overline{R}}\left(x,t;\xi,\tau\right)\frac{\overline{W_bC_b}\overline{T_a}+\overline{Q_m}}{\rho C}\cdot\exp\left(\frac{\overline{W_bC_b}}{\rho C}\tau\right)d\xi+ \tag{92}$$
$$\int_0^L G_{\overline{R}}\left(x,t;\xi,\tau\right)\Big|_{\tau=0}T_0(\xi)d\xi$$

Substituting Equation (92) into Equation (85) leads to the mean temperature $\overline{T(x,t)}$. The above solution is applicable to any transient environmental temperature $T_f(t)$ and space-dependent metabolic heat generation $\overline{Q_m(x,t)}$. For the fluctuation temperature, the solving process is as follows. Using the similar transformation as Equation (85)

$$T'(x,t)=R'(x,t)\cdot\exp\left(-\frac{\overline{W_bC_b}}{\rho C}t\right) \tag{93}$$

Equations (76) and (82-84) can be respectively converted to

$$\frac{\partial R'}{\partial t}=\alpha\frac{\partial^2 R'}{\partial x^2}+\frac{W_b'C_b(\overline{T_a}-\overline{T})+\overline{W_bC_b}T_a'+Q_m'}{\rho C}\cdot\exp\left(\frac{\overline{W_bC_b}}{\rho C}t\right) \tag{94}$$

$$K\frac{\partial R'}{\partial x} = h_f R', \quad x = 0 \tag{95}$$

$$-K\frac{\partial R'}{\partial x} = 0, \quad x = L \tag{96}$$

$$R' = 0, \quad t = 0 \tag{97}$$

Following the same procedure described above, the Green's function of this equation set is:

$$G_{R'}(x,t;\xi,\tau) = G_{\overline{R}}(x,t;\xi,\tau) \tag{98}$$

where, $G_{\overline{R}}(x,t;\xi,\tau)$ was given in Equation (90).

Consequently, the fluctuation variable R' in Equation (94) can be obtained as

$$R'(x,t) = \int_0^t d\tau \int_0^L G_{R'}(x,t;\xi,\tau) \frac{W_b'C_b(\overline{T_a} - \overline{T}) + \overline{W_b}C_bT_a' + Q_m'}{\rho C} \cdot \exp\left(\frac{\overline{W_b}C_b}{\rho C}\tau\right) d\xi \tag{99}$$

where, the mean temperature $\overline{T(x,t)}$ was given by Equation (85). Substituting Equation (99) into Equation (93), the temperature fluctuation $T'(x,t)$ can be obtained. As illustration, two calculation examples using the above analytical solutions were presented in this section to investigate the temperature fluctuation phenomena in living tissues. In the following illustration calculations, the typical values for tissue properties were taken from [27]. For clarity, only effect of the pulsative blood perfusion W_b' alone will be considered while the pulsative arterial temperature and metabolic heat generation were not taken into account, namely, $T_a' = 0$, $Q_m' = 0$, and a constant mean blood perfusion $\overline{W_b}$ was assumed. Here, the pulsative blood perfusion W_b' was far less than the mean value $\overline{W_b}$, and its simplest form can be expressed as a periodic quantity with constant frequency and oscillation amplitude such as $W_{bm}\cos\omega t$ (where ω is frequency, W_{bm} the amplitude, and $W_{bm} \ll \overline{W_b}$). However, to be more general, a stochastic pulsative perfusion form as $W_b' = 0.1\overline{W_b}(0.5 - ran)$ (where, ran is random number generated by Fortran function) was adopted for calculations.

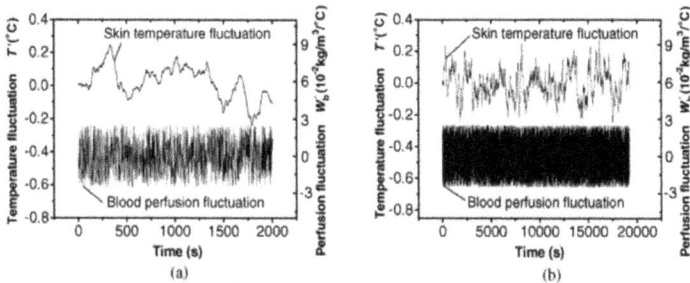

Figure 4. Temperature fluctuation due to pulsative blood perfusion ($\overline{W_b} = 0.5 kg/m^3$ °C)

Fig. 4 depicted both the skin surface temperature fluctuation (the mean perfusion $\overline{W_b} = 0.5 kg/m^3 \,°C$) and the blood perfusion perturbation W_b' causing this behavior. As a comparison, Fig. 4(a) and Fig. 4(b) gave temperature fluctuations spanning two different time scope. It can be seen that small perturbation on blood perfusion resulted in an evident and observable temperature fluctuation in the living tissues. This result accords with the commonly observed phenomenon that the measured tissue temperature appears as fluctuating even when the measured animal is under physiologically basal state. It can also be found that the frequency of temperature fluctuation appears much smaller than that of the blood perfusion fluctuation, which implies that intravital biological tissue tends to keep its temperature stable. This result indicates that the stochastic fluctuation of blood perfusion in intravital biological tissue may also contribute to the tissue temperature oscillations, and the internal relations between blood perfusion fluctuation and the temperature oscillation need further clarification.

In this section, the perturbation model for characterizing the temperature fluctuation in living tissues was illustrated and its exact analytical solution was obtained which has wide applicability. One of the most important results in this section is perhaps that small perturbation in blood perfusion result in evidently observable temperature fluctuation in the living tissues. And the larger blood perfusion, the more liable for the living tissues to keep its temperature stable. This model provides a new theoretical foundation for better understanding the thermal fluctuation behavior in living tissues.

4. Bioheat transfer problems with phase change

4.1. Analytical solutions to 3-D phase change problems during cryopreservation

Derivation such solutions was based on the moving heat source method, in which all the thermal properties were considered as constants, and phase transition was assumed to occur in a single temperature [28]. The density, specific heat and heat conductivity of solid phase were considered to be the same as those of the liquid phase, respectively. To simplify the problem, only computation in a regular geometry characterized by Cartesian coordinates was considered, as shown in Fig. 5. According to the geometrical symmetry, only 1/8 of the whole cubic tissue was chosen as the study object, whose center is set as the origin point, and l, s_1, s_2 represent the distances between the origin and the boundaries of the tissue along x, y, z directions, respectively.

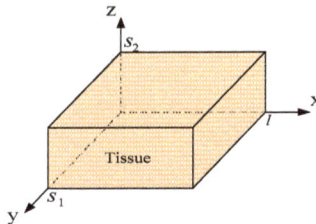

Figure 5. Schematic of 1/8 cuboidal tissue subject to cryopreservation [29]

The energy equations for different phase regions were then written. For the liquid phase:

$$k\nabla^2 T_l(x,y,z,t) = \rho c \frac{\partial T_l(x,y,z,t)}{\partial t}, \quad \text{in region} \quad R_l, \quad t>0 \tag{100}$$

For the solid phase:

$$k\nabla^2 T_s(x,y,z,t) = \rho c \frac{\partial T_s(x,y,z,t)}{\partial t}, \quad \text{in region} \quad R_s, \quad t>0 \tag{101}$$

where k, ρ, c denote thermal conductivity, density and specific heat of tissue, respectively; T_s, T_l are the temperature distributions for solid and liquid phase, respectively; t is time; R_l, R_s denote solid and liquid region; the subscript l, s represent the liquid and solid phase, respectively. To obtain an analytical solution, all the parameters were assumed as uniform and remain constant.

It should be pointed out that the physical properties for the biological tissues would change during the phase change process. Therefore it may cause certain errors when assuming both the frozen and unfreezing regions take the same physical parameters. According to existing measurements, the density changes little and thus can be used as a constant. However, the other parameters, especially the thermal conductivity and the specific heat, would change significantly. For such case, one can choose to adopt an equivalent physical property to represent the original parameter, i.e. the parameters could take into concern contributions from both frozen and unfreezing phase, such that $k = \bar{k} = (k_l + k_s)/2$ and $c = \bar{c} = (c_l + c_s)/2$. This will minimize the errors via a simple however intuitive way. The solid and liquid phases are separated by an obvious interface, which can be expressed as

$$S(x_0,y_0,z_0,t) = z_0 - s(x_0,y_0,z_0) = 0 \tag{102}$$

where, s denotes the moving solid-liquid interface, (x_0,y_0,z_0) represents position of any point in this specific interface and each of them is time dependent.

In the solid-liquid interface, conservation of energy and continuum of temperature read as

$$k\frac{\partial T_s}{\partial n} - k\frac{\partial T_l}{\partial n} = \rho L \frac{\partial s}{\partial t}\bigg|_n \quad \text{at} \quad S(x,y,z,t)=0, \quad t>0 \tag{103}$$

$$T_s = T_l = T_m \quad \text{at} \quad S(x,y,z,t)=0, \quad t>0 \tag{104}$$

where n is the unit normal vector; L is the latent heat of tissue; T_m is the phase change temperature; $v_n(t) = \dfrac{\partial s}{\partial t}\bigg|_n$ represents the velocity of the interface. Based on the moving heat source method [28, 29], the above phase change problem can be equivalently combined to a heat conduction problem with an interior moving heat source term, i.e.

$$\nabla^2 T + \frac{\rho L}{k}\frac{\partial s(x,y,t)}{\partial t}\delta(z-z_0) = \frac{1}{\alpha}\frac{\partial T}{\partial t}, \quad \text{in region} \quad R_s + R_l, \, t > 0 \tag{105}$$

where, $\alpha = k/\rho c$ is the thermal diffusivity; z_0 is z coordinate of the arbitrary point p in the interface; $\delta(z-z_0)$ is the delta function. For brief, Equation (105) can be rewritten as

$$\nabla^2 T + \frac{q_r(x,y,z,t)}{k} = \frac{1}{\alpha}\frac{\partial T}{\partial t} \tag{106}$$

where, a generalized volumetric heat source has been expressed as

$$q_r(x,y,z,t) = \rho L \frac{\partial s(x,y,t)}{\partial t}\delta(z-z_0) \tag{107}$$

The typical cooling situations most encountered in a cryopreservation [8, 10] include the following cases: (a) convective cooling at all boundaries by liquid nitrogen; (b) fixed temperature cooling at all boundaries through contacting to copper plate with very low temperature; (c) fixed temperature cooling at upside and underside surface of tissues and convective cooling at side faces; (d) convective cooling at upside and underside surface of tissues and fixed temperature cooling at side faces. Usually, the boundary types as (c) and (d) were adopted to increase the cooling rate. In this section, the analytical solutions will be presented according to the above four cooling cases, respectively.

Case (a): Convective cooling at all boundaries

Clinically, one of the most commonly used cooling approaches is to immerse the processed tissue into liquid nitrogen and frequently shift it up and down so as to enhance the heat exchange between the tissue and the liquid. Such boundary conditions can be defined as

$$\begin{aligned}
\frac{\partial T}{\partial x} &= 0 \quad at \quad x = 0; & -k\frac{\partial T}{\partial x} &= h(T - T_f) \quad at \quad x = l; \\
\frac{\partial T}{\partial y} &= 0 \quad at \quad y = 0; & -k\frac{\partial T}{\partial y} &= h(T - T_f) \quad at \quad y = s_1; \\
\frac{\partial T}{\partial z} &= 0 \quad at \quad z = 0; & -k\frac{\partial T}{\partial z} &= h(T - T_f) \quad at \quad z = s_2;
\end{aligned} \tag{108}$$

where, T_f is temperature of the cooling liquid; h is the convective heat transfer coefficient. Here, the three planes, i.e. $x = 0$, $y = 0$, $z = 0$ are defined as adiabatic boundaries, and the other three planes, i.e. $x=l$, $y=s_1$, $z=s_2$, are subjected to forced convective cooling conditions. Without losing any generality, the initial temperature is defined as

$$T(x,y,z,0) = T_0(x,y,z) \tag{109}$$

Using transformation $\theta = T - T_f$, Equations (106), (108) and (109) can be rewritten as

$$\nabla^2\theta + \frac{q_r}{k} = \frac{1}{\alpha}\frac{\partial\theta}{\partial t} \tag{110}$$

$$
\begin{array}{llllll}
\dfrac{\partial\theta}{\partial x} = 0 & at & x = 0; & -k\dfrac{\partial\theta}{\partial x} = h\theta & at & x = l; \\[3mm]
\dfrac{\partial\theta}{\partial y} = 0 & at & y = 0; & -k\dfrac{\partial\theta}{\partial y} = h\theta & at & y = s_1; \\[3mm]
\dfrac{\partial\theta}{\partial z} = 0 & at & z = 0; & -k\dfrac{\partial\theta}{\partial z} = h\theta & at & z = s_2;
\end{array}
\tag{111}
$$

$$\theta(x,y,z,0) = F(x,y,z) = T_0(x,y,z) - T_f \tag{112}$$

To solve for the Green's function of the above equations, the following auxiliary problem needs to be considered for the same region:

$$\nabla^2 G = \delta(x-\xi)\delta(y-\eta)\delta(z-\lambda)\delta(t-\tau) + \frac{1}{\alpha}\frac{\partial G}{\partial t} \tag{113}$$

$$
\begin{array}{llllll}
\dfrac{\partial G}{\partial x} = 0 & at & x = 0; & -k\dfrac{\partial G}{\partial x} = hG & at & x = l; \\[3mm]
\dfrac{\partial G}{\partial y} = 0 & at & y = 0; & -k\dfrac{\partial G}{\partial y} = hG & at & y = s_1; \\[3mm]
\dfrac{\partial G}{\partial z} = 0 & at & z = 0; & -k\dfrac{\partial G}{\partial z} = hG & at & z = s_2;
\end{array}
\tag{114}
$$

$$G(x,y,z,0) = 0 \tag{115}$$

The final expression for the Green's function of Equations (110) and (111) can be obtained as:

$$G(x,y,z,t;\xi,\eta,\lambda,\tau) = -\sum_{k=0}^{\infty}\sum_{n=0}^{\infty}\sum_{j=0}^{\infty}\frac{\cos(\beta_j x)\cos(\gamma_n y)\cos(\mu_k z)\cos(\beta_j\xi)\cos(\gamma_n\eta)\cos(\mu_k\lambda)}{N(\beta_j)N(\gamma_n)N(\mu_k)}\times$$
$$H(t-\tau)\exp[-\alpha(\beta_j^2 + \gamma_n^2 + \mu_k^2)(t-\tau)] \tag{116}$$

$$j = 0,1,2; \qquad n = 0,1,2; \qquad k = 0,1,2,$$

where, $N(\beta_j) = \dfrac{l[\beta_j^2 + (h/k)^2] + h/k}{2[\beta_j^2 + (h/k)^2]}$, β_j are positive roots of the equation $\beta_j\tan(\beta_j l) = h/k$;

$N(\gamma_n) = \dfrac{s_1[\gamma_n^2 + (h/k)^2] + h/k}{2[\gamma_n^2 + (h/k)^2]}$, γ_n are positive roots of equation $\gamma_n\tan(\gamma_n s_1) = h/k$; and

$N(\mu_k) = \dfrac{s_2[\mu_k^2 + (h/k)^2] + h/k}{2[\mu_k^2 + (h/k)^2]}$, μ_k are positive roots of $\mu_k\tan(\mu_k s_2) = h/k$.

Finally, according to the above results and expression for the heat source term in Equation (107), the analytical solution to the temperature field under totally convective cooling conditions can then be obtained as:

$$\theta(x,y,z,t) = T(x,y,z,t) - T_f = -\int_0^t d\tau \int_0^l \int_0^{s_1} \int_0^{s_2} \frac{q_r}{k} G(x,y,z,t;\xi,\eta,\lambda,\tau)d\xi d\eta d\lambda -$$
$$\int_0^l \int_0^{s_1} \int_0^{s_2} \frac{1}{\alpha} G(x,y,z,t;\xi,\eta,\lambda,0) \cdot F(\xi,\eta,\lambda)d\xi d\eta d\lambda \tag{117}$$

From the first term containing time in the above analytical solution, it can be seen that the thermal diffusivity α appears in the exponential term, i.e. $\exp[-\alpha(\beta_j^2 + \gamma_n^2 + \mu_k^2)(t-\tau)$, which indicates that the time for the temperature to reach the thermal equilibrium state depends exponentially on the thermal diffusivity.

Case (b): Fixed temperature cooling at all boundaries

Clinically, direct cooling the tissues through contacting it to copper plate pre-cooled by liquid nitrogen has been proved to be more effective than cooling by convection [28]. Therefore, it is very essential to get the temperature field of the tissue under totally fixed temperature cooling boundary conditions. For this problem, the form of the control equations still remain the same, so did the solution procedures of the Green's function method, since only the boundary conditions were slightly changed. Assuming that T_p is the temperature of the cooling plate, using transformation $\theta = T - T_p$, and solving the auxiliary problem via the similar way as that used in the convective cooling case, one can obtain the Green's function solution for the present case as:

$$G(x,y,z,t;\xi,\eta,\lambda,\tau) = -\sum_{k=0}^{\infty}\sum_{n=0}^{\infty}\sum_{j=0}^{\infty} \frac{8\cos(\beta_j x)\cos(\gamma_n y)\cos(\mu_k z)\cos(\beta_j \xi)\cos(\gamma_n \eta)\cos(\mu_k \lambda)}{s_1 s_2 l} \times$$
$$H(t-\tau)\exp[-\alpha(\beta_j^2 + \gamma_n^2 + \mu_k^2)(t-\tau)] \tag{118}$$
$$j = 0,1,2; \qquad n = 0,1,2; \qquad k = 0,1,2,$$

Then the transient temperature field can be constructed as:

$$\theta(x,y,z,t) = T(x,y,z,t) - T_p = -\int_0^t d\tau \int_0^l \int_0^{s_1} \int_0^{s_2} \frac{q_r}{k} G(x,y,z,t;\xi,\eta,\lambda,\tau)d\xi d\eta d\lambda$$
$$-\int_0^l \int_0^{s_1} \int_0^{s_2} \frac{1}{\alpha} G(x,y,z,t;\xi,\eta,\lambda,0) \cdot F(\xi,\eta,\lambda)d\xi d\eta d\lambda \tag{119}$$

where, $F(x,y,z) = \theta(x,y,z,0) = T_0(x,y,z) - T_p$.

In practice, demanded by certain specific cooling rate and mechanical factors, sometimes one has to apply different cooling strategies on each side of the tissue surfaces. Thus, it is essential to take into account the complex hybrid boundary conditions. For brief, we assume that the temperature of the cooling plate T_p is equal to that of the cooling fluid T_f. Then the Green's function solutions for the following two boundaries can be obtained accordingly.

Case (c): Fixed temperature cooling at upside and underside surface and convective cooling at side faces

$$G(x,y,z,t;\xi,\eta,\lambda,\tau) = -\sum_{k=0}^{\infty}\sum_{n=0}^{\infty}\sum_{j=0}^{\infty} \frac{2\cos(\beta_j x)\cos(\gamma_n y)\cos(\mu_k z)\cos(\beta_j \xi)\cos(\gamma_n \eta)\cos(\mu_k \lambda)}{N(\beta_j)N(\gamma_n)s_2} \times$$
$$H(t-\tau)\exp[-\alpha(\beta_j^2 + \gamma_n^2 + \mu_k^2)(t-\tau)] \qquad (120)$$
$$j = 0,1,2; \qquad n = 0,1,2; \qquad k = 0,1,2,$$

Case (d): Convective cooling at upside and underside and fixed temperature cooling at side faces

$$G(x,y,z,t;\xi,\eta,\lambda,\tau) = -\sum_{k=0}^{\infty}\sum_{n=0}^{\infty}\sum_{j=0}^{\infty} \frac{4\cos(\beta_j x)\cos(\gamma_n y)\cos(\mu_k z)\cos(\beta_j \xi)\cos(\gamma_n \eta)\cos(\mu_k \lambda)}{N(\mu_k)s_2 l} \times$$
$$H(t-\tau)\exp[-\alpha(\beta_j^2 + \gamma_n^2 + \mu_k^2)(t-\tau)] \qquad (121)$$
$$j = 0,1,2; \qquad n = 0,1,2; \qquad k = 0,1,2,$$

Considering that expressions for the transient temperature field for the above two cases still remain similar to that of Equation (117), they have not been rewritten here for brief.

It should be pointed that there still exist many difficulties to calculate the exact temperature field from the above analytical solutions. However, the solution forms can still be flexibly applied to analyze certain special problems. As indicated in [28], in the freezing or warming process there must exist a maximum cooling or warming rate at some places of the tissue, which is varying with the time. Theoretically, this transient position can be predicted by using Equations (117) and (119) or other equations for corresponding processes. For example, one can obtain $\dfrac{\partial}{\partial x_i}\left[\dfrac{\partial T(x,y,z,t)}{\partial t}\right]$ (where x_i may represent x, y, z direction) easily by utilizing any of the above Green's function solutions. After making it equal to zero, one can solve for x_i which is just the position where the maximum cooling or rewarming rate occurs. Knowing such position is of importance for the operation of a successful cryopreservation procedure, since the maximum cooling or warming rate is the crucial factor to cause injury to biological materials. Here, computation of $\partial T(x,y,z,t)/\partial t$ can eliminate the time integral term in Equations (117) and (119), which would simplify the solution form. Overall, the present analytical method in virtue of its straightforward form is of great significance to evaluate the phase change problem in cryobiology.

4.2. Analytical solutions to 3-D phase change problems during cryosurgery

Cryosurgery is very different from cryopreservation, since living tissue has to be considered. Consequently, it must take into account the effects of blood perfusion and metabolic heat generation into bioheat equation. Here, the Pennes equation is applied to characterize the heat transfer process in the living tissue. To avoid the complex boundary conditions, the calculation tissue domain is chosen as a whole cuboid as shown in Fig. 6, where a cryoprobe with length l_1 was settled in the center. Then the location of the cryoprobe is at $x_0 = l / 2$, $y_0 = s_1 / 2$, and the range for z coordinate is $0.5(s_2 - l_1) \leq z \leq 0.5(s_2 + l_1)$.

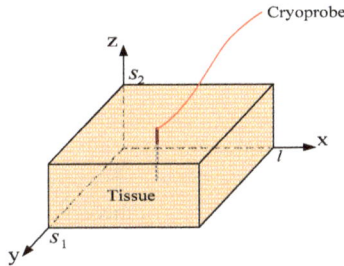

Figure 6. Schematic of the living tissue domain subject to cryosurgery

The energy equations for the tissue before and after it was frozen are respectively as:

For the liquid phase

$$\rho c \frac{\partial T_l}{\partial t} = k\nabla^2 T_l + \omega_b \rho_b c_b (T_a - T_l) + Q_m + Q_c, \text{ in region } R_l, \ t > 0 \qquad (122)$$

For the solid phase

$$\rho c \frac{\partial T_s(x,y,z,t)}{\partial t} = k\nabla^2 T_s(x,y,z,t) + Q_c, \text{ in region } R_s, \ t > 0 \qquad (123)$$

where, ρ_b, c_b are density, specific heat of blood; ω_b is blood perfusion; T_a is supplying arterial temperature; Q_m is volumetric metabolic heat; Q_c is heat sink, and the other parameters represent the same meanings as before.

The control equations in the solid-liquid interface are the same as before. The above phase change problem can then be equivalently transformed to a heat conduction problem, i.e.

$$\rho c \frac{\partial T}{\partial t} = k\nabla^2 T + Q_r \qquad (124)$$

where, Q_r is the moving heat source term, which consists of three parts in a cryosurgical process: the metabolic heat source term Q_m, the heat sink term Q_c and the phase change heat source term q_r, i.e.

$$Q_r = q_r + Q_m + Q_c = \rho L \frac{\partial s(x,y,t)}{\partial t} \delta(z-z_0) + q'_m H(z-z_0) + Q_c \tag{125}$$

where, $q'_m = w_b c_b \rho_b (T_a - T) + q_m$, reflecting contributions of the blood heat transfer and the metabolic heat generation in unfrozen region. Other parameters and functions have the same definitions with those of cryopreservation.

To simplify the problem, the cryoprobe inserted into the deep tissue is treated as a linear heat sink [29] and assumed to supply a constant cold amount Q_c, i.e. $Q_c = B$, which can also include many different discrete terms representing cooling effects from multiple cryoprobes and expressed as $Q_c = \sum Q_i \delta(x-x_i, y-y_i, z-z_i)$ (which reflects the amount of cold at the location of (x_i, y_i, z_i), where i is the sequence number of the cryoprobe; Q_i is the amount of cold released by the ith probe; δ is the Delta function). The solution expressed below indicated that the Green's function method can deal with the multi-probe freezing problem. This is however not easy to be dealt with even by numerical computation. In this side, the analytical solution embodies its particular theoretical significance.

Equation (124) can be rewritten as:

$$\frac{\partial T}{\partial t} = \alpha \nabla^2 T - \frac{\omega_b c_b \rho_b}{\rho c} H(z-z_0)T + g(z,t) \tag{126}$$

where, $g(z,t) = [(\omega_b c_b \rho_b T_a + q_m)H(z-z_0) + q_r + Q_c] / \rho c$. In a cryosurgical procedure, the cryoprobe is inserted into the target tissue, which will subject to a specific temperature drop due to cooling of heat sink effect of the probe. Since the skin surface is exposed to the ambient environment, the boundary conditions can thus be treated as:

$$
\begin{aligned}
-k\frac{\partial T}{\partial x} = 0 \quad &at \quad x = 0; & -k\frac{\partial T}{\partial x} = 0 \quad &at \quad x = l \\
-k\frac{\partial T}{\partial y} = 0 \quad &at \quad y = 0; & -k\frac{\partial T}{\partial y} = 0 \quad &at \quad y = s_1 \\
-k\frac{\partial T}{\partial z} = 0 \quad &at \quad z = 0; & -k\frac{\partial T}{\partial z} = h(T - T_f) \quad &at \quad z = s_2
\end{aligned}
\tag{127}
$$

As shown in Fig. 6, the origin (x=0, y=0, z=0) is defined differently from the cryopreservation. The adiabatic boundaries are applied on the side surfaces along the x, y axis. This is because the side surfaces are far from the heat sink and are not influenced by the heat sink in the deep tissue and the external convective conditions. Finally, the initial condition is defined as $T(x,y,z,0) = T_0(x,y,z)$, and every boundary is defined as adiabatic at this initial state. Although the control equations and boundary conditions in cryosurgery are very different from that in cryopreservation, the solution procedures still remain similar if certain transformations are introduced. In the following, only those steps different from the above will be addressed. To make solution of the problem feasible, we have adopted before the following specific transformation:

$$T(x,y,z,t) = T_0(x,y,z) + W(x,y,z,t)\exp[-\frac{\omega_b \rho_b c_b}{\rho c}H(z-z_0)t] \tag{128}$$

Substituting it into Equation (126), one obtains

$$\frac{\partial W}{\partial t} = \alpha\nabla^2 W + g_1(z,t)\exp[\frac{\omega_b \rho_b c_b}{\rho c}H(z-z_0)t] \tag{129}$$

where, $g_1(z,t) = \{[\omega_b c_b \rho_b (T_a - T_0) + q_m]H(z-z_0) + q_r + Q_c\}/\rho c$. To simplify the equation, the volumetric moving heat source term in Equation (129) can be expressed as

$$q_r^* = g_1(z,t)\exp[\frac{\omega_b \rho_b c_b}{\rho c}H(z-z_0)t] \tag{130}$$

The boundary conditions are rewritten as:

$$-k\frac{\partial W}{\partial x} = 0 \quad at \quad x = 0; \quad -k\frac{\partial W}{\partial x} = 0 \quad at \quad x = l$$

$$-k\frac{\partial W}{\partial y} = 0 \quad at \quad y = 0; \quad -k\frac{\partial W}{\partial y} = 0 \quad at \quad y = s_1 \tag{131}$$

$$-k\frac{\partial W}{\partial z} = 0 \quad at \quad z = 0; \quad -k\frac{\partial W}{\partial z} = h[W - f(t)] \quad at \quad z = s_2$$

where, $f(t) = (T_f - T_0)\exp[\frac{\omega_b \rho_b c_b}{\rho c}H(z-z_0)t]H(t)$, and the initial condition is defined as $W(x,y,z,t) = 0$. Applying the same theoretical strategies as used in solving cryopreservation into the above problem, one can obtain the Green's function expression for a 3-D cryosurgical process as:

$$G(x,y,z,t;\xi,\eta,\lambda,\tau) = \frac{1}{s_1 s_2 l}\sum_{k=0}^{\infty}\sum_{n=0}^{\infty}\sum_{j=0}^{\infty}R_1 R_2 \frac{2[\mu_k^2 + (h/k)^2]}{[\mu_k^2 + (h/k)^2] + h/k}\exp[-\alpha(\beta_j^2 + \gamma_n^2 + \mu_k^2)(t-\tau)]\times$$
$$H(t-\tau)\cos(\beta_j x)\cos(\gamma_n y)\cos(\mu_k z)\cos(\beta_j \xi)\cos(\gamma_n \eta)\cos(\mu_k \lambda) \tag{132}$$
$$j = 0,1,2; \qquad n = 0,1,2; \qquad k = 0,1,2,$$

where, $R_1 = \begin{cases} 1, & j=0 \\ 2, & j=1,2,3,\cdots \end{cases}$; $R_2 = \begin{cases} 1, & n=0 \\ 2, & n=1,2,3,\cdots \end{cases}$; $\beta_j = \frac{j\pi}{l}$; $\gamma_n = \frac{n\pi}{s_1}$; μ_k is the positive

root of $\mu_k \tan(\mu_k s_2) = h/k$. Taking into account of the moving source term q_r^* and the boundary conditions, i.e. Equation (131), one can obtain the transient temperature field of the tissue corresponding to Equation (129) as:

$$W(x,y,z,t) = -\int_0^t d\tau \int_0^l \int_0^{s_1} \int_0^{s_2} \frac{q_r^*}{k} G(x,y,z,t;\xi,\eta,\lambda,\tau) d\xi d\eta d\lambda -$$

$$\int_0^t d\tau \int_0^l \int_0^{s_1} \frac{1}{k} G(x,y,z,t;\xi,\eta,\lambda,\tau)\Big|_{\lambda=s_2} \cdot h \cdot f(t) d\xi d\eta - \qquad (133)$$

$$\int_0^l \int_0^{s_1} \int_0^{s_2} \frac{1}{\alpha} G(x,y,z,t;\xi,\eta,\lambda,0) F(\xi,\eta,\lambda) d\xi d\eta d\lambda$$

The above procedures illustrate the basic strategy to exactly solve the three dimensional phase change problem of biological tissues in vivo, which involves the blood perfusion and metabolic heat etc. However, the integral equation is so complex due to moving phase change front inherited in the integral term, that calculating the equation based on the above analytical expressions is still a challenge. This requests certain development of the applied mathematics. However, a simplified form for the present solution can be utilized to analyze some specific one dimensional heat transfer problems.

4.3. Analytical solutions to 3-D temperature distribution in tissues embedded with large blood vessels during cryosurgery

From the viewpoint of heat transfer, a large blood vessel (also termed a thermally significant vessel) denotes a vessel larger than 0.5 mm in diameter [30]. Anatomically, tumors are often situated close to or embedded with large blood vessels, since a tumor's quick growth ultimately depends on nutrients supplied by its blood vessel network. During cryosurgery, the blood flow inside a large vessel represents a source which heats the nearby frozen tissues and, thereby, limits freezing lesions during cryosurgery. Under this condition, a part of the vital tumor cells may remain in the cryolesion and lead to recurrence of tumors after cryosurgical treatment. More specifically, tumor cell survival in the vicinity of large blood vessels is often correlated with tumor recurrence after treatment [30]. Consequently, it is difficult to implement an effective cryosurgery when a tumor is contiguous to a large blood vessel. To better understand the effect of blood flow to the temperature distribution of living tissues subject to freezing, a conceptual model for characterizing the heat transfer in 3-D cylindrical tissues embedded with a single blood vessel was illustrated in this section. And a closed form analytical solution to this model was provided to explore different factors' thermal influences to the freezing mechanism of living tissues.

The geometry used for the analysis is depicted as Fig. 7, which is consisted of three distinct concentric cylinders: the most interior region representing a large blood vessel, the intermediate for unfrozen liquid-phase tissue and the outer the frozen tissue. In Fig. 7, symmetrical condition in θ direction can be applied. Then the 3-D bioheat transfer will degenerate to a 2-D problem. The cylinder is long enough so that its end effects to the heat transfer can be neglected. For simplicity in analytical solution, only steady state temperature field was assumed in both the vessel and the surrounding tissue. And the same constant thermal properties for different tissue area were considered. Therefore heat conduction in the regions of both liquid and frozen tissue can be described by only a single equation.

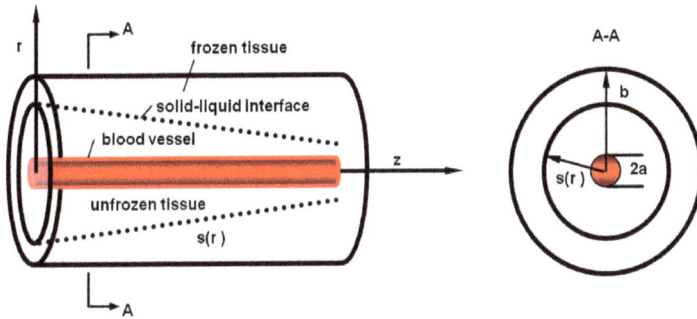

Figure 7. Schematic of cylindrical tissues embedded with a single blood vessel [30]

To carry out the theoretical analysis, additional assumptions for tissues were made as follows: there is no heat flow across the boundaries at z=0 and z=L. In reality, if the vessel length is long enough, these adiabatic boundary conditions can be closely satisfied. And the cylindrical tissue surface (r=b) was assumed to be immersed into a cooling medium or subjected to a circular cryoprobe with constant freezing temperature T_0. This situation also reflects the cooling of an interior tissue inside the biological body, which is frozen on the surface. After introducing the dimensionless parameter $T_t^* = (T_t - T_0)/(T_a - T_0)$, where T_a stands for the body core temperature usually fixed at 37°C, the temperature distribution for the cylindrical tissue area can be described by the following equations [30]

$$\begin{cases} \dfrac{1}{r}\dfrac{\partial}{\partial r}(r\dfrac{\partial T_t^*(r,z)}{\partial r}) + \dfrac{\partial T_t^*(r,z)}{\partial z^2} = 0 \\ T_t^* = 0 & at \quad r = b \\ T_t^* = f(z) & at \quad r = a \\ \dfrac{\partial T_t^*(r,z)}{\partial z} = 0, & at \quad z = 0, L \end{cases} \tag{134}$$

where, f(z) is temperature along the vessel wall, which is unknown here and needs to be calculated later using temperature continuity condition on the vessel wall.

Blood flow velocity profile in the vessel can be obtained as $u = u_0(1 - \dfrac{r^2}{a^2})$ by solving the Navier-Stokes equation for an incompressible steady-state fluid, where a stands for the radius of blood vessel, and u_0 the average blood flow velocity over the vessel cross-section. Both temperature and heat flux on the vessel wall ($r = a$) satisfy continuity conditions. Defining the following dimensionless parameters: $T_b^* = \dfrac{T_b - T_0}{T_a - T_0}, T_c^* = \dfrac{T_c - T_0}{T_a - T_0}$, the governing equations for the vessel read as

$$\begin{cases} \dfrac{K_b}{\rho_b C_b}\dfrac{1}{r}\dfrac{\partial}{\partial r}(r\dfrac{\partial T_b^*}{\partial r})) = u_0(1-\dfrac{r^2}{a^2})\dfrac{dT_c^*}{dz} \\ \dfrac{\partial T_b^*}{\partial r} = 0 \quad , \qquad at \ \ r=0 \\ T_b^* = f(z) \quad , \qquad at \ \ r=a \\ T_b^* = 1 \quad , \qquad at \ \ z=0 \end{cases} \tag{135}$$

where, T_c^* is the vessel bulk temperature within cylindrical symmetry, and defined as

$$T_c^* = \frac{4}{a^2}\int_0^a T_b^*(1-\frac{r^2}{a^2})r dr \tag{136}$$

In the above equations, subscript t, b designate tissue and blood respectively, * means the dimensionless parameters. The variable separation method was applied to solve the Equation (134), whose solution can be rewritten as product of an axial term $Z(z)$ and a radial one $Ro(r)$, i.e. $T_t^*(r,z) = R_0(r)Z(z)$, with $Z(z)$ satisfying

$$\begin{cases} \dfrac{d^2 Z(z)}{dz^2}+\beta_m^2 Z(z)=0 \qquad (0<z<L) \\ \dfrac{dZ(z)}{dz}=0 \qquad\qquad (z=0,L) \end{cases} \tag{137}$$

and $Ro(r)$ satisfying

$$\begin{cases} \dfrac{d^2 R_0(r)}{dr^2}+\dfrac{1}{r}\dfrac{dR_0(r)}{dr}-\beta_m^2 R_0(r)=0 \qquad (a<r<b) \\ R_0=0 \qquad\qquad\qquad (r=b) \end{cases} \tag{138}$$

The solution to Equation (137) is thus obtained as $Z(z)=C_1\cos(\beta_m z)+C_2\sin(\beta_m z)$. β_m is the positive roots of $\sin(\beta L)=0$. Then

$$\beta_m = \frac{m\pi}{L}, m=0,1,2\cdots \tag{139}$$

Therefore

$$Z(z)=\cos(\beta_m z) \tag{140}$$

with normal as

$$N(\beta_m)=\begin{cases} L/2 \quad , & m=1,2,3,\cdots \\ L \quad , & m=0 \end{cases} \tag{141}$$

The solution to Equation (138) can be obtained as

$$R_0 = \begin{cases} C_3[-\dfrac{K_0(\beta_m b)}{I_0(\beta_m b)}I_0(\beta_m r) + K_0(\beta_m r)], \beta_m \neq 0 \\ C_4 \ln(r / b), \qquad\qquad\qquad \beta_0 = 0 \end{cases} \tag{142}$$

Then the solution to Equation (134) can be expressed as

$$T_t^*(r,z) = \sum_{m=1}^{\infty} A_m[-\frac{K_0(\beta_m b)}{I_0(\beta_m b)}I_0(\beta_m r) + K_0(\beta_m r)]Z(\beta_m,z) + A_0 \ln(r / b)Z(\beta_0,z) \tag{143}$$

where, A_m, A_0 are coefficients. Substituting Equation (134) into Equation (143) leads to

$$f(z) = \sum_{m=1}^{\infty} A_m[-\frac{K_0(\beta_m b)}{I_0(\beta_m b)}I_0(\beta_m a) + K_0(\beta_m a)]Z(\beta_m,z) + A_0 \ln(r / b)Z(\beta_0,z), \quad 0 < z \le L \tag{144}$$

where, A_m, A_0 are obtained as

$$A_m = \frac{1}{[-\dfrac{K_0(\beta_m b)}{I_0(\beta_m b)}I_0(\beta_m a) + K_0(\beta_m a)]N(\beta_m)}\int_0^L Z(\beta_m,z)f(z)dz \tag{145}$$

$$A_0 = \frac{1}{\ln(a / b)N(\beta_0)}\int_0^L Z(\beta_m,z)f(z)dz \tag{146}$$

Substituting Equations (145-146) into Equation (143) leads to

$$T_t^*(r,z) = (T_a - T_0)\left(\sum_{m=1}^{\infty} \frac{2[-\dfrac{K_0(\beta_m b)}{I_0(\beta_m b)}I_0(\beta_m r) + K_0(\beta_m r)]}{L[-\dfrac{K_0(\beta_m b)}{I_0(\beta_m b)}I_0(\beta_m a) + K_0(\beta_m a)]}\cos(\beta_m z)\int_0^L \cos(\beta_m z)f(z)dz \right.$$
$$\left. + \frac{\ln(r / b)}{L\ln(a / b)}\int_0^L f(z)dz \right) + T_0 \tag{149}$$

As to the exact temperature profile within blood vessel, if defining $U = \dfrac{u_0 \rho_b C_b}{4K_b}\dfrac{dT_c^*}{dz}$, one has from Equation (135)

$$T_b^* = U[r^2 - \frac{r^4}{4a^2} + C_5(z)\ln r + C_6(z)] \tag{148}$$

Applying Equation (135) to this equation, $C_5(z) = 0$ was obtained. Substituting Equation (148) into Equation (136) leads to

$$T_c^* = [\frac{7a^2}{24} + C_6(z)]\frac{u_0 \rho_b C_b}{4K_b}\frac{dT_c^*}{dz}$$

(149)

Further, it can be derived as

$$T_c^* = T_{c0}^* \int_0^z \frac{1}{[\frac{7a^2}{24} + C_6(z)]\frac{u_0 \rho_b C_b}{4K_b}} dz$$

(150)

where, T_{c0}^* is defined as $T_{c0}^* = \frac{4}{a^2}\int_0^a T_{b0}^*(1 - \frac{r^2}{a^2})r dr$, T_{b0}^* is the vessel inlet temperature at z=0. From Equation (135), one obtains $T_{b0}^* = 1$. Therefore

$$T_{c0}^* = 1$$

(151)

Substituting Equation (151) into Equation (150) leads to

$$T_b^*(r,z) = \frac{[r^2 - \frac{r^4}{4a^2} + C_6(z)]}{4[\frac{7a^2}{24} + C_6(z)]} e^{\int_0^z \frac{1}{[\frac{7a^2}{24} + C_6(z)]\frac{u_0 \rho_b C_b}{4K_b}} dz}$$

(152)

At r=a, one has

$$f(z) = \frac{[\frac{3}{4}a^2 + C_6(z)]}{4[\frac{7a^2}{24} + C_6(z)]} e^{\int_0^z \frac{1}{[\frac{7a^2}{24} + C_6(z)]\frac{u_0 \rho_b C_b}{4K_b}} dz}$$

(153)

Using the continuity condition for heat flux on the blood vessel wall, one approximately has

$$k_t \frac{\partial T_t}{\partial r}\Big|_{r=a} = k_b \frac{\partial T_b}{\partial r}\Big|_{r=a}$$

(154)

The two terms in this equation can further be obtained from Equation (147) and Equation (152), respectively, which are

$$\left.\frac{\partial T_t(r,z)}{\partial r}\right|_{r=a} = (T_a - T_0)\left\{\sum_{m=1}^{\infty}\frac{2\beta_m[-\frac{K_0(\beta_m b)}{I_0(\beta_m b)}I_1(\beta_m a) + K_1(\beta_m a)]}{L[-\frac{K_0(\beta_m b)}{I_0(\beta_m b)}I_0(\beta_m a) + K_0(\beta_m a)]}\cos(\beta_m z)\int_0^L \cos(\beta_m z)f(z)dz\right.$$

$$\left.+\frac{1}{La\ln(a/R)}\int_0^L f(z)dz\right\}$$

(155)

and

$$\left.\frac{\partial T_b}{\partial r}\right|_{r=a} = \frac{a(T_a - T_0)}{4[\frac{7a^2}{24} + C_6(z)]}e^{\int_0^z\frac{1}{[\frac{7a^2}{24}+C_6(z)]\frac{\mu_0\rho_b C_b}{4K_b}}dz}$$

(156)

Exact calculation on $C_6(z)$ from Equation (154) appears extremely difficult. However, if treating C_6 as a constant, and substituting Equation (155) and Equation (156) into Equation (154), then integrating it from $z = 0$ to L, one has

$$\overline{C}_6 = a^2\left[\frac{K_b}{K_t}\ln(\frac{a}{b}) - \frac{3}{4}\right]$$

(157)

This constant $\overline{C}_6$ does not exactly satisfy Equation (154) indeed. However it is a simple and intuitive approximation. According to former calculation [30], it can be found that heat fluxes calculated using temperature from both sides of the blood vessel wall were almost identical. Therefore, an approximate estimation on C_6 can also be made as follows. In Equation (154), if the axial position z is fixed, C_6 can be obtained through numerical iteration. And the relative error $[C_6(z) - \overline{C}_6]/\overline{C}_6$ was very small along the axial direction. For the case of aorta, the largest error is less than 1%, while for the case of terminal branches, it is 6%. Overall, the smaller vessel diameter, the larger error in the estimated value $\overline{C}_6$. However, since the present discussion was mainly focused on the case of large blood vessel (for tissues with extremely small vessel, a collective model such as Pennes equation is often applicable), therefore treating $C_6(z)$ as a constant was an acceptable approximation. But for more complex situations where the above approximation cannot hold, exactly solving the Equation (157) for $C_6(z)$ is still very necessary in the future.

When analyzing the thermal effect of large blood vessel in cryosurgery, an important issue is when the blood vessel begins to freeze and how to control cryoprobe's temperature to completely freeze the target tumor. Substituting Equation (157) into Equation (152), one can set up the relation for the blood vessel temperature $T_b(a,L)$ to reach the phase change point T_m, i.e.

$$
T_0 = \frac{\dfrac{\frac{3}{4}a^2 + \bar{C}_6}{4(\frac{7}{24}a^2 + \bar{C}_6)} e^{\dfrac{L}{4(\frac{7}{24}a^2 + \bar{C}_6)(\frac{u_0 \rho_b C_b}{4 K_b})}} T_a - T_m}{\dfrac{\frac{3}{4}a^2 + \bar{C}_6}{4(\frac{7}{24}a^2 + \bar{C}_6)} e^{\dfrac{L}{4(\frac{7}{24}a^2 + \bar{C}_6)(\frac{u_0 \rho_b C_b}{4 K_b})}} - 1} \tag{158}
$$

Up to now, the above analysis was based on a steady state assumption. And only a single blood vessel was considered for the sake of analytical solution, although it does provide certain important information for understanding the phase change heat transfer in living tissues with blood vessel. Clinically, knowledge on the transient temperature response is still very necessary for the successful operation of a cryosurgery. However, such non-steady state problem cannot be dealt with by the present method. This needs further efforts in the near future using numerical approach.

5. Conclusion

This chapter has presented an overview on several typical closed form analytical solutions to 3-D bioheat transfer problems with or without phase change as developed before in the authors' laboratory. In these solutions, relatively complex boundary conditions and heating/cooling on skin surface or inside biological bodies were addressed. In addition, the theoretical strategies towards analytically solving the complex 3-D bioheat transfer problems were outlined by the mathematical transformation, the Green's function method, and the moving heat source model etc.

The analytical solutions introduced in this chapter can be used to predicate the evolution of temperature distribution inside the target tissues during tumor hyperthermia, cryosurgery, cryopreservation, thermal diagnostics, thermal comfort analysis, brain hypothermia resuscitation, and burn injury evaluation. Through fitting the predicted with the experimentally measured temperatures at the skin surface, some thermal parameters of biological tissues such as blood perfusion, thermal conductivity, and heat capacity, can be estimated non-invasively. Moreover, based on the requirements for freezing/heating necrosis temperature of tissue, an approach to optimize the parameters of cryosurgical/hyperthermic treatment can be obtained using the presented analytical solutions. Therefore, the presented analytical solutions are very useful for a variety of thermal-oriented biomedical studies. However, it should be pointed out that although such analytical solutions have some versatility in dealing many bioheat transfer problems, numerical approaches are still needed for more complex situations. In fact, the relation between analytical and numerical solutions should be complementary. On one hand, numerical approach can deal with more complex problem than analytical way. On the other hand, the analytical results can serve as benchmark solutions for numerical analyses on complex situations. In summary, it is believed that even the applications with some simplified conditions do not affect the applicability of the present analytical solutions.

Nomenclature

a	Radius of blood vessel [m]
b	Radius of tissue [m]
c	Specific heat of tissue [$J/kg°C$]
c_b	Specific heat of blood [$J/kg°C$]
$C_1,...,C_6$	Coefficients
f_1	Surface heat flux
f_2	Temperature of the cooling medium
h_0	Heat convection coefficient [$W/m^2°C$]
h_f	Heat convection coefficient [$W/m^2°C$]
k	Thermal conductivity of tissue [$W/m·°C$]
L	Distance between skin surface and body core [m]
s_1	Width of the tissue domain in y direction [m]
s_2	Width of the tissue domain in z direction [m]
P_1	Point heating power [W/m^3]
R	Transformed temperature [$°C$]
Q_m	Metabolic rate of tissue [W/m^3]
Q_r	Spatial heating [W/m^3]
t , τ	Time [s]
T	Tissue temperature [$°C$]
T_a	Artery temperature [$°C$]
T_c	Body core temperature [$°C$]
T_f	Fluid temperature [$°C$]
u_0	Average blood flow velocity [m/s]
W	Transformed temperature [$°C$]
x, y, z	Coordinate [m]
X	Location
$\xi, \vartheta, \varsigma$	Coordinate [m]

α Thermal diffusivity of tissue [m^2/s]

η Scattering coefficient [$1/m$]

ω_b Blood perfusion [$ml / s / ml$]

ρ Density of tissue [kg/m^3]

ρ_b Density of blood [kg/m^3]

Ω Spatial domain

Γ Temporal averaging period [s]

Secondary parameters

A $A = \omega_b \rho_b c_b / k$

W_b $W_b = \rho_b \omega_b$

Subscripts

b Blood

l Liquid phase

m Phase change point or metabolism

n Unit normal vector

s Solid phase

t Tissue

Superscripts

—   Mean value

'   Fluctuation value

*   Dimensionless parameters

Author details

Zhong-Shan Deng
Technical Institute of Physics and Chemistry, Chinese Academy of Sciences, Beijing, China

Jing Liu
Technical Institute of Physics and Chemistry, Chinese Academy of Sciences, Beijing, China
Department of Biomedical Engineering, School of Medicine, Tsinghua University, Beijing, China

Acknowledgement

Part of the researches as presented in this chapter has been supported by the National Natural Science Foundation of China under grants Grant Nos. 51076161 and 81071255, the Specialized Research Fund for the Doctoral Program of Higher Education, and Research Fund from Tsinghua University under Grant No. 523003001.

6. References

[1] Liu J, Wang CC (1997) Bioheat Transfer (in Chinese). Beijing: Science Press. 435 p.

[2] Liu J, Deng ZS (2008) Physics of Tumor Hyperthermia (in Chinese). Beijing: Science Press. 381 p.

[3] Liu J, Chen X, Xu LX (1999) New Thermal Wave Aspects on Burn Evaluation of Skin Subjected to Instantaneous Heating. IEEE Transactions on Biomedical Engineering 46: 420-428.

[4] Lv YG, Liu J, Zhang J (2006) Theoretical Evaluation on Burn Injury of Human Respiratory Tract due to Inhalation of Hot Gas at the Early Stage of Fires. Burns 32: 436-446.

[5] Liu J (2007) Cooling Strategies and Transport Theories for Brain Hypothermia Resuscitation. Frontiers of Energy and Power Engineering in China l: 32-57.

[6] Deng ZS, Liu J (2004) Computational Study on Temperature Mapping Over Skin Surface and Its Implementation in Disease Diagnostics. Computers in Biology and Medicine 34: 2004.

[7] Fanger PO (1970) Thermal Comfort - Analysis and Applications in Environmental Engineering. New York: McGraw-Hill. 244 p.

[8] Liu J (2007) Principles of Cryogenic Biomedical Engineering (in Chinese). Beijing: Science Press. 338 p.

[9] Diller KR (1992) Modeling of Bioheat Transfer Processes at High and Low Temperatures. Advances in Heat Transfer 22: 157-357.

[10] Hua TC, Ren HS (1994) Cryogenic Biomedical Technology (in Chinese). Beijing: Science Press. 412 p.

[11] Roemer RB (1999) Engineering Aspects of Hyperthermia Therapy. Annual Review of Biomedical Engineering. 1: 347-376.

[12] Gage AA, Baust J (1998) Mechanism of Tissue Injury in Cryosurgery. Cryobiology. 37: 171-186.

[13] Deng ZS, Liu J (2002) Analytical Study on Bioheat Transfer Problems with Spatial or Transient Heating on Skin Surface or Inside Biological Bodies. ASME Journal of Biomechanical Engineering. 124: 638-649.

[14] Pennes HH (1948) Analysis of Tissue and Arterial Blood Temperatures in the Resting Human Forearm. Journal of Applied Physiology. 1: 93-122.

[15] Liu J (2006) Bioheat Transfer Model. In: Akay M, editor. Wiley Encyclopedia of Biomedical Engineering. John Wiley & Sons. pp. 429-438.

[16] Wulff W (1974) The Energy Conservation Equation for Living Tissues. IEEE Transactions on Biomedical Engineering. 21: 494-497.

[17] Chen MM, Holmes KR (1980) Microvascular Contributions in Tissue Heat Transfer. Annals of the New York Academy of Sciences. 335: 137-150.

[18] Weinbaum S, Jiji LM (1985) A New Simplified Bioheat Equation for the Effect of Blood Flow on Local Average Tissue Temperature. ASME Journal of Biomechanical Engineering. 107: 131-139.

[19] Liu J, Deng ZS (2009) Numerical Methods for Solving Bioheat Transfer Equations in Complex Situations. In: Minkowycz WJ, Sparrow EM, Abraham JP, editors. Advances in Numerical Heat Transfer (vol.3). Taylor & Francis. pp. 75-120.

[20] Deng ZS, Liu J (2002) Monte Carlo Method to Solve Multi-Dimensional Bioheat Transfer Problem. Numerical Heat Transfer, Part B: Fundamentals. 42: 543-567.

[21] Vyas R, Rustgi ML (1992) Green's Function Solution to the Tissue Bioheat Equation. Medical Physics. 19: 1319-1324.

[22] Gao B, Langer S, Corry PM (1995) Application of the Time-Dependent Green's Function and Fourier Transforms to the Solution of the Bioheat Equation. International Journal of Hyperthermia. 11: 267-285.

[23] Durkee JW, Antich PP, Lee CE (1990) Exact-Solutions to the Multiregion Time-Dependent Bioheat Equation - Solution Development. Physics in Medicine and Biology. 35: 847-867.

[24] Durkee JW, Antich PP (1991) Exact-Solution to the Multiregion Time-Dependent Bioheat Equation with Transient Heat-Sources and Boundary-Conditions. Physics in Medicine and Biology. 36: 345-368.

[25] Kou HS, Shih TC, Lin WL (2003) Effect of the Directional Blood Flow on Thermal Dose Distribution during Thermal Therapy: An Application of a Green's Function Based on the Porous Model. Physics in Medicine and Biology. 48: 1577-1589.

[26] Liu J, Zhu L, Xu LX (2000) Studies on the Three-Dimensional Temperature Transients in the Canine Prostate during Transurethral Microwave Thermal Therapy. ASME Journal of Biomechanical Engineering. 122: 372-379.

[27] Deng ZS, Liu J (2001) Blood Perfusion Based Model for Characterizing the Temperature Fluctuation in Living Tissues. Physica A: Statistical Mechanics and Its Applications. 300: 521-530.

[28] Liu J, Zhou YX (2002) Analytical Study on the Freezing and Thawing Processes of Biological Skin with Finite Thickness. Heat and Mass Transfer. 38: 319-326.

[29] Li FF, Liu J, Yue K (2009) Exact Analytical Solution to Three-Dimensional Phase Change Heat Transfer Problems in Biological Tissues Subject to Freezing. Applied Mathematics and Mechanics. 30: 63-72.

[30] Zhang YT, Liu J, Zhou YX (2002) Pilot Study on Cryogenic Heat Transfer in Biological Tissues Embedded with Large Blood Vessels. Forschung im Ingenieurwesen - Engineering Research. 67: 188-197.

Chapter 9

The Effects of Hall and Joule Currents and Variable Properties on an Unsteady MHD Laminar Convective Flow Over a Porous Rotating Disk with Viscous Dissipation

Abdus Sattar and Mohammad Ferdows

Additional information is available at the end of the chapter

1. Introduction

Heat transfer from convection in a rotating body is of theoretical as well as practical importance in the thermal analysis of rotating components of various types of mechanical devises. The rotating disk is one of a number of such geometrical configurations of rotating bodies which is of primary interest. Many practical systems can be modeled in terms of disk rotating in an infinite environment or in a housing. The importance of heat transfer from a rotating body can thus be ascertained in cases of many types of machineries, for example computer disk drives(Herrero *et al* 1994), rotating disk reactors for bio-fuel production and gas or marine turbines(Owen and Rogers 1989).

Heat transfer from a rotating disk by convection has been investigated theoretically by Wagner(1948), Millsaps and Pohlhausen(1952), Kreith and Taylor(1956), Kreith, Taylor and Chong(1959) and Sparrow and Gregg(1959). The theory thus established predicts that in the laminar flow regime heat and mass transfer coefficients are uniform over the entire surface of a rotating disk. Following pioneer treatment of von Karman(1921) for a flow over a rotating disk, an exact solution of complete Navier-Stokes' and energy equations was obtained by Millsaps and Pohlhausen for laminar convective flow. The rate of heat and mass transfer from a rotating disk at various speeds in an infinite environment in both laminar and turbulent flows were measured by Kreith, Taylor and Chong(1959). On the other hand Popiel and Boguslawski(1975) measured the heat transfer coefficient at a certain location over a disk rotating at different angular speeds.

The applied magnetic field effects on a steady flow due to the rotation of a disk of infinite or finite extend were studied by El-Mistikaway and Attia(1990) and El-Mistikaway et al. (1991). Aboul-Hassan and Attia(1997) also studied steady hydrodynamic flow due to an infinite disk rotating with uniform angular velocity in the presence of an axial magnetic field with Hall current. Attia(1998) separately studied the effects of suction as well as injection in the presence of a magnetic field on the unsteady flow past a rotating porous disk. It was observed by him that strong injection tend to destabilize the laminar boundary layer but when magnetic field works even with strong injection, it stabilizes the boundary layer.

The heat transfer phenomenon along with magneto-hydrodynamic effect on an unsteady incompressible flow due to an infinite rotating disk were studied by Maleque and Sattar(2003) using implicit finite difference scheme of Crank-Nicolson method. Later Maleque and Sattar(2005) investigated numerically the steady three-dimensional MHD free convective laminar incompressible boundary layer flow due to an infinite rotating disk in an axial uniform magnetic field taking into account the Hall current.

In classical treatment of thermal boundary layers, fluid properties such as density, viscosity, and thermal conductivity are assumed to be constant. But experiments indicate that the assumption of constant fluid property only makes sense if temperature does not change rapidly as far as application is concerned. To predict the flow behavior accurately, it may be necessary to take into account these properties as variables. It is of course known that these physical properties may change significantly with the change of temperature of the flow. Zakerullah and Ackroyd(1979) taking into account the variable properties analyzed the laminar natural convection boundary layer flow on a horizontal circular disk. Herwig(1985) analyzed the influence of variable properties on a laminar fully developed pipe flow with constant heat flux across the wall. He showed how the exponents in the property ratio method depend on the fluid properties. Herwig and Wikeren(1986) made a similar analysis in case of a wedge flow. In case of a fully developed flow in a concentric annuli, the effects of the variable properties have been investigated by Herwig and Klamp(1988). Maleque and Sattar(2002) , however, studied the effects of variable viscosity and Hall current on an unsteady MHD laminar convective flow due to a rotating disk. Similar unsteady hydromagnetic flow due to an infinite rotating disk was studied by Attia(2006) takig into account the temperature dependent viscosity in a porous medium with Hall and ion-slip currents. The effects of variable properties(density, viscosity and thermal conductivity) on the steady laminar convective flow due to a rotating disk were shown by Maleque and Sattar(2005a) while Maleque and Sattar(2005b) further investigated the same problem in presence of Hall current. Osalusi and Sibanda(2006) revisited the problem of Maleque and Sattar(2005b), considering magnetic effect. Osalusi et al.(2008) , however, considered an unsteady MHD flow over a porous rotating disk with variable properties in the presence of Hall and Ion-slip currents. Rahman(2010) recently made a similar study on the slip-flow with variable properties due to a porous rotating disk.

Most of the above studies were in cases of steady flows accept few. The reason is that the theoretical treatment of unsteady problems is a difficult task. However, one can rely on the sophisticated numerical tools such as finite difference or finite element methods to solve the unsteady problems but the solutions such obtained are non-similar. Problem of course arises when one tries to obtain similarity solutions of an unsteady flow. A similarity technique for unsteady boundary layer problems was thus fathered by Sattar and Hossain(1992), which has been incorporated here to investigate the effects of Hall and Joule currents on an unsteady MHD laminar convective flow due to a porous rotating disk with viscous dissipation.

2. Nomenclature

a, b, c arbitrary exponents

B_0 magnetic flux density

F dimensionless radial velocity

G dimensionless vertical velocity

H dimensionless tangential velocity

J_h Joule heating parameter

k thermal conductivity

k_∞ uniform condition of thermal conductivity

m Hall current

M magnetic parameter

p pressure of the fluid

p_∞ uniform condition of pressure

P_r Prandtl number

N_u Nusselt number

$\mathbf{q}$ velocity vector

q_w rate of heat transfer

r radial axis

R_{ex} magnetic Reynolds number

R_e rotational Reynolds number

T temperature of the fluid

T_w uniform surface temperature

T_∞ free stream temperature

(u,v,w) velocity components along (r,φ,z) coordinates

U_0 mean velocity of the fluid

w_w uniform suction/injection velocity

z vertical axis

2.1. Greek letters

δ a time dependent length scale

φ tangential axis

γ relative temperature difference parameter

λ a constant

θ dimensionless temperature

σ electrical conductivity

η similarity variable

τ_r, τ_t radial and tangential stresses

v kinematic coefficient of viscosity

v_∞ uniform condition of kinematic coefficient of viscosity

μ viscosity

μ_∞ uniform condition of viscosity

ρ density of the fluid

ρ_∞ uniform condition of density

Ω angular velocity

3. Physical model

Let us consider a disk which is placed at $z = 0$ in a cylindrical polar coordinate system (r, φ, z) where z is the vertical axis and r and φ are the radial and tangential axes respectively. The disk is assumed to rotate with an angular velocity Ω and the fluid occupies the region $z > 0$. Let the components of the flow velocity $\mathbf{q} = (u, v, w)$ be in the directions of increasing (r, φ, z) respectively. Let p be the pressure, ρ be the density and T be the temperature of the fluid while the surface of the rotating disk is maintained at a uniform temperature T_w. For away from the wall, free stream is kept at a constant temperature T_∞ and at a constant pressure p_∞. The fluid is assumed to be Newtonian, viscous and electrically conducting. An external magnetic field is applied in the z-direction having a constant magnetic flux density B_0 which is assumed unchanged by taking small magnetic Reynolds number $(R_{ex} \langle\langle 1 \rangle)$. The electron-atom collision frequency is assumed to be relatively high, so that the Hall effect is assumed to exist. Geometry of the physical model is shown below.

The fluid properties viscosity (μ), thermal conductivity (k) and the density (ρ) are taken as functions of temperature alone and obey the following laws [Jayaraj (1995)]

$$\mu = \mu_\infty \left[T \Big/ T_\infty \right]^a , k = k_\infty \left[T \Big/ T_\infty \right]^b , \rho = \rho_\infty \left[T \Big/ T_\infty \right]^c \tag{1}$$

where a, b and c are arbitrary exponents and $\mu_\infty, k_\infty, \rho_\infty$ are the uniform conditions of viscosity, thermal conductivity and the density. For the present analysis the fluid considered in the flue gas. For flue gases the values of the exponent are a = 0.7, b = 0.83 and c = -1.0 (ideal gas).

Based on the above features , the Navier Stokes equations and Energy equation , which are the governing equations of the problem, due to unsteady axially symmetric, compressible MHD laminar flow of a homogenous fluid take the following form:

$$\frac{\partial}{\partial t}(\rho r)+\frac{\partial}{\partial r}(\rho ru)+\frac{\partial}{\partial z}(\rho rw)=0 \tag{2}$$

$$\rho\left(\frac{\partial u}{\partial t}+u\frac{\partial u}{\partial r}-\frac{v^2}{r}+w\frac{\partial u}{\partial z}\right)=-\frac{\partial p}{\partial r}+\frac{\partial}{\partial r}\left(\mu\frac{\partial u}{\partial r}\right)+\frac{\partial}{\partial r}\left(\mu\frac{u}{r}\right)+\frac{\partial}{\partial z}\left(\mu\frac{\partial u}{\partial z}\right)-\frac{\sigma B_0^2}{1+m^2}(u-mv) \tag{3}$$

$$\rho\left(\frac{\partial v}{\partial t}+u\frac{\partial v}{\partial r}-\frac{uv}{r}+w\frac{\partial v}{\partial z}\right)=\frac{\partial}{\partial r}\left(\mu\frac{\partial v}{\partial r}\right)+\frac{\partial}{\partial r}\left(\mu\frac{v}{r}\right)+\frac{\partial}{\partial z}\left(\mu\frac{\partial v}{\partial z}\right)-\frac{\sigma B_0^2}{1+m^2}(v+mu) \tag{4}$$

$$\rho\left(\frac{\partial w}{\partial t}+u\frac{\partial w}{\partial r}+w\frac{\partial z}{\partial z}\right)=-\frac{\partial p}{\partial z}+\frac{\partial}{\partial r}\left(\mu\frac{\partial w}{\partial r}\right)+\frac{1}{r}\frac{\partial}{\partial r}(\mu w)+\frac{\partial}{\partial z}\left(\mu\frac{\partial w}{\partial z}\right) \tag{5}$$

$$\frac{\partial T}{\partial t}+u\frac{\partial T}{\partial r}+w\frac{\partial T}{\partial z}=\frac{\partial}{\partial r}\left(k\frac{\partial T}{\partial r}\right)+\frac{k}{r}\frac{\partial T}{\partial r}+\frac{\partial}{\partial z}\left(k\frac{\partial T}{\partial z}\right)+\frac{\sigma B_0^2}{1+m^2}(u^2+v^2)+$$

$$+\mu\left[\left(\frac{\partial u}{\partial z}\right)^2+\left(\frac{\partial v}{\partial z}\right)^2\right] \tag{6}$$

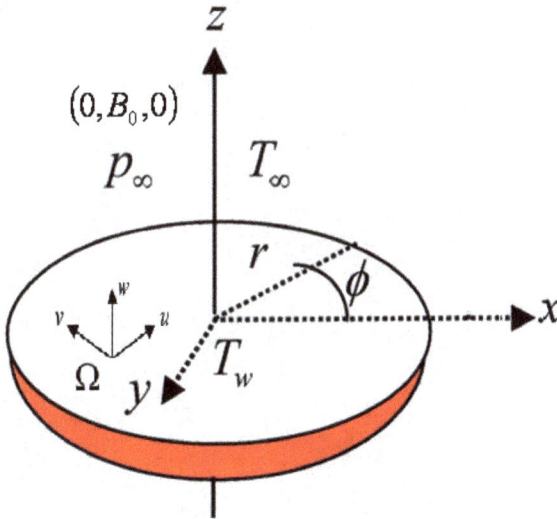

Scheme 1. Geometry of the physical model

In the above equations (2)-(6), m represents the Hall current and in equation (6) the last two terms respectively represent magnetic and viscous dissipation terms.

The appropriate boundary conditions of the flow induced by the infinite disk $(z = 0)$ which is started impulsively into steady rotation with constant angular velocity Ω and a uniform suction/injection velocity w_w through the disk are given by

$$u = 0 \ , \ v = \Omega r \ , \ w = w_w \ , \ T = T_w \text{ at } z = 0 \ \ \}$$ (7)

$$u \to 0 \ , \ v \to 0 \ , \ T \to T_\infty \ , \ p \to p_\infty \text{ as } z \to \infty .$$

4. Similarity transformations

In order to tackle the unsteady character of the motion unlike other approaches for example that of Chamkha & Ahmed(2011), a new similarity parameter taken as a function of time is introduced as $\delta = \delta(t)$. Here δ is a time dependent length scale and is a new parameter that has been fathered by Sattar & Hossain(1992).

Hence to obtain similarity solutions of the above governing equations the following similarity transformations which are little deviated from the usual von-Karman transformations are introduced in terms of the similarity parameter δ :

$$\eta = \frac{z}{\delta} , u = r\Omega F(\eta), v = r\Omega H(\eta), w = r\Omega G(\eta), T = T_\infty + \Delta T\theta(\eta), p = p_\infty + 2\mu_\infty \Omega P(\eta),$$ (8)

where Ω is a constant angular velocity and $\Delta T = T_w - T_\infty$ and T_w is the temperature of the disk wall.

Following the laws in (1) the unsteady governing partial differential equations (2)-(6) are then transformed respectively to the following set of dimensionless nonlinear ordinary differential equations through the introduction of the transformations in (8):

$$c\gamma(1+\gamma\theta)^{-1}\left[R_eH\theta' - \frac{\delta}{\upsilon_\infty}\frac{d\delta}{dt}\eta\theta'\right] + 2R_eF + R_eH' = 0$$ (9)

$$(1+\gamma\theta)^{c-a}\left[R_e(F^2 - G^2 + HF') - \frac{\delta}{\upsilon_\infty}\frac{d\delta}{dt}\eta F'\right] = \gamma a(1+\gamma\theta)^{-1}F'\theta' + F'' -$$

$$\frac{M}{1+m^2}(F - mG)(1+\gamma\theta)^{-a}$$ (10)

$$(1+\gamma\theta)^{c-a}\left[R_e(2FG + HG') - \frac{\delta}{\upsilon_\infty}\frac{d\delta}{dt}\eta G'\right] = \gamma a(1+\gamma\theta)^{-1}G'\theta' + G'' -$$

$$\frac{M}{1+m^2}(G+mF)(1+\gamma\theta)^{-a} \tag{11}$$

$$\frac{\delta}{\upsilon_\infty}\frac{d\delta}{dt}(1+\gamma\theta)^{c-a}\left[H-H^{'}\right]+P^{'}=\gamma a(1+\gamma\theta)^{-1}H^{'}\theta^{'}+H^{''} \tag{12}$$

$$P_r(1+\gamma\theta)^{c-a}\left[R_eH\theta^{'}-\frac{\delta}{\upsilon_\infty}\frac{d\delta}{dt}\eta\theta^{'}\right]=\theta^{''}+\gamma b(1+\gamma\theta)^{-1}\theta^2+\frac{M}{1+m^2}J_hP_r(G^2+F^2)+$$

$$J_hP_r(1+\gamma\theta)^{-a}(F^{'2}+G^{'2}) \tag{13}$$

where $M=\dfrac{\sigma B_0^2\delta^2}{\mu_\infty}$ is the magnetic parameter, $P_r=\dfrac{\rho_\infty\upsilon_\infty c_p}{k_\infty}$ is the Prandtl number,

$R_e=\dfrac{\Omega\delta^2}{\upsilon_\infty}$ is the rotational Reynolds number, $J_h=\dfrac{r^2\Omega^2}{c_p\Delta T}$ is the Joule heating parameter and

$\gamma=\dfrac{\Delta T}{T_\infty}$ is the relative temperature difference parameter which is positive for heated surface

and negative for cooled surface and zero for uniform properties.

The equations (10) to (14) are similar in time accept for the term $\left(\dfrac{\delta}{\upsilon_\infty}\dfrac{d\delta}{dt}\right)$ where t appears

explicitly. Thus the similarity conditions requires that $\left(\dfrac{\delta}{\upsilon_\infty}\dfrac{d\delta}{dt}\right)$ must be a constant. Hence

following the work of Sattar & Hossain(1992) one can try a class of solutions of equations
(10)-(14) by assuming

$$\frac{\delta}{\upsilon_\infty}\frac{d\delta}{dt}=\lambda \text{ (a constant)}. \tag{14}$$

Thus introducing (14) with the conditions that $\delta=0$ when $t=0$, one obtain

$$\delta=\sqrt{2\lambda\upsilon t} . \tag{15}$$

It thus appears from (15) that the length scale δ is consistent with the usual length scale
considered for various non-steady flows(Schlichting,1958) . Since δ is a scaling factor as
well as a similarity parameter, any value of λ in equation (15) would not change the nature
of the solutions except that the scale would be different.

Now making a realistic choice of λ to be equal to 2 in equation (15), equations (9) to (13)
finally become

$$c\gamma(1+\gamma\theta)^{-1}\left[R_eH\theta^{'}-2\eta\theta^{'}\right]+2R_eF+R_eH^{'}=0 \tag{16}$$

$$(1+\gamma\theta)^{c-a}\left[R_e(F^2-G^2+HF')-2\eta F'\right]=\gamma a(1+\gamma\theta)^{-1}F'\theta'+F''-\frac{M}{1+m^2}(F-mG)(1+\gamma\theta)^{-a} \quad (17)$$

$$(1+\gamma\theta)^{c-a}\left[R_e(2FG+HG')-2\eta G'\right]=\gamma a(1+\gamma\theta)^{-1}G'\theta'+G''-\frac{M}{1+m^2}(G+mF)(1+\gamma\theta)^{-a} \quad (18)$$

$$2(1+\gamma\theta)^{c-a}\left[H-H'\right]+P'=\gamma a(1+\gamma\theta)^{-1}H'\theta'+H'' \quad (19)$$

$$P_r((1+\gamma\theta)^{c-b}\left[R_eH\theta'-2\eta\theta'\right]=\theta''+\gamma b(1+\gamma\theta)^{-1}\theta'^2+\frac{M}{1+m^2}J_hP_r(G^2+F^2)+$$

$$+J_hP_r(1+\gamma\theta)^{-a}(F'^2+G'^2) \ . \quad (20)$$

With reference to the transformations (8), the boundary conditions (7) transform to

$$F(0)=0, G(0)=0, H(0)=W_S, \theta(0)=1$$
$$F(\infty)=0, G(\infty)=0, p(\infty)=0, \theta(\infty)=0 \quad (21)$$

where, $W_S=\dfrac{w_w}{\delta\Omega}$ represents a uniform suction ($W_S\langle 0$) or injection $(W_S\rangle 0)$ at the surface.

The quantities which are of physical interest relevant to our problem are the local skin-friction coefficients (radial and tangential) and the local Nusselt number.

Now since the radial (surface) and tangential stresses are respectively given by

$$\tau_r=\left[\mu\left(\frac{\partial u}{\partial z}+\frac{\partial w}{\partial r}\right)\right]_{z=0} \text{ and } \tau_t=\left[\mu\left(\frac{\partial v}{\partial z}+\frac{1}{r}\frac{\partial w}{\partial\varphi}\right)\right]_{z=0},$$

the dimensionless radial and tangential skin-friction coefficients are respectively obtained as

$$\frac{U_0^2\delta}{v_\infty\gamma\Omega}(1+\gamma\theta)^{-a}=F'(0) \quad (22)$$

$$\frac{U_0^2\delta}{v_\infty\gamma\Omega}(1+\gamma\theta)^{a-c}=G'(0) \quad (23)$$

where U_0 is taken to be a mean velocity of the flow.

Again the rate of heat transfer from the disk surface to the fluid is given by

$$q_w=-\left(k\frac{\partial T}{\partial z}\right)_{Z=0}.$$

Hence the Nusselt number defined by

$$N_u = \frac{\delta q_w}{k\Delta T}$$

is obtained as

$$N_u = -\theta'(0).$$

5. Numerical method

The nonlinear coupled ordinary differential equations (16) to (20) with the boundary conditions (21) have been solved numerically applying Natchtsheim-Swigert(1965) iteration technique(for detailed discussion of the method see Maleque and Sattar(2002)) along with sixth-order Runge-Kutta integration scheme. A step size of $\Delta\eta = .01$ was selected to be satisfactory for a convergence criteria of 10^{-7}. The value of η_∞ was found to each iteration loop by the statement $\eta_\infty = \eta_\infty + \Delta\eta$. The maximum value of η_∞ was determined when the value of the unknown boundary conditions at $\eta = 0$ does not change to successful loop with an error less than 10^{-7}.

6. Steady case

When the flow is steady, δ is no longer a function of time rather can be considered to be a characteristic length scale such as L. Thus in equations (9) to (13) we can take

$$\frac{d\delta}{dt} = \frac{dL}{dt} = 0.$$

Thus putting $\frac{d\delta}{dt} = 0$ in equations (9) to (13) we obtain the following equations:

$$H' + 2F + c\gamma(1+\gamma\theta)^{-1}H\theta = 0 \tag{24}$$

$$F'' + a\gamma(1+\gamma\theta)^{-1}F'\theta' - \left[R_L(F^2 - G^2 + HF')\right](1+\gamma\theta)^{c-a} - \frac{M}{1+m^2}(F - mG)(1+\gamma\theta)^{-a} = 0 \tag{25}$$

$$G'' + a\gamma(1+\gamma\theta)^{-1}G'\theta' - \left[R_L(2FG + HG')\right](1+\gamma\theta)^{c-a} - \frac{M}{1+m^2}(G + mF)(1+\gamma\theta)^{-a} = 0 \tag{26}$$

$$H' + \gamma a(1+\gamma\theta)^{-1}H'\theta' - P' = 0 \tag{27}$$

$$\theta'' + b\gamma(1+\gamma\theta)^{-1}\theta'^2 - P_r R_L(1+\gamma\theta)^{c-a} + \frac{M}{1+m^2}J_h P_r(G^2 + F^2) + J_h P_r(1+\gamma\theta)^{-a}(F'^2 + G'^2) = 0. \tag{28}$$

In the above equations

$$R_L = \frac{\Omega L^2}{v_\infty}.$$

The above equations exactly correspond to those of Maleque and Sattar(2005a) , therefore the solutions to the above equations have not been explored here for brevity. However, numerical values of the radial , tangential and rate of heat transfer coefficients for three different values of the relative temperature difference parameter γ is presented in Table-1 and compared with those of Maleque and Sattar(2005a).

γ	$F'(0)$	Maleque and Sattar(2005a)	$-G'(0)$	Maleque and Sattar(2005a)	$-\theta'(0)$	Maleque and Sattar(2005a)
-0.5	0.457	0.468	2.084	2.086	0.868	0.867
0.0	0.371	0.372	1.234	1.233	0.721	0.720
0.5	0.167	0.168	0.624	0.622	0.559	0.559

Table 1. Values of $F'(0), -G'(0)$ and $-\theta'(0)$ for various values of γ when $R_L = 1$, $J_h = 0$, $\lambda = 0$, $m = M = 0.1$.

Although the comparison should show exact values, due to the differences in the present code and that of Maleque and Sattar(2005a) there are differences in the calculated values of $F'(0)$, $-G'(0)$, and $-\theta'(0)$. Percentage wise differences have therefore been calculated and found to be maximum 2.35% and minimum 0.0% w.r.t three decimal places of the calculated values which shows a good agreement between our calculated results and that of Maleque and Sattar(2005a).

7. Unsteady solutions

As a result of the numerical calculations the radial, tangential and axial velocity profiles and temperature profiles are displayed in Figures 1-24 for various values of the governing parameters. In the analysis the fluid considered is flue gas for which $P_r = 0.64$ and the values of the exponents a, b and c are taken to be as $a = 0.7$, $b = 0.83$ and $c = -1.0$.

Variation of the radial, tangential and axial velocity profiles and temperature profiles under the influence of γ are shown in Figures 1-4. From Fig.1 it can be seen that due to the existence of the centrifugal force the radial velocity attains maximum values $0.23_{\gamma=-0.5}$, $0.13_{\gamma=0.0}$ and $0.10_{\gamma=0.5}$ close to the surface of the disk(approximately at $\eta = 0.75$). Thus at $\eta = 0.75$ the boundary layer thickness of the surface of the disk is reduced due to the increase in γ. From Fig-2, it is seen that the tangential velocity profile decreases in the interval $\eta \in [0,0.75]$, but for

$\eta\rangle 0.75$ this situation breaks down and the consequence is that the tangential velocity increases with the increase of γ. From Fig-3, it is seen that close to the disk surface γ has a tendency to reduce the motion and induce more flow far from the boundary indicating that there is a

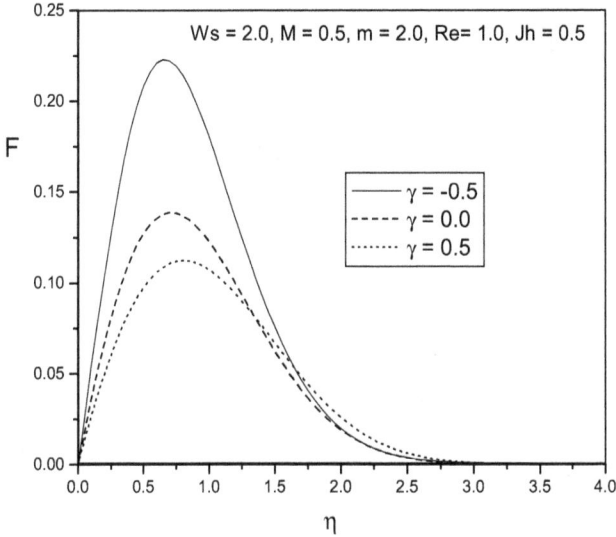

Figure 1. The dimensional radial velocity profiles against η for different values of γ

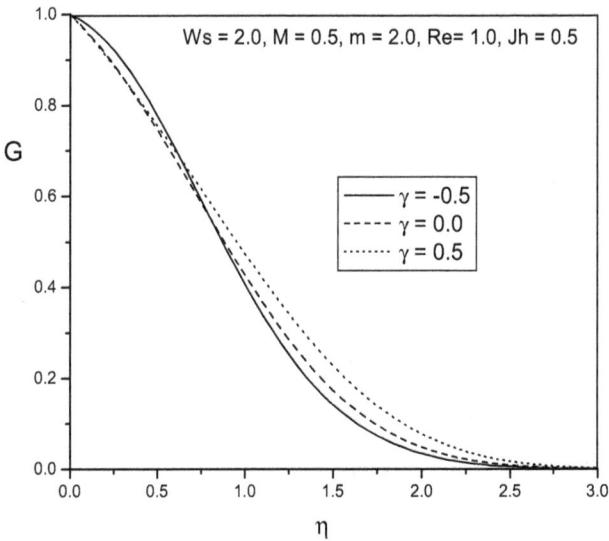

Figure 2. The dimensional tangential velocity profiles against η for different values of γ

separation flow and is detected at $\eta = 0.75$ (approximately). On the other hand, from Fig-4 it is observed that there is a small rate of decrease of the temperature close to the surface and then the temperature distributions starts increasing with increasing γ. This means that the thermal boundary layer induces more flow far from the surface of the disk.

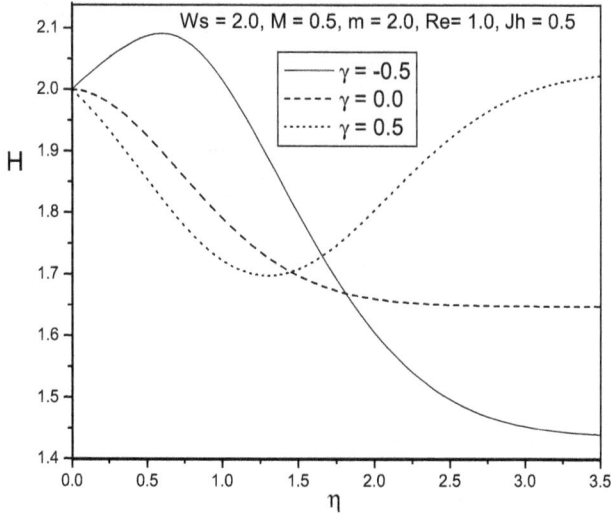

Figure 3. The dimensional axial velocity profiles against η for different values of γ

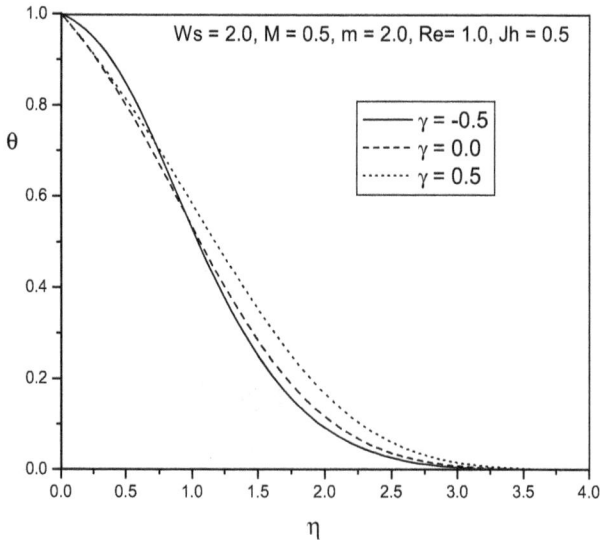

Figure 4. The dimensional temperature profiles against η for different values of γ

Figures 5-8, present the effects of uniform suction as well as injection(W_s) on the flow, which characterizes the flow behavior. It is evident from these figures that the boundary layer is increasingly blown away from the disk to form an interlayer between the injection and the outer flow regions. Also all flow profiles increase monotonically with increasing W_s. From Fig.5 it is apparent that radial velocity in this case attains maximum values approximately $0.19|_{W_s=4,\eta=1}, 0.11|_{W_s=2,\eta=0.75}, 0.05|_{W_s=0,\eta=0.5}, 0.03|_{W_s=-2,\eta=0.25}$. This implies that the momentum boundary layer thickness decreases due to an increase in the values of Ws in different regions like $\eta = 1, 0.75, 0.5, 0.25$. Thus reduced flows are observed for increase in injection ($W_s < 0$) and induced flows are observed for increase in suction ($W_s > 0$) in the total flow behavior.

The effects of the magnetic parameter M on the radial, tangential and axial velocities and temperature profiles are depicted in Figures 9-12. We see that the radial velocity increases with the increase in M. It can also be seen that at each value of M there exists local maxima in radial velocity distributions. The maximum values of velocities are approximately $0.13|_{M=1,\eta=1}, 0.11|_{M=0.5,\eta=1}, 0.08|_{M=0,\eta=1}$. Thus we can say that at $\eta = 1$ the boundary layer thickness increases. From Fig.-10, we see that tangential velocity decreases with increasing M. On the other hand, axial velocity decreases with increasing M and seen that at each value of M there exists local minima in the profiles. The effect of M is found to be almost not significant for temperature distributions.

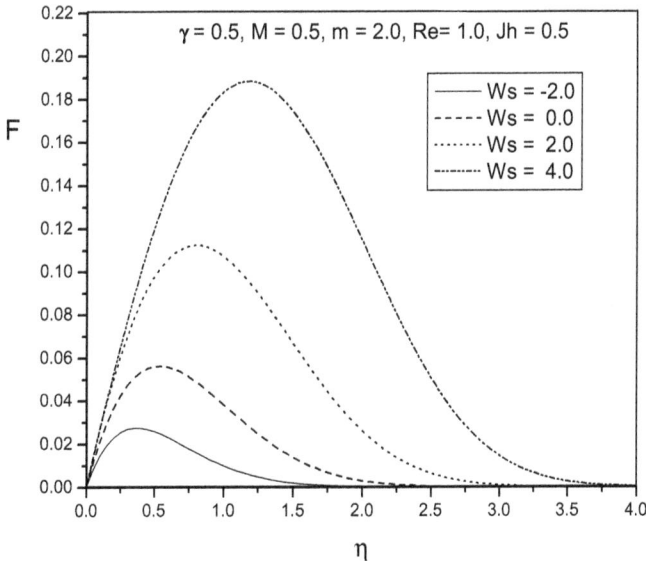

Figure 5. Fig-5: The dimensional radial velocity profiles against η for different values of W_s.

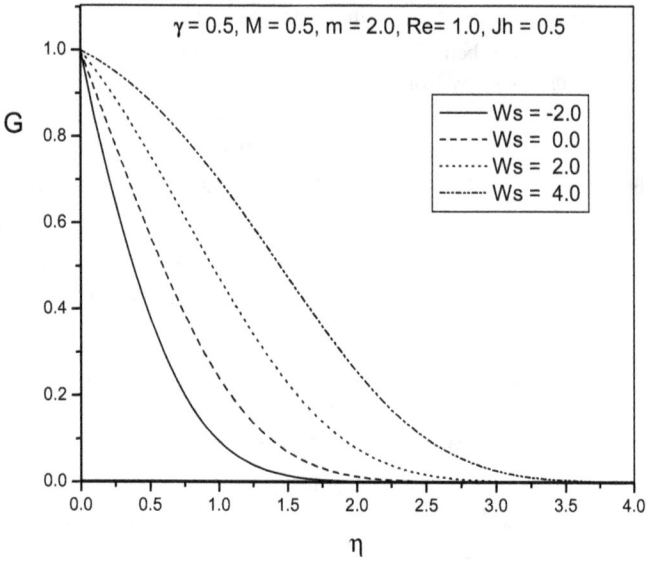

Figure 6. The dimensional tangential velocity profiles against η for different values of W_s .

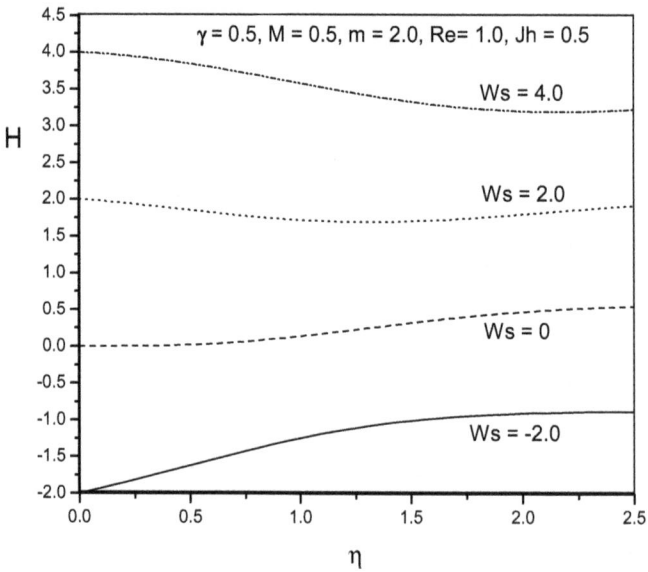

Figure 7. The dimensional axial velocity profiles against η for different values of W_s.

The Effects of Hall and Joule Currents and Variable Properties on an Unsteady MHD
Laminar Convective Flow Over a Porous Rotating Disk with Viscous Dissipation

257

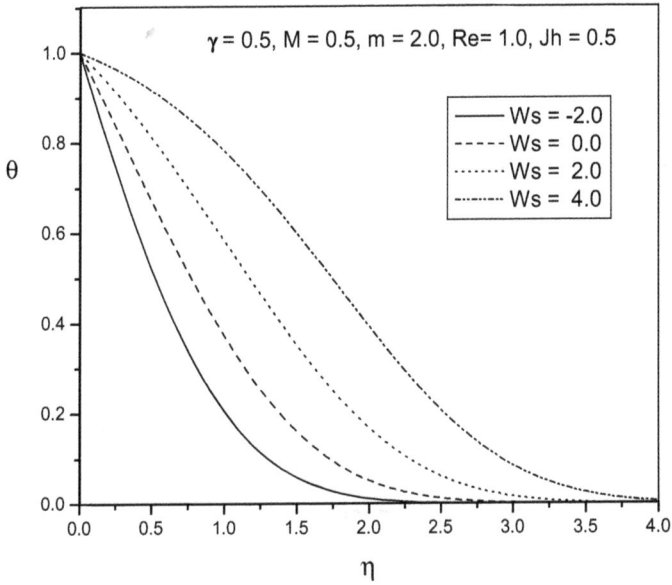

Figure 8. The dimensional temperature profiles against η for different values of W_s.

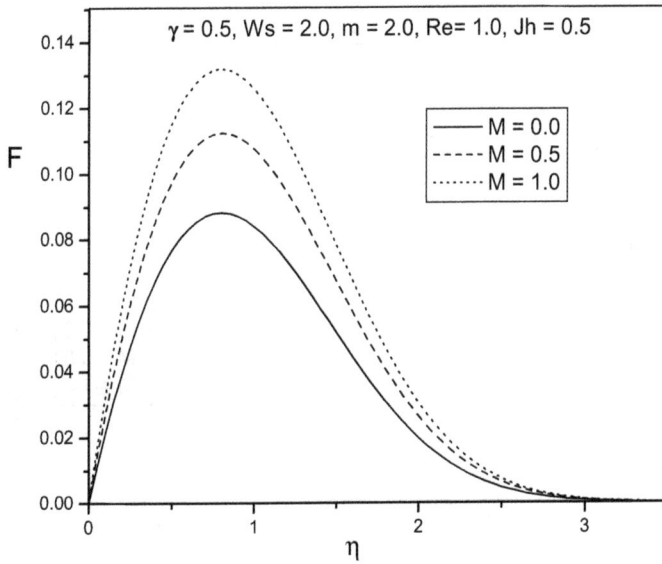

Figure 9. The dimensional radial velocity profiles against η for different values of M.

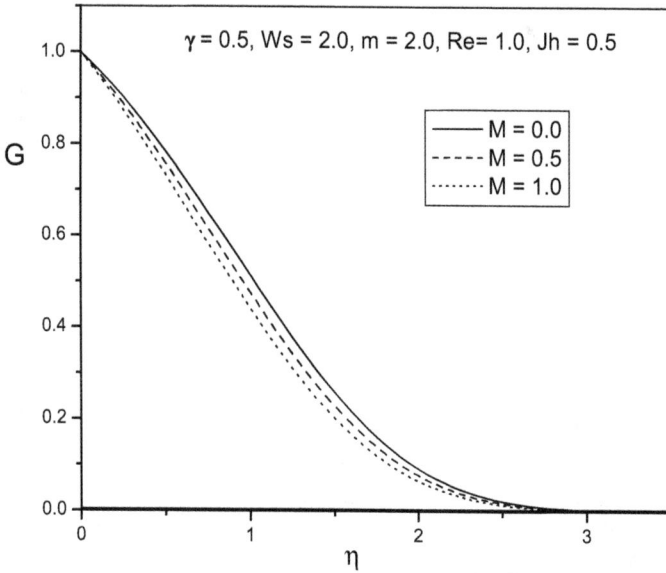

Figure 10. The dimensional tangential velocity profiles against η for different values of M.

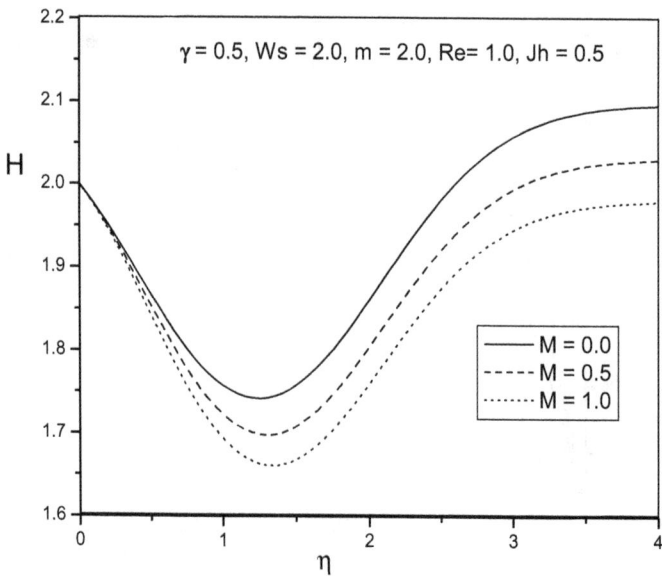

Figure 11. The dimensional axial velocity profiles against η for different values of M.

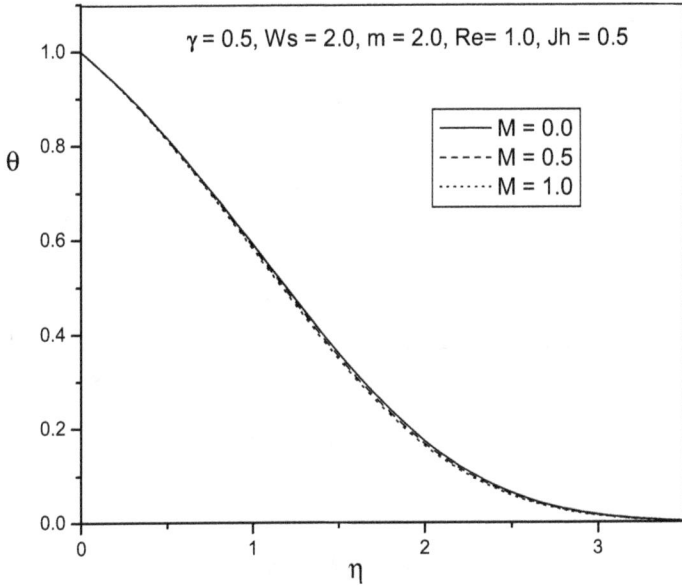

Figure 12. The dimensional temperature profiles against η for different values of M.

The effects of varying the Hall current parameter m on the flow distributions are shown in Figures 13-16. It can be seen that the radial velocity distributions increases with increasing m. This is due to the fact that for large values of m, the term $\dfrac{1}{1+m^2}$ is very small and hence the resistive effect of the magnetic field is diminished. This phenomenon for small and large values of m has been effectively explained by Hassan and Attia(1997). The maximum velocities are approximately $0.12|_{m=50,\eta=1}, 0.10|_{m=5,\eta=1}, 0.9|_{m=2,\eta=1}, 0.05|_{m=0,\eta=1}$. Thus at $\eta=1$ the boundary layer thickness increases due to the increase in m. Tangential distribution increases with the increase of m. On the other hand, the axial velocity decrease with increasing m and shows local minima indicating that the boundary layer thickness decreases. Like the magnetic parameter M the effect of m is also found to be not much significant in case of temperature distributions.

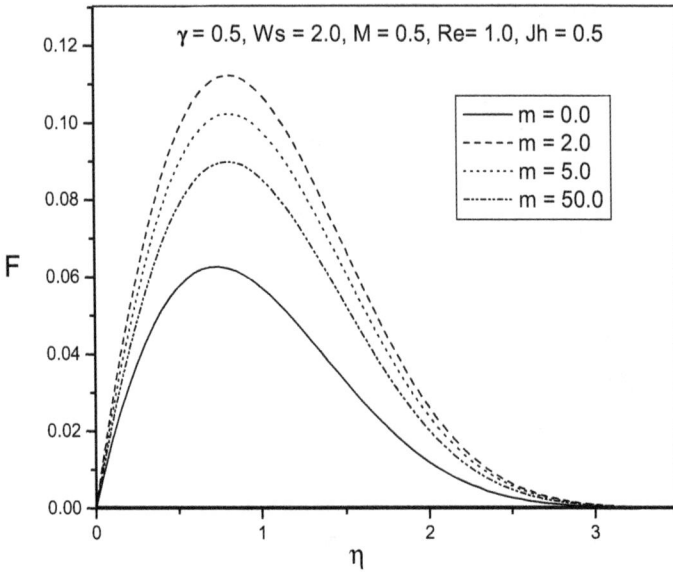

Figure 13. The dimensional radial velocity profiles against η for different values of m.

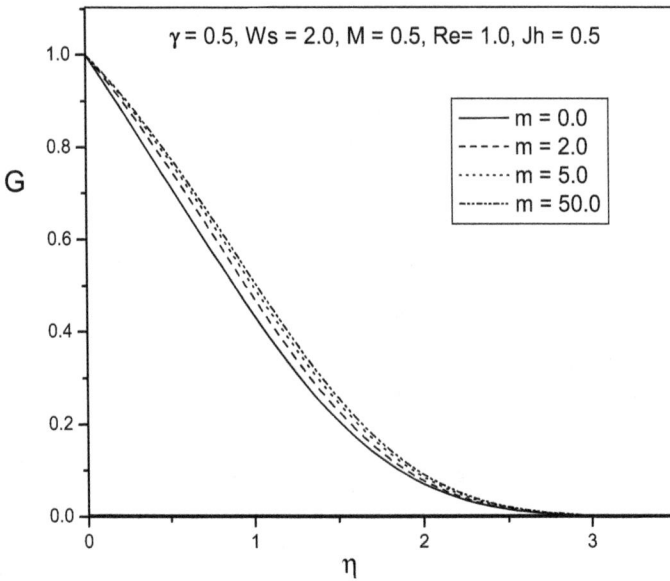

Figure 14. The dimensional tangential velocity profiles against η for different values of m.

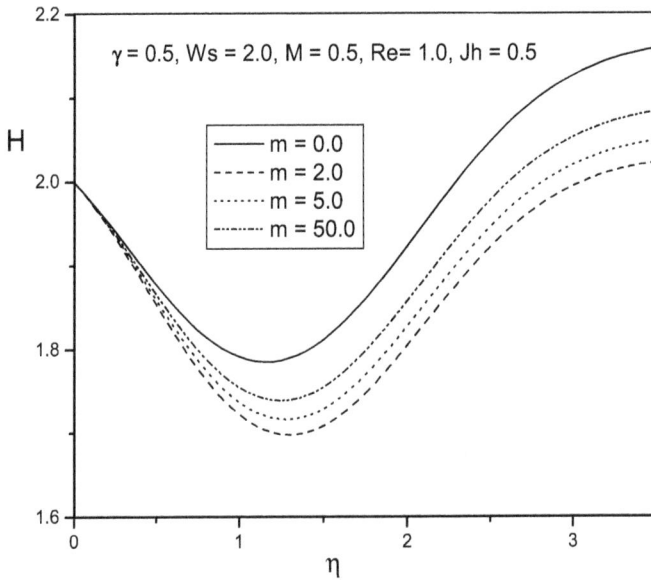

Figure 15. The dimensional axial velocity profiles against η for different values of m.

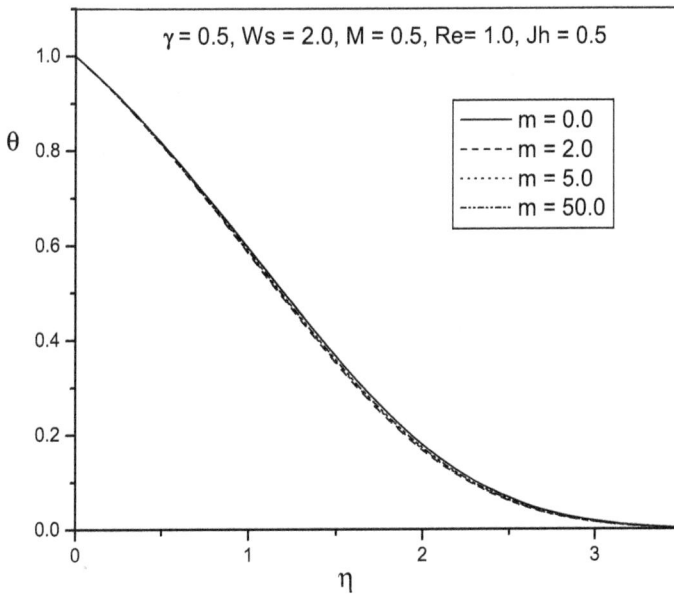

Figure 16. The dimensional temperature profiles against η for different values of m.

Figures 17-20 present the effects of the rotational Reynolds number, R_e, on the flow behavior. It is seen that Reynolds number accelerates the fluid motion in radial and tangential velocity profiles and temperature profiles. However the behavior of the axial velocity profiles decreases with increasing R_e.

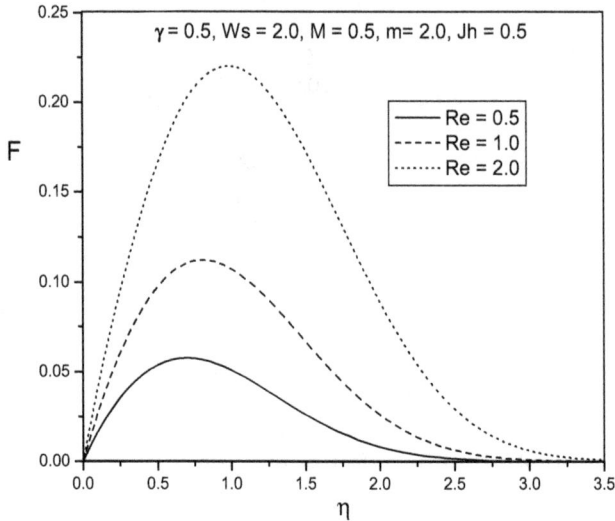

Figure 17. The dimensional radial velocity profiles against η for different values of R_e.

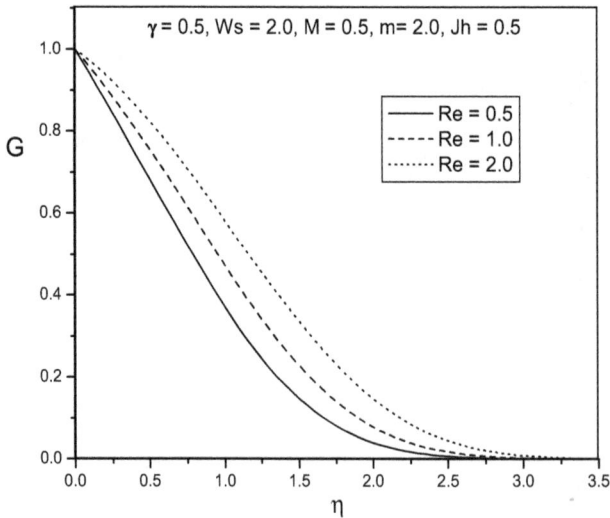

Figure 18. The dimensional tangential velocity profiles against η for different values of R_e.

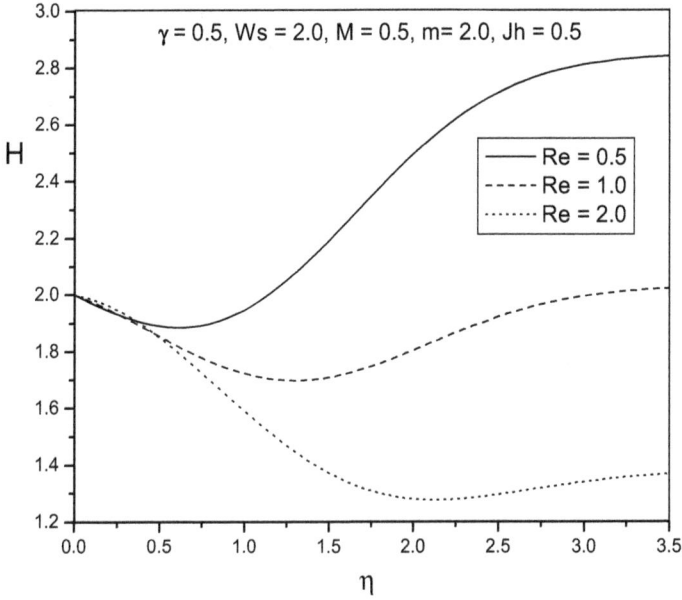

Figure 19. The dimensional axial velocity profiles against η for different values of R_e.

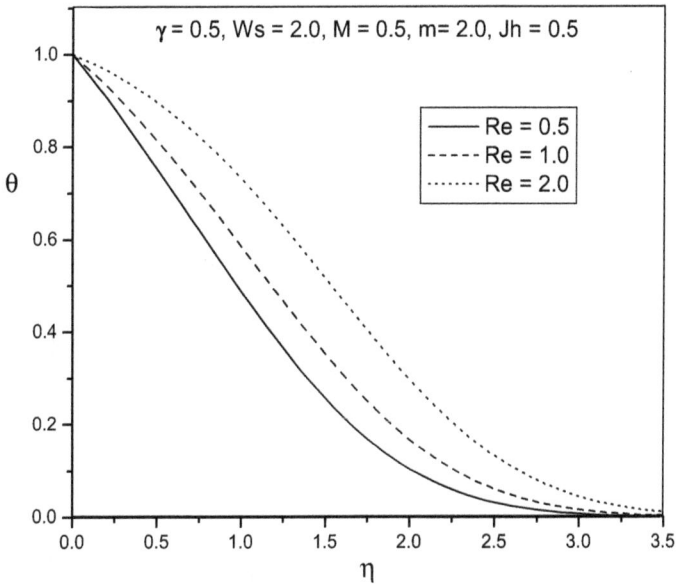

Figure 20. The dimensional temperature profiles against η for different values of R_e.

Figures 21-24 present the effects of the Joule heating parameter, J_h, on the flow behavior. It is seen that the axial velocity profiles and temperature profiles increase with increasing J_h while velocity profiles is generally much smaller between 0.1 and 0.5 except for large values of J_h, when it increases above 0.5, which is expectable on physical basis. The radial and tangential velocity profiles have no significant impact.

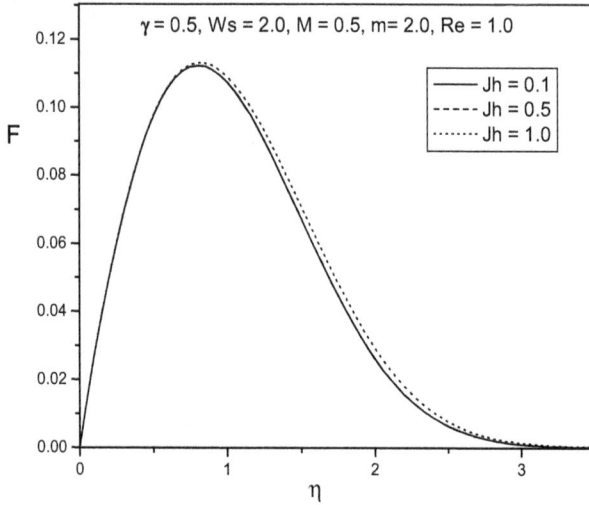

Figure 21. The dimensional radial velocity profiles against η for different values of J_h.

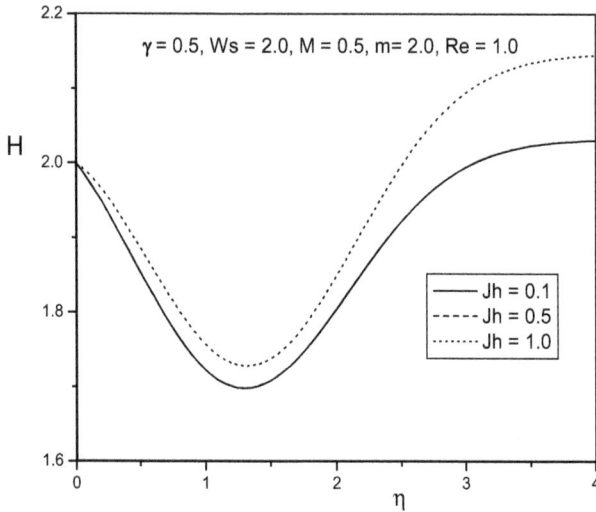

Figure 22. The dimensional tangential velocity profiles against η for different values of J_h.

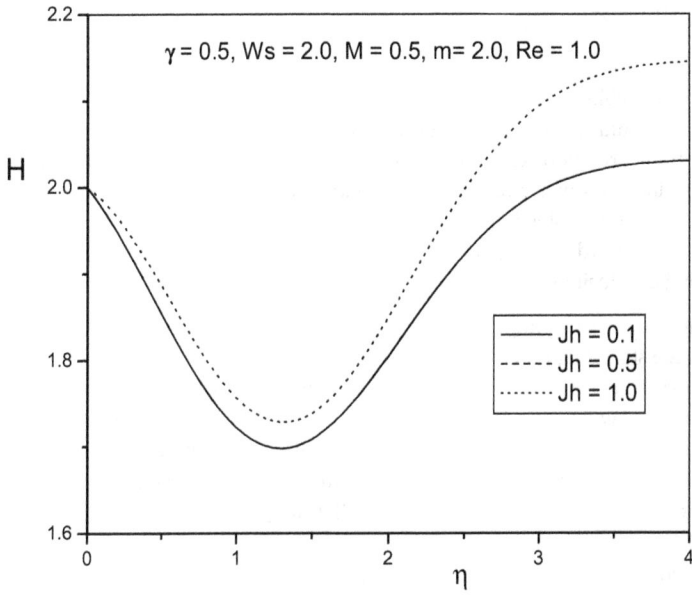

Figure 23. The dimensionless axial velocity profiles against η for different values of J_h.

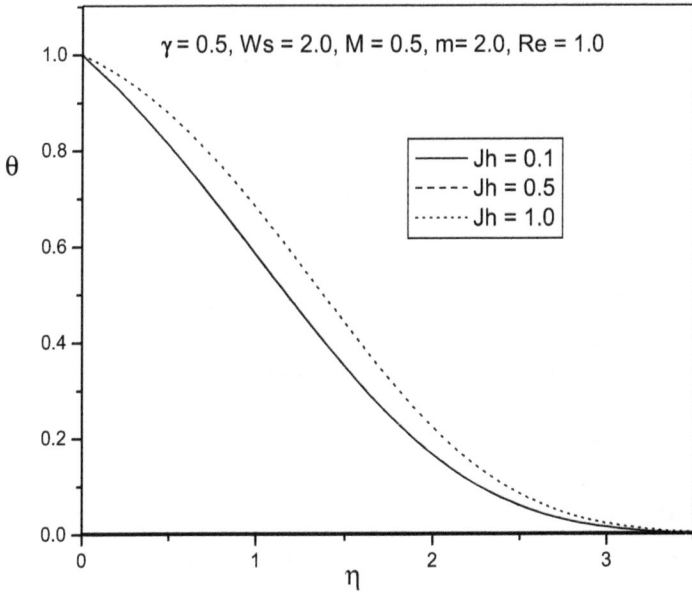

Figure 24. The dimensional temperature profiles against η for different values of J_h.

8. Concluding remarks

In this study, we have investigated numerically the heat transfer phenomenon along with the effects of variable properties for a 2-D unsteady hydrodynamic flow past a rotating disk taking into account viscous dissipation, Joule and hall currents. Using a new class of similarity transformation close to von-Karman, the governing equations have been transformed into non-linear ordinary differential equations that are locally similar. These equations have been solved using the Nachtsheim-Swigert shooting iteration technique along with a sixth-order Runge-Kutta integration scheme. Based on the resulting solutions the following conclusions can be drawn:

1. Similarity approach adopted in the analysis has the advantage that one can separately obtain the steady and unsteady solutions.
2. A comparison of the steady results for the radial and tangential stresses and the rate of heat transfer with those from the available literature leads credence to the numerical code used and hence to the results obtained in the unsteady case.
3. The relative temperature difference parameter γ taken as the variable properties parameter has marked effects on the radial and axial velocity profiles. Close to the surface of the disk tangential velocities and temperature slow down but shortly after they increase with the increasing values of γ.
4. As an influence of the relative temperature difference parameter γ, the thermal boundary layer induces more flow far from the surface of the disk.
5. Separation of flow was detected in different regimes of the momentum and thermal boundary layers.
6. Reduced flows have been observed for increase in injection $(W_s < 0)$ while induced Flows were observed for increase in suction $(W_s > 0)$.
7. Local maxima and local minima have been observed in the cases of radial and axial velocities for M, m and J_h.
8. As R_e increases, axial and tangential velocity profiles and temperature profiles increase while radial profiles decrease.

Author details

Abdus Sattar
North South University, Bashundhara Baridhara, Dhaka, Bangladesh

Mohammad Ferdows
Department of Mathematics, Dhaka University, Dhaka, Bangladesh

9. References

Aboul-Hassan, A. L. and Attia, H. A. (1997). " Flow due to a rotating disk with Hall currents", *Phys. Lett. A.*, vol. 228, pp. 286-290.

Attia, H. A. (1998). "Unsteady MHD flow near a rotating porous disk with uniform suction or injection", *Fluid Dynamics Research*, vol. 23, pp. 283-290.

Attia, H. A. (2006). "Unsteady flow and heat transfer of viscous incompressible fluid with temperature dependent viscosity due to a rotating disk in a porous medium", *J. Phys. A. Math. Gen.*, vol. 39, pp.979-991.

Chamkha, A. J. and Ahmed, S. A. (2011). "Similarity solution for unsteady MHD flow near a stagnation point of three dimensional porous body with heat and mass transfer, heat generation/absorption and chemical reaction", *J. Appl. Fluid Mech.*, vol. 4(2), pp.87-94.

EL-Mistikawy, T. M. A. and Attia, H. A. (1990). " Rotating disk flow in the presence of strong magnetic field", *Proc. Third Int. Cong. Fluid Mech., Cairo Egypt*, 2-4 January 1990, vol. 3, pp. 1211-1222.

El-Mistikawy, T. M. A. , Attia, H. A. and Megahed, A. A. (1991). "The rotating disk flow in the presence of weak magnetic field", *Proc. Fourth Conf. Theoret. Appl. Mec., Cairo, Egypt*, 5-7 Nov. 1991, pp. 69-82.

Hassan, L. A and Attia, H. A. (1997). "Flow due to a rotating disk with Hall effects", *Physics letter*, vol. A228, pp. 246-290.

Herrero, J. , Humphrey, J. A. C. and Giralt. F. (1994). "Comperative analysis of coupled flow and heat transfer between co-rotating disks in rotating and fixed cylindrical enclosures", *ASME Heat Transfer Div.* vol. 300, pp. 111-121.

Herwig, H. (1985). " The effect of variable properties on momentum and heat transfer in a tube with constant heat flux across the wall", *Int. J. Heat Mass Transfer*, vol. 28, pp. 424-441.

Herwig, H. and Wickern, G. (1986). " The effect of variable properties on laminar boundary layer flow", *Warme und Stoffubertragung*, vol. 20, pp. 47-57.

Herwig, H. and Klemp, K. (1988), " Variable property effects of fully developed laminar flow in concentric annuli", *ASME J. Heat Transfer*, vol. 110, pp. 314- 320.

Kreith, F. and Taylor, J. H. (1956). "Heat transfer from a rotating disk in turbulent flow", *ASME paper no. 56-A-146*.

Kreith, F. , Taylor, J. H. and Chong, J. P. (1959). "Heat and mass transfer from a rotating disk", *Journal of Heat Transfer*, vol. 18, pp. 95-105.

Maleque, Kh. A. and Sattar, M. A. (2002). "The effects of Hall current and variable viscosity on an unsteady MHD laminar convective flow due to a rotating disc", *J. Energy, Heat and Mass Transfer*, vol. 24, pp. 335-348.

Maleque, Kh. A. and Sattar, M. A. (2003). "Transient convective flow due to a rotating disc with magnetic field and heat absorption", *J. Energy, Heat and Mass Transfer*, vol. 25, pp. 279-291.

Maleque, Kh. A. and Sattar, M. A. (2005). "MHD convection flow due to a rotating disk with hall effect", *J. Energy, Heat and Mass Transfer*, vol. 27, pp. 211-228.

Maleque , Kh. A. and Sattar, M. A. (2005a). "The effects of variable properties and Hall current on steady MHD laminar convective fluid due to a porous rotating disk', *Int. J. Heat and Mass Transfer*, vol. 48, pp. 4963-4972.

Maleque, Kh. A. and Sattar, M. A. (2005b). "Steady laminar convective flowwith variable properties due to a porous rotating disk", *ASME J. Heat Transfer*, vol. 127, pp.1406-1409.

Millsaps, K. and Pohlhausen, K. (1952). "Heat transfer by laminar flow from a rotating plate", *J. Aeronautical Sciences*, vol. 19. Pp. 120-126.

Nachtsheim, P. R. and Swigert, P. (1965). "Satisfaction of the asymptotic boundary conditions in numerical solution of non-linear equations of boundary layer type", *NASA TND*-3004.

Osalusi, E. and Sibanda, P. (2006). "On variable laminar convective flow properties due to a porous rotating disk in a magnetic field", Romanian J. Phys., vol. 51, pp.933-944.

Osalusi, E., Side, J. and Harris, R. (2008). "Thermal-diffusion and diffusion thermo effects on combined heat and mass transfer of a steady MHD convective and slip flow due to a rotating disk with viscous dissipation and Ohming heating", *Int. Commu. Heat and mass Transfer*, vol. 35., pp. 908-915.

Owen, J. M. and Rogers, R. H. (1989). "Flow and heat transfer in a rotating disc system", *Rotor-Stator System*, vol. 1, Research Studies Press, Taunton, U.K. and John Wiley, NY.

Popiel, C. Z. O. and Boguslawski, L. (1975). "Local heat transfer coefficients on the rotating disk in still air', *Int. J. Heat and Fluid Flow*, vol. 18, pp.167-170.

Rahman, M. M. (2010). "Convective Hydromagnetic slip flow with variable properties due to a porous rotating disk", *SQU Journal of Science*, vol. 15, pp. 55-79.

Sattar, M. A. and Hossain, M. M. (1992). "Unsteady hydromagnetic free convection flow with hall current and mass transfer along an accelerated porous plate with time dependent temperature and concentration", *Canadian Journal of Physics*, Vol. 70(5), pp. 67-72.

Schlichting, H. (1958). Boundary Layer Theory, McGraw Hill, Sixth Edition.

Sparrow, E. M. and Gregg, J. L. (1959). "Heat transfer from a rotating disk to fluids of any Prandtle number", *ASME J. Heat Transfer*, vol. 81, pp.249-251.

Jayaraj, S. (1995). " Thermophoresis in laminar flow over cold inclined plates with variable properties", Heat Mass Transfer, Vol. 30, pp. 167-173.

von. Karman, T. (1921). "Uber laminare und turbulente reibung", *ZAMM*, vol.1, pp.233-255.

Wagner, C. (1948). "Heat transfer from a rotating disk to ambient air", *Journal of Applied Physics*, vol. 19, pp. 837-839.

Zakerullah, M. and Ackroyd, J. A. D. (1979). "Laminar natural convection boundary layers on Horizontal Circular discs", *J. Appl. Math. Phys.* , vol.30, pp. 427-435.

Chapter 10

Stochastic Analysis of Heat Conduction and Thermal Stresses in Solids: A Review

Ryoichi Chiba

Additional information is available at the end of the chapter

1. Introduction

In general, accurately predicting the thermal or mechanical loads acting on structural components is very difficult. There is no question that their material properties are random variables and thus are usually stated in terms of average values with attached uncertainties. Taking these factors into consideration, the *factor of safety* was introduced into structural design. However, current design methods based on the factor of safety cannot quantitatively estimate the safety of structures. In order to circumvent such an issue, the probability theory and mathematical statistics have been applied to many engineering problems. This allows us to determine the safety both quantitatively and objectively on the basis of the concept of reliability.

Currently, the application of probabilistic methods to engineering problems, which stems from the random vibration theory, has been broadened to the field of heat transfer. As structures subjected to extreme thermal load currently hold a prominent position in industries, the stochastic analysis of heat conduction and related thermal stresses in solids has drawn attention. In addition, the stochastic analyses of heat conduction in not only homogeneous but also nonhomogeneous bodies are being carried out more frequently because of the fabrication of advanced heat-resistant materials characterized by nonhomogeneity in recent years owing to advances in material manufacturing technology.

This article reviews research achievements for the stochastic analysis of heat conduction and related thermal stresses in solids. The objective of this review is to provide researchers and engineers, mainly in the field of heat transfer and thermoelasticity, with basic information useful for assessing the reliability of high-temperature apparatus. It is beyond the scope of this article to provide basic knowledge about the theory of probability and random processes, which is necessary to describe randomness mathematically. Readers not familiar with this discipline are recommended to refer to textbooks related to stochastic modeling.

2. Overview

Table 1 summarizes existing studies that used probabilistic methods for heat conduction and thermal stress analysis. The existing studies are organized according to the type of parameters considered as stochastic quantities; the classification also distinguishes (i) homogeneity/nonhomogeneity of object materials to be analyzed and (ii) presence/absence of the analysis of thermal stress fields (including displacement fields due to thermal deformation). Note that the analysis of thermal stress fields includes the analysis of heat conduction as a prerequisite. Table 1 indicates that among the studies for homogeneous bodies, many treated heat conduction problems only, but studies that also investigated the effects of random parameters on thermal stresses (or thermal deformation) are limited. Moreover, studies that focused on nonhomogeneous bodies are far fewer than those that targeted homogeneous bodies, although the former have gradually increased since the early 1990s, coupled with the emergence of functionally graded materials (FGMs) [1].

Samuels [2] was the first to conduct a seminal study on heat conduction analysis using probabilistic methods. He analyzed a plate and sphere with randomly fluctuating surface temperature and spatiotemporally random internal heat generation. Parkus [3] was the first to study random thermal stresses; he successfully analyzed the thermoelastic problem of a semi-infinite body using probabilistic methods.

With regard to parameters considered as stochastic quantities, many papers have presented the analysis of problems where the surface temperature of an object or the temperatures of its surrounding media are regarded as stochastic quantities, i.e., random heating problems. This is probably because random heating problems are strongly related to the design of thermal insulating systems for equipment sensitive to temperature changes, for example. Moreover, quite a few studies considered the material properties of analysis objects as stochastic quantities. This is because the fact that any materials show variability in their properties to a greater or lesser extent has become public knowledge. An FGM, which is a typical nonhomogeneous material, includes more factors to produce large variability in the material properties, as compared to other materials. From early on, Poterasu et al. [4] focused on the large variability in the material properties of FGMs and attempted a stochastic analysis of thermal stresses in consideration of their randomness. However, their study unfortunately remained confined to a formulation based on the stochastic finite element method. It was ten years before full-scale studies on thermal stresses in nonhomogeneous bodies whose material properties were assumed to be stochastic quantities began to be conducted.

In the rest of this article, existing papers related to this topic are classified into six groups according to the type of random or uncertain parameters considered in the analysis, and an extensive literature review is presented for each group. The review gives special emphasis to analytical methods used in the respective papers.

Random parameter	Homogeneous bodies		Nonhomogeneous bodies	
	Heat conduction	Thermal stresses	Heat conduction	Thermal stresses
Surface temperature (or ambient temperature)	[2, 5-23]	[3, 24-37]		[38-41]
Heat flux	[23]			
Initial temperature	[11, 15, 21, 22, 42-47]	[48]		[49]
Material properties	[13, 21, 23, 47, 50-68]	[37, 69-74, 127]	[75-77]	[4, 78-84]
Heat transfer coefficients	[11, 12, 14, 21, 47, 56, 65, 85, 86]	[33, 35, 36, 87-89]	[90]	[91]
Emissivity	[23, 61, 65]			
Heat generation rate	[2, 12, 15, 43, 46, 57, 66, 92-94]	[95]		
Geometry	[86, 96]	[70, 73, 97, 98]		[78]

* There are also some review articles [99-102].

Table 1. Summary of stochastic heat conduction/thermal stress studies*

3. Case of random surface temperature or ambient temperature

Recently, as reliability and safety gain increasing importance in the design phase of high-temperature apparatuses or heat-resistant structures, conventional deterministic thermal stress analysis alone is not sufficient; analysis that considers uncertainties included in the analysis objects themselves and/or thermal environments (e.g., temperature of the surrounding media) is required. In general, accurately predicting the thermal or mechanical loads acting on structural components is very difficult [103]. Representative examples of such situations are random high-cycle temperature fluctuations observed at the upper core structure of fast-breeder reactors [34] and random variations in heat transfer coefficients (HTCs) around the stator vanes of gas turbines [87]. When uncertain factors are involved in thermal environments, the temperature and thermal stresses in objects should be evaluated stochastically.

We focused on existing studies that examined cases of random thermal boundary conditions. Samuels [2] analyzed the temperature field of a plate and sphere whose surface temperature fluctuated randomly; this was a pioneering work on heat conduction analysis using probabilistic methods. He applied the theory of random processes to determine the mean square temperature of bodies under the random heat conditions. Hung [6] analyzed the heat conduction of straight and circular fins whose root temperatures fluctuate randomly, and Yoshimura et al. [10] analyzed the temperature field for a rectangular fin whose ambient temperature fluctuates; the former acknowledged the approach adopted by Samuels [6].

Heller [7] derived the frequency response function for the temperature of an infinite plate subjected to random heating, from which the standard deviation of the temperature was estimated. Using a similar method, Heller also addressed the stochastic analysis for two-dimensional non-axisymmetric heat conduction of an infinite multilayered cylinder subjected to random heating at the outer surface [9]. Novichkov *et al.* [8], with the aid of the Monte Carlo method, evaluated the correlation function and spectral density function of the temperature for a double-layered infinite plate subjected to temporally random-varying heating. Gaikovich [16, 17] analytically obtained the covariance functions of temperature and *brightness* temperature in a homogeneous semi-infinite body for which the boundary temperature is given by a stationary random process. Subsequently, he studied the same problem in terms of spectral densities and presented significantly simpler results for the statistics [18, 19]. Nicola *et al.* [20] developed a variance propagation algorithm to investigate the effect of ambient temperature modeled as a random process on the variability of the steady-state temperature in a complex-shaped body cooled by convection. Chantasiriwan [22] solved the stochastic heat conduction problem under random boundary and initial conditions using a meshless method—the multiquadric collocation method. He demonstrated that the value of "the shape parameter" strongly influences the solution accuracy.

In contrast, studies on thermal stresses for randomly heated bodies originated from Parkus' work [3] on semi-infinite bodies. Heller [27] analyzed thermal stresses for a steel pipe with a concrete cylinder as a core material for which the surface temperature is expressed as a narrow-band random process. Lenyuk *et al.* [28] investigated the non-Fourier heat conduction and related thermal stresses in a semi-infinite body in contact with a medium whose temperature is a random process. The same analytical method was applied to the stochastic thermal stress problem of infinite cylinders [29]. Miller [31] derived the power spectrum of stress intensity factors from the temperature spectrum through the response function of temperature while supposing that the temperature variation in the analysis object can be modeled as a stationary Gaussian random process; he calculated the mean growth rate of a crack due to thermal fatigue using a statistical method. Singh *et al.* [30] evaluated the characteristics of random thermal stresses in a long hollow cylinder whose temperature at the outer surface is a random process, on the basis of the concept of a complex frequency response function.

Amada [32] stochastically analyzed the temperature and thermal stresses in an infinite plate, a solid sphere, and a solid cylinder where the surface temperature was assumed to be a stationary process. Consider an infinite plate of thickness h, where the temperature of one side of the surfaces T_∞ fluctuates randomly. The initial temperature of the plate is assumed to be zero. The heat conduction equation is expressed in a dimensionless form by Eq. (1).

$$\frac{\partial \theta}{\partial \tau} = \frac{\partial^2 \theta}{\partial \xi^2} \qquad (1)$$

where $\theta = T / T_0$, $\tau = \kappa t / h^2$, $\xi = x / h$, T_0 denotes a reference temperature, κ is the thermal diffusivity, t is the time, and x is the through-thickness coordinate. The initial and boundary conditions are given by Eq. (2a) and Eqs. (2b), (2c), respectively.

$$\theta = 0 \quad \text{at } \tau = 0 \qquad \text{(a)}$$
$$\theta = \theta_\infty(\tau) \text{ at } \zeta = 0 \qquad \text{(b)} \qquad (2)$$
$$\theta = 0 \quad \text{at } \zeta = 1 \qquad \text{(c)}$$

where $\theta_\infty = T_\infty / T_0$. An analytical solution to Eqs. (1) and (2) is obtained in the form of Eq. (3).

$$\theta(\xi,\tau) = 2\pi \sum_{n=1}^{\infty} n \sin(n\pi\xi) \int_0^\tau \theta_\infty(\tau-\eta) \exp[-(n\pi)^2 \eta] d\eta \qquad (3)$$

If θ_∞ is a stationary process, the autocorrelation function of the temperature, R_θ, is given by Eq. (4).

$$R_\theta(\xi,\lambda) = \langle \theta(\xi,\tau) \cdot \theta(\xi,\tau+\lambda) \rangle =$$
$$4\pi^2 \sum_{m=1}^{\infty} \sum_{n=1}^{\infty} mn \sin(m\pi\xi)\sin(n\pi\xi) \int_0^\infty \int_0^\infty R_{\theta_\infty}(\lambda+\mu-\eta)\exp[-(m\pi)^2\mu - (n\pi)^2\eta]d\eta d\mu \qquad (4)$$

where < > denotes the expectation operator and R_{θ_∞} and λ represent the autocorrelation function of θ_∞ and an arbitrary nondimensionalized time interval, respectively. The spectral density of the temperature, S_θ, is expressed as Eq. (5).

$$S_\theta(\xi,\omega) = 4\pi^2 \sum_{m=1}^{\infty} \sum_{n=1}^{\infty} mn \sin(m\pi\xi)\sin(n\pi\xi) \int_0^\infty \int_0^\infty S_{\theta_\infty}(\omega)\exp\{-[(m\pi)^2 - i\omega]\mu - [(n\pi)^2 + i\omega]\eta\}d\eta d\mu \qquad (5)$$

where S_{θ_∞} denotes the spectral density of θ_∞; ω, an angular frequency; and i, the imaginary number. For a solid sphere of radius a whose surface temperature is a stationary process, T_∞, the autocorrelation function of the temperature is derived as Eq. (6).

$$R_\theta(\rho,\lambda) =$$
$$\frac{4\pi^2}{\rho^2} \sum_{m=1}^{\infty} \sum_{n=1}^{\infty} mn(-1)^{m+n} \sin(m\pi\rho)\sin(n\pi\rho) \int_0^\infty \int_0^\infty R_{\theta_\infty}(\lambda+\mu-\eta)\exp[-(m\pi)^2\mu - (n\pi)^2\eta]d\eta d\mu \qquad (6)$$

where $\rho = r/a$ and r denotes the radial coordinate. For a solid cylinder of radius a whose surface temperature is a stationary process, T_∞, R_θ is obtained in the form of Eq. (7).

$$R_\theta(\rho,\lambda) = 4\sum_{m=1}^{\infty} \sum_{n=1}^{\infty} \gamma_m \gamma_n \frac{J_0(\gamma_m\rho)}{J_1(\gamma_m)} \frac{J_0(\gamma_n\rho)}{J_1(\gamma_n)} \int_0^\infty \int_0^\infty R_{\theta_\infty}(\lambda+\mu-\eta)\exp[-\gamma_m^2\mu - \gamma_n^2\eta]d\eta d\mu \qquad (7)$$

where $J_0(\ )$ and $J_1(\ )$ are Bessel functions of the first kind of order 0 and 1, respectively. γ_n is the n-th positive root of the transcendental Eq. (8).

$$J_0(\gamma_n) = 0, \quad n = 1, 2, 3,...,\infty \qquad (8)$$

The mean square temperature is obtained by substituting $\lambda = 0$ into Eqs. (4), (6), and (7).

Tanaka *et al.* [34] proposed an analytical method for obtaining not only the probability distribution of the residual life of an infinite plate with cracks but also the statistical properties of the crack length from the power spectral density of random surface temperature variations. The temperature variation is modeled as a narrow band stationary Gaussian process.

With regard to nonhomogeneous bodies, Sugano et al. analyzed the stochastic thermal stress problems of a nonhomogeneous plate [38] and disk [39] with randomly fluctuating surface temperature. They derived analytical solutions to the statistics of temperature and thermal stresses, assuming that the material properties of the objects vary in a certain way along one direction. Consider a nonhomogeneous annular disk of inner radius r_0 and outer radius r_1 with zero initial temperature, as shown in Figure 1. At the inner radius, the disk is subjected to non-axisymmetric heating due to the boundary temperature $T_\infty \cdot f(\phi)$, which varies randomly with respect to time and is symmetric about the x-axis. At the outer radius, heat dissipates to the surrounding medium of zero temperature via an HTC h_1. Given that the specific heat c and density d of the disk are assumed to be constant, the heat conduction equation, initial condition, and boundary conditions for this nonhomogeneous disk are expressed by Eq. (9a), Eq. (9b), and Eqs. (9c), (9d), respectively.

$$\frac{1}{r}\frac{\partial}{\partial r}\left[rK(r)\frac{\partial T}{\partial r}\right] + \frac{K(r)}{r^2}\frac{\partial^2 T}{\partial \phi^2} = dc\frac{\partial T}{\partial t} \quad \text{(a)}$$

$$T = 0 \text{ at } t = 0 \quad \text{(b)}$$

$$T = T_\infty \cdot f(\phi) \text{ at } r = r_0 \quad \text{(c)}$$

$$K(r)\frac{\partial T}{\partial r} + h_1 T = 0 \text{ at } r = r_1 \quad \text{(d)}$$

$$\text{(9)}$$

where the thermal conductivity K is given by a power function of r as Eq. (10).

$$K(r) = K_0 \left(\frac{r}{r_0}\right)^\beta \quad \text{(10)}$$

If T_∞ is a stationary process, the autocorrelation function of the temperature, R_T, is given by Eq. (11).

$R_T(r,\phi,\lambda) =$

$$\left(\frac{\kappa_0^*}{\pi}\right)^2 \left(\frac{r}{r_0}\right)^{-\beta} \sum_{m,n=0}^{\infty} \sum_{k,l=1}^{\infty} \varepsilon_m \varepsilon_n \frac{L_{mk}(r)L_{nl}(r)}{H_{mk}H_{nl}} \Theta_m \Theta_n \cos m\phi \cos n\phi \int_0^\infty \int_0^\infty R_{T_\infty}(\lambda + p - q)\exp\left[-\frac{\kappa_0^*}{4}(w_{mk}^2 q + w_{nl}^2 p)\right]dpdq \quad \text{(11)}$$

where $\kappa_0^* = (K_0 / dc)r_0^{-\beta}$, $\varepsilon_0 = 1$, $\varepsilon_1 = \varepsilon_2 = \cdots\cdots = 2$, $\Theta_n = \int_0^\pi f(\phi)\cos n\phi\, d\phi$, and $L_{nl}(r)$ and H_{nl} are an explicit function of r and a constant (without going into detail) determined for each set of n and l, respectively. Furthermore, w_{nl} denotes the l-th positive root satisfying a transcendental equation determined for each value of n. For more details, see [39]. The mean square temperature is obtained by substituting $\lambda = 0$ into Eq. (11).

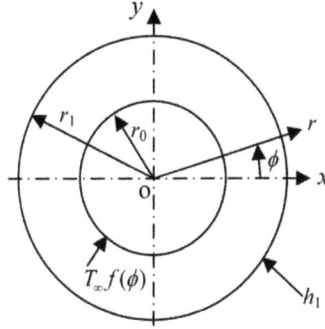

Figure 1. Schematic of a nonhomogeneous annular disk subjected to temporally random heating [39]

As a numerical example, Sugano *et al.* considered the case where the autocorrelation function of T_∞ and function f are given by Eqs. (12a) and (12b), respectively.

$$R_{T_\infty}(\lambda) = \exp(-V \mid \lambda \mid) \qquad \text{(a)}$$

$$f(\phi) = \begin{cases} 1 - \phi^2 / \alpha^2 & \text{for } 0 \le \phi \le \alpha \\ 0 & \text{for } \alpha \le \phi \le \pi - \alpha \qquad \text{(b)} \\ 1 - (\phi - \pi)^2 / \alpha^2 & \text{for } \pi - \alpha \le \phi \le \pi \end{cases} \qquad (12)$$

Figure 2 shows the spatial distribution of the mean square temperature for $V = 1$, $\alpha = \pi/6$, $r_1/r_0 = 2$, and $h_1 r_0 / K_0 = 0.1$.

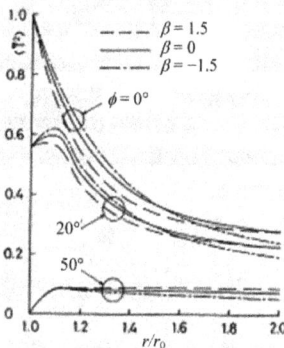

Figure 2. Effects of location-dependent thermal conductivity on mean square temperature [39]

Chiba *et al.* [41] addressed the stochastic thermal stress problem of functionally graded plates convectively heated from the surrounding medium whose temperature fluctuates randomly. The material properties of the plates are allowed to vary arbitrarily along the thickness direction. Under the condition that the surrounding medium temperature is expressed as a stationary process, they derived analytical solutions of statistics for this problem.

4. Case of random initial temperature

The initial temperature of structures is often uncertain (or random) in a real environment. For example, when space planes or space shuttles reenter the atmosphere, the initial temperature distribution of the fuselages is always uncertain [104]. Moreover, the temperature distribution in high-temperature apparatus, such as gas turbines, at the time of the resumption of operation is an uncertain factor in the design phase because of the time elapsed from shutdown and heat transfer from/to the surrounding media, such as a working fluid [103]. In order to investigate the effects of such randomness included in the initial temperature on the temperature and thermal stresses, stochastic analysis is absolutely imperative.

Thus far, stochastic studies on the heat conduction and thermal stress problems of solids with a random initial temperature have been limited. Ahmadi [42] studied the temperature field of an infinite plate and a semi-infinite body for which the initial temperature is a random field and showed that the randomness in the temperature diminishes over time. Subsequently, Grigorkiv *et al.* [44] conducted a similar study from the viewpoint of non-Fourier heat conduction. Tasaka [45] studied the convergence of statistical finite element solutions to one-dimensional heat conduction under a random initial condition. He also presented three different approaches for obtaining the statistics of temperature for the one-dimensional heat conduction problem in which the initial temperature is a random field and the internal heat generation is spatiotemporally random (i.e., a random wave) [46]: an analytical solution, a semi-analytical solution, and a numerical solution based on the finite difference method (FDM). He compared the numerical results of these different approaches. Nicolai *et al.* [47] investigated the transient behavior of temperature variance in bodies with random initial temperature using the stochastic finite element method (FEM). Scheerlinck *et al.* [21] analyzed the coupled heat and mass transfer problem in which the material properties, initial condition, and boundary conditions are random fields. In [21], a first-order perturbation algorithm based on the Galerkin finite-element discretization of Luikov's heat and mass transfer equations for capillary porous bodies was developed.

However, very few existing studies have dealt with the thermal stress problems for a random initial temperature. Chiba *et al.* [48] extended the work of Ahmadi [42] to thermal stress fields; they analytically obtained the autocorrelation functions of temperature and thermal stresses in seven simple-shaped bodies with the initial temtperature modeled as a homogeneous random field. These include an infinite plate, an infinite strip, a hollow sphere, an infinite body with a spherical hole, an infinite hollow cylinder, an annular disk, and an infinite body with a circular cylindrical hole. As a numerical example, the mean square temperature and the mean square thermal stresses were also calculated for when the initial temperature is given by white noise. The transient behavior of these statistics was graphically represented, as shown in Figure 3. Note that τ denotes the Fourier number; ζ, the dimensionless radial coordinate (= r/r_0); and $\bar{b}$, the ratio of the inner and outer radii (= r_1/r_0). The temperatures at the inner and outer radii are both considered to be deterministic; therefore, the mean square temperatures at the surfaces are zero.

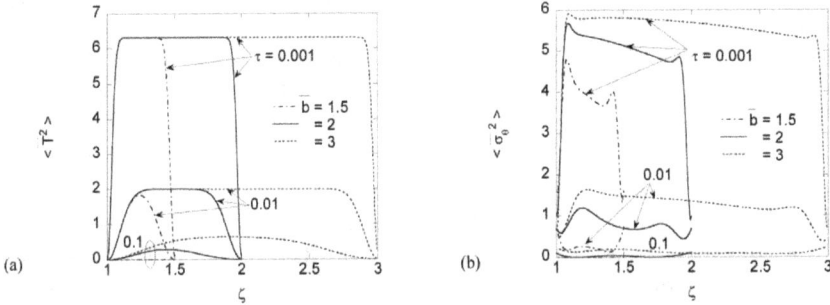

Figure 3. Transient behavior of (a) mean square temperature and (b) mean square tangential stress in a hollow sphere with random initial temperature modeled as white noise [48]

Sugano *et al.* [49] used probability theory to analyze the heat conduction and thermal stress problems of functionally graded plates with the initial temperature assumed to be a homogeneous random field. The material properties of the plates were allowed to vary arbitrarily along the thickness direction. The autocorrelation functions of temperature and thermal stresses were derived in an explicit form. Numerical calculations were performed for when the initial temperature is white noise or a Markov random field. The relationships between the through-thickness variation in the material composition and the statistics of the temperature and the thermal stresses were discussed.

5. Case of random material properties

As can be found in Table 1, there are a relatively large number of stochastic studies on the heat conduction and thermal stress problems of objects with random material properties. Some examples of studies that employed analytical (mathematical) methods are as follows: Chen *et al.* [50] analyzed the temperature field of a semi-infinite body with random thermal diffusivity by a perturbation method; Keller [54], Ahmadi [51, 52], Fox *et al.* [53], and Tzou [58] analyzed the heat conduction of approximately homogeneous bodies with random thermal conductivity; and Barrett [78] and Tzou [75] analyzed nonhomogeneous bodies with random thermal conductivity. Srivastava *et al.* [63] analyzed the one-dimensional steady heat conduction for thermal conductivity given by a random field and obtained analytical solutions to the mean and variance of the temperature in Earth's crust. Subsequently, Srivastava [67] presented analytical solutions to the mean and variance of heat flux for the same problem; numerical results demonstrated that a decrease in the correlation length scale of the thermal conductivity increases the variability in the heat flux. Kotulski [69] investigated the thermoelastic wave propagation in a solid with a random coefficient of thermal expansion.

Examples that employed numerical methods are given below. Nakamura *et al.* [13] analyzed the heat conduction when the thermal properties of analysis objects and ambient temperature are random, and Emery [65] treated the case where the thermal conductivity, HTC, and emissivity are random. The stochastic FEM [105] was used in both works. Manolis *et al.* [77] investigated the stochastic steady-state heat conduction in a nonhomogeneous solid whose thermal conductivity varies linearly along one direction at the macroscopic level but is spatially random at the microscopic level (see Figure 4). They mathematically described the random thermal conductivity with Eq. (13).

$$k(x,\gamma) = k_0 + [k_1 + \varepsilon k_2(\gamma)]x \qquad (13)$$

where k_0 and k_1 are constants, ε is a small parameter ($\varepsilon \ll 1$), k_2 is a zero-mean random field, and γ is a random variable. Note that the slope of the thermal conductivity consists of a constant plus a zero-mean random part. In their study, stochastic analysis of the heat conduction was carried out using a boundary integral equation approach.

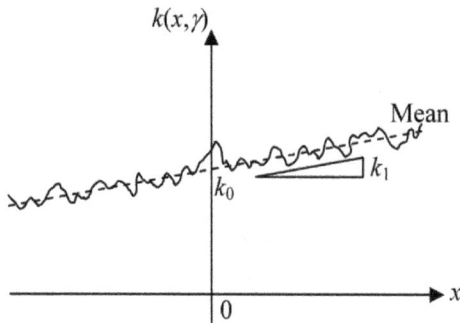

Figure 4. Linearly varying thermal conductivity [77]

Hien *et al.* [62] analyzed, using the stochastic FEM, the nonlinear heat conduction in solids whose thermal conductivity or specific heat is temperature-dependent and is considered to be a random field. Emery *et al.* [61] also analyzed the nonlinear heat conduction, which is attributed to not only random material properties but also random boundary conditions. Kaminski *et al.* [64] studied the heat conduction in composite materials whose thermal conductivity and specific heat are given as a random field by using the stochastic FEM. In [64], numerical results offered an interesting insight that whereas the randomness of the thermal conductivity monotonically increases the variability of temperature with time, the randomness of the specific heat increases the variability of temperature in the beginning but decreases it after a certain period of time. Liu [66] presented an analytical solution to the relative variability of temperature in bodies for when the thermal properties and internal heat generation are random. Although this is not an exact statistic, it is useful to roughly understand the extent to which the temperature distribution is influenced by random input parameters. Nakamura *et al.* [23] performed stochastic heat conduction analysis of atmospheric reentry vehicles. They considered the aerodynamic heat flux, heat shield emissivity, and insulator thermal conductivity to be random variables. They analyzed the transient behavior of the statistics of temperature at several locations by using the FEM and Monte Carlo simulation.

Aida *et al.* [72] estimated the probability of crack initiation due to thermal stresses by using an FEM-based Monte Carlo simulation method where the elastic modulus and the tensile and compressive strengths have uncertainties. Nakamura *et al.* [37] conducted a similar study for the case of uncertain Young's modulus, tensile and compressive strengths, and ambient temperature by using the first-order approximation theory. In addition, an interesting study on the stochastic finite element analysis of a thermal deformation problem was reported for a carbon fiber reinforced plastic (CFRP)-laminated plate whose fiber orientation angle is random [71]. Using the stochastic FEM, Sluzalec [73] analyzed the thermoelastic deflection of a rectangular plate subjected to a thermal and mechanical load concentrated at the center, where the plate has material properties and a thickness given by a two-dimensional random field.

With regard to the stochastic analysis of FGMs, which considers the uncertainties of material properties, Poterasu *et al.* [4] formulated the stochastic FEM for the thermoelastic problem of FGMs whose thermomechanical properties are homogeneous random fields. Ferrante *et al.* [80] considered the volume fraction and porosity to be random fields with spatial correlation in an FGM plate having linearly varying material composition distribution and analyzed the steady-state thermal stresses using Monte Carlo simulation. The analysis of the results showed that deviations in the ceramic/metal volume fraction produce significant randomness in the thermal stress and safety factor distribution of the plate. Chiba *et al.* [79, 82] stochastically analyzed the transient heat conduction and thermal stress problems of infinite FGM plates with an uncertain thermal conductivity and coefficient of thermal expansion. The FGM plates were assumed to have arbitrary thermal and mechanical nonhomogeneities along the thickness direction. Two methods were used for the analysis: the direct Monte Carlo simulation method [79] and a perturbation method [82]. Sugano *et al.* [81] analyzed the thermoelastic problem of nonhomogeneous plates with a random thermal conductivity and coefficient of thermal expansion.

Hosseini *et al.* [74] analyzed thermoelastic waves in a thick hollow cylinder for which some material properties are independently random. The statistics of temperature, stresses, and displacement were obtained by combining a hybrid numerical method, which consists of the Galerkin FEM and the Newmark FDM, and Monte Carlo simulation. Their numerical results demonstrated that the peak positions of the variances of the temperature and displacement progress with time in response to the progression of the heat wave front. Unfortunately, the sample size for the Monte Carlo simulation was unspecified (150 samples according to histograms shown therein); therefore, the degree of reliability of the presented numerical results is unclear. Using the same method, Hosseini *et al.* also conducted the thermoelastic wave analysis of an FGM thick hollow cylinder with random material properties [83]. Fairly recently, the same research group carried out this stochastic analysis using the meshless local Petrov-Galerkin method accompanied with Monte Carlo simulation [84]. This approach does not require the functionally graded cylinder to be assumed to be a multilayered cylinder with different material properties in each layer.

6. Case of random heat transfer coefficient

In real thermal environments, the HTCs of object surfaces are known to vary both temporally and spatially. For example, in the spinning process of light fiber wires, unsteady gas flow in furnaces has been reported to vary the spatial distribution of the HTCs of the fiber wire surface, which results in the variability of the wire diameter [106]. However, accurately predicting this spatial distribution of the HTCs is very difficult. In addition, the HTCs of turbine disk surfaces are influenced by many factors: the disk rotational speed, the presence or absence of shroud and neighboring disks, the distance from them, the velocity of cooling air, and its flow pattern. To make matters worse, these influences are nonlinear and change rapidly. Thus, accurate prediction of the HTCs is quite difficult. Moreover, there seems to be a measurement uncertainty of over 50% for the overall heat transfer coefficients of heat transfer surfaces of heat exchangers [107]. As long as the predicted values of HTCs include the uncertainties described above, a quantitative evaluation of the statistics of temperature and thermal stresses is needed to maintain an appropriate level of product quality or structure reliability. Hence, the temperature and thermal stresses in objects should be analyzed on the basis of the probability theory.

Stochastic studies on the heat conduction and thermal stress problems that consider spatial or temporal randomness in HTCs are scarce. Using a stochastic boundary element method, Drewniak [56] analyzed the steady-state heat conduction in solids for which the thermal conductivity or HTC is modeled as a random field. Madera [14] and Emery [65] analyzed the stochastic heat conduction problems of a rectangular fin for which they expressed the HTCs of the heat transfer surfaces as Gaussian white noise and a random field, respectively. The former derived partial differential equations for the expected value and covariance of temperature to present analytical solutions to these statistics under steady conditions. The latter used a higher-order perturbation method to analyze the problem and concluded that the use of first-order estimation for the standard deviation of temperature and second-order estimation for the mean response is preferable. Furthermore, the scale of correlation has been shown to have a strong effect on the statistics of the response. Kuznetsov [85] analyzed the stochastic heat conduction problem of an infinite strip for which the HTC is not a random field but is spatially random. Chiba [90] analytically obtained the second-order statistics, i.e., the mean and standard deviation, of the temperature for axisymmetrically heated FGM annular disks for which the surface HTCs are random fields. However, the abovementioned studies did not consider thermal stresses induced by the temperature changes.

In contrast, Mori *et al.* [87] numerically analyzed the statistics of steady thermal stresses produced in the stator vane of a gas turbine heated from the surroundings via random HTCs by using the stochastic FEM. Klevtsov *et al.* [33] investigated random thermal stress fluctuations in steam generation pipes, which are induced by randomly fluctuating HTCs/ambient temperature. Subsequently, the same authors [35, 36] estimated the variability of temperature and thermal stresses in a pipe for which the HTC of the inner surface and the temperature of the medium flowing in the pipe are random processes. They proposed a high-accuracy technique for predicting the material fatigue life that reflects the estimated variability. Ishikawa [88] used the FDM to analyze the coupled thermoelastic problem of a beam having HTCs described by a random field.

Chiba [89] analytically derived the second-order statistics of temperature and thermal stresses by a perturbation method in homogeneous annular disks for which the HTCs of the major surfaces are random fields. He assumed that the disks are subjected to a deterministic axisymmetric heat load. Numerical calculations were performed for the case where the surface HTCs are band-limited white noise random fields. The mean $E[\theta]$ and standard deviation $S[\theta]$ of the dimensionless temperature in the disks supposed as annular fins are shown in Figure 5. The ratio of the inner and outer radii of the disks is 0.2, and the coefficient of variation of the random HTCs is 0.1. In Figure 5, τ denotes a dimensionless time (Fourier number), and m is a nondimensionalized HTC (Biot number). The results are shown for the two cases in which the HTC mean is uniform throughout the disk surfaces and varies linearly along the radial direction.

Chiba [91] then analyzed the second-order statistics of the temperature and thermal stresses in FGM annular disks of variable thickness via Monte Carlo simulation; the HTCs of the disk surfaces were considered to be random fields and the disks were subjected to axisymmetric heat loads at the inner and outer radii.

7. Case of random internal heat generation/sink

Samuels [2] obtained analytical solutions of the mean and mean square value of the temperature field in a plate and sphere with randomly fluctuating surface temperature and spatiotemporally random internal heat generation. Becus [43] presented a solution to the heat conduction problem with a random heat source and random initial and boundary conditions. He also conducted a series of analytical studies on random heat conduction [108-111], which contributed greatly to the subsequent growth of this field. Vasseur *et al.* [92] analytically obtained relationships between the autocorrelation functions of heat generation and heat flow in the three-dimensional steady heat conduction of a homogeneous body, where the internal heat generation is expressed as a two-dimensional homogeneous random field. Nielsen [57] extended this work to address the case where thermal conductivity and heat generation are given by a random field and a cross-correlation exists between them. Val'kovskaya *et al.* [15] analyzed the stochastic temperature field in a two-layer solid disk subjected to heat sources for which power is a random function of time and radial coordinates. They also considered the randomness included in the initial temperature and ambient temperature. Ishikawa [93] analyzed the one-dimensional non-Fourier heat conduction in a solid with internal heat generation of white noise. Srivastava *et al.* [94] analyzed the one-dimensional steady-state heat conduction in solids with internal heat generation described by a random field. They derived exact solutions for the mean and variance of the temperature.

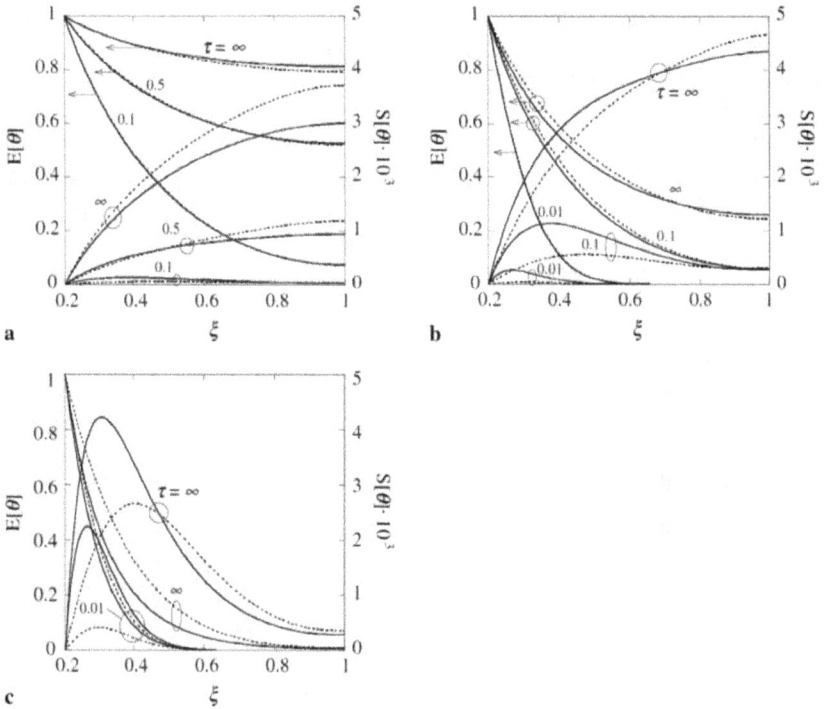

Figure 5. Transient distribution of the mean and standard deviation of the dimensionless temperature for uniform HTCs (solid curves) and linearly varying HTCs (broken curves) where (a) $m = 0.01$, (b) $m = 0.1$, and (c) $m = 1$ [89]

8. Case of random geometry

Shvets *et al.* [97] analytically evaluated the temperature field and thermal stresses in cylinders for which the radius fluctuates randomly along the circumferential direction. A perturbation method and Laplace transform were used for the analysis. Smith [70] discussed the structural reliability of hollow cylinders or hollow spheres; their inner and outer radii and material properties were random, and they were heated by the surroundings. Clarke [96] analytically solved the heat conduction problem of a layered plate for which the stacking sequence of layers and the layer thickness are random. This problem was later re-examined by Willis *et al.* [112]. Mori *et al.* [98] discussed the effects of the shape irregularity and thickness variability at a welded joint all around the body on stresses produced under internal pressure in a large-sized, thin-walled pressure vessel. The stochastic FEM was used to solve this random geometry problem.

Emery *et al.* [86] estimated the standard deviations of the heat fluxes in an object for which the heated surface roughness or HTC at the heated surface was assumed to be a random field. They adopted the FEM and direct Monte Carlo simulations. In order to treat this class of stochastic heat conduction problems, the problem of a stochastic region was converted to one in which the conductivity is stochastic through a coordinate transformation. Figure 6 shows the standard deviations for the x and y components of the heat flux in a slab where heat is mainly transferred along the width direction (x-direction) and the heated surface roughness is a random field with a certain spatial correlation length.

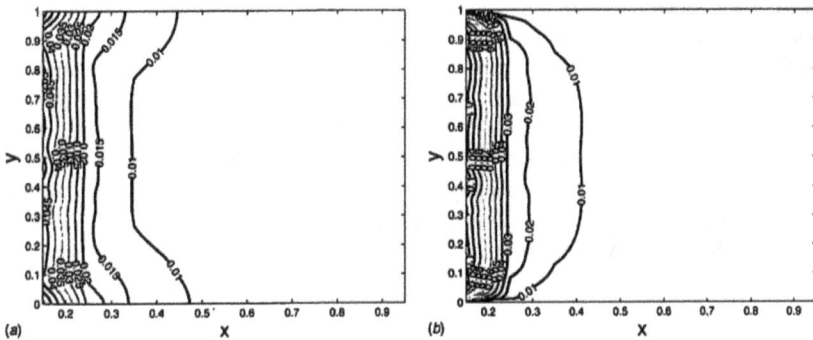

Figure 6. Contours of the standard deviations of the heat fluxes for an edge roughness of 2% and a correlation length of 5% of the slab mean width: (a) $\sigma(q_x)$ and (b) $\sigma(q_y)$ [86]

9. Numerical methods for stochastic heat conduction problems

In this section, we present numerical methods proposed thus far for the stochastic analysis of heat conduction. Case *et al.* [113] used the stochastic FEM to obtain the statistics of displacement due to thermal deformation in a solid for which the temperature distribution is expressed by simple functions of random variables. The Young's modulus and the coefficient of thermal expansion were assumed to depend on temperature. Nicolai *et al.* [11] proposed a numerical method for the computation of the statistics of the transient temperature field in heated foods with random initial and boundary condition parameters. This method is based on space discretization by the FEM and the stochastic systems theory. The same authors also developed a method for computing the statistics of the temperature field in objects with random thermophysical parameters; this method is based on incorporating perturbations of the thermal parameters in the FEM [60].

Madera [12, 114] obtained partial differential equations for the expectation and correlation of the stochastic temperature field for a solid subjected to random heating expressed by Gaussian white noise. These equations were solved by analytical methods (e.g., Green's function method) or numerical methods (e.g., control volume method). Madera [76] also developed a numerical method for determining three-dimensional transient temperature fields described by stochastic heat conduction equations with random coefficients and by stochastic initial and boundary conditions. This method is based on stochastic mathematical model discretization by the FEM or the FDM and on the solution of the Volterra stochastic integral equations.

Nicolai et al. [115] extended the perturbation algorithm of Sluzalec [59] and Fadale et al. [116] for linear heat conduction problems with random field parameters to the algorithm used for the analysis of nonlinear heat conduction under random conditions. This extended algorithm can consider the effects of (i) random field initial conditions, (ii) correlated thermophysical parameters, and (iii) correlated boundary condition parameters. In addition, Nicolai et al. [117] presented a variance propagation algorithm for heat conduction problems with parameters involving random fluctuations in time. This algorithm is based on the FEM and involves the numerical solution of Sylvester and Lyapunov differential systems. This algorithm was also applied to calculate the mean and variance of the temperature at arbitrary positions in heated cylinders with random wave parameters [118]. Nicolai et al. [119] described how to use variance propagation algorithms for calculating the statistical characteristics of the stochastic temperature field of heated solids in detail. Moreover, they developed an extended variance propagation algorithm for stochastic coupled heat and mass transfer problems under random process boundary conditions [120]. The numerical results of their study demonstrated that the random fluctuations of the process conditions may cause considerable variability in the temperature and moisture content in solids undergoing a drying process.

Liu et al. [121] extended the stochastic FEM to analyze the heat conduction problems that simultaneously consider randomness in the material properties and heat load conditions. Le Maitre et al. [122] developed a computationally efficient numerical technique for solving two-dimensional stochastic diffusion equations, in which a multigrid technique is applied to the system of equations arising from the polynomial chaos and the FDM.

Xiu et al. [123] solved the two-dimensional transient heat conduction subject to random inputs by generalized polynomial chaos expansion. This study is a natural extension of their earlier work on stochastic steady-state diffusion problems [124]. Xiu et al. [125] also treated diffusion problems in domains with rough boundaries considered as random fields. They proposed a novel computational framework based on the use of stochastic mappings to transform the original problem in a random domain into a stochastic problem in a deterministic domain and solved the transformed stochastic problem using a stochastic Galerkin method and Monte Carlo simulations.

Saleh et al. [68] analyzed stochastic one-dimensional heat conduction with random heat capacity or random thermal conductivity, which they modeled as a random field. They employed the stochastic FEM based on the Karhunen-Loeve decomposition and the projection of the solution on chaos polynomials. Recently, Nicolai et al. [126] extended the interval and fuzzy FEMs to nonlinear heat conduction problems with uncertain parameters and verified the efficiency of the proposed methodology with two case studies from food processing engineering. In [102], some of the above-mentioned algorithms developed by Nicolai *et al.* for stochastic heat conduction analysis were outlined and illustrated by some simple examples from thermal food process engineering.

10. Concluding remarks

This article reviewed the historical progress in the stochastic analysis of heat conduction and related thermal stresses. Existing papers related to this topic were classified into six groups according to the type of random or uncertain parameters considered in the analysis, and an extensive literature review was presented for each group. The overview indicates that among the studies for homogeneous bodies, many treated heat conduction problems only, but only a limited number also investigated the effects of random parameters on thermal stresses (or thermal deformation). Studies on nonhomogeneous bodies are far fewer than those targeting homogeneous bodies, although the former are increasing in number. With regard to parameters considered as stochastic quantities, a number of studies have analyzed problems in which the surface temperatures of an object or ambient temperatures are regarded as stochastic quantities, i.e., random heating problems. Furthermore, quite a few studies assumed the material properties of analysis objects to be stochastic quantities.

Some future research directions related to this topic are suggested as follows:

i.   In studies on the heat conduction and thermal stress analysis of bodies subjected to random heating, only stationary random processes have been targeted so far. Thus, when time functions included in thermal boundary conditions are nonstationary random processes, such as earthquake vibration, we need to divide the whole time interval into several intervals, conduct the analysis for a stationary process in each interval, and finally join the analysis results for the respective intervals. In an actual operation environment, structures may often be subjected to heat loads that are difficult to regard as stationary random processes. Therefore, the above analyses treated in the framework of the theory of a stationary random process need to be extended to nonstationary random processes.

ii. Uncertainties included in the material properties of "materials with microstructure," including particle-dispersed composite materials, are attributed to the variability in the microstructure (or microstructural morphology) as well as the variability in the material properties of the constituents. Hence, stochastic studies based on micromechanics, which consider various parameters at the microscale (e.g., the material properties of constituents and the shape and size of dispersed particles) as stochastic quantities, would be another potential research topic.

iii. All the existing stochastic studies on thermal stresses in solids have focused on the elastic range. Structures in which advanced heat-resistant materials are used are supposed to undergo extremely large temperature gradients; therefore, they may undergo partial plastic deformation. Thus, in order to extend the discussion from the elastic range to the plastic range, the same type of stochastic studies on the thermo-elastic-plastic behavior is another topic remaining to be addressed.

The stochastic analyses of this field lead directly to the reliability evaluation of high-temperature structures that require higher safety, such as space planes, hot gas turbines, and atomic reactors. Because these analyses are required in a variety of fields— e.g., food engineering and geophysical science—, they have wide applicability.

Author details

Ryoichi Chiba
Department of Mechanical Systems Engineering, Asahikawa National College of Technology, Asahikawa, Japan

11. References

[1] Miyamoto Y, Kaysser WA, Rabin BH, Kawasaki A, Ford RG (1999) Functionally graded materials: design, processing and applications. Boston: Kluwer Academic.

[2] Samuels JC (1966) Heat conduction in solids with random external temperatures and/or random internal heat generation. Int J Heat Mass Tran. 9:301-14.

[3] Parkus H (1962) Thermal stresses in bodies with random surface temperature. ZAMM. 42:499-507.

[4] Poterasu VF, Tanaka K, Sugano Y (1995) Stochastic FEM formulation of thermoelastic behavior in functionally gradient materials. Memoirs Tokyo Metro Inst Tech. 8:83-8.

[5] Moskalenko VN (1968) Random temperature fields in plates and shells. Int Appl Mech. 4:6-9.

[6] Hung HM (1969) Heat Transfer of Thin Fins with Stochastic Root Temperature. J Heat Tran. 91:129-34.

[7] Heller RA (1976) Temperature response of an infinitely thick slab to random surface temperatures. Mech Res Comm. 3:379-85.

[8] Novichkov YN, Butko AM (1979) Investigation of random temperature fields in two-layer plates with the use of the Monte Carlo method. Int Appl Mech. 15:160-5.

[9] Heller RA, Singh MP (1980) Temperature Distribution in Randomly and Asymmetrically Heated Cylindrical Structures. High Temp - High Pressures. 12:457-63.

[10] Yoshimura T, Campo A (1981) Extended surface heat rejection accounting for stochastic sink temperatures. AIAA Journal. 19:221-5.

[11] Nicolai BM, Baerdemaeker JD (1992) Simulation of heat transfer in foods with stochastic initial and boundary conditions. Trans IChemE Part C. 70:78-82.

[12] Madera AG (1993) Modelling of stochastic heat transfer in a solid. Appl Math Model. 17:664-8.

[13] Nakamura H, Hamada S (1994) The finite element analysis of heat conduction for the concrete structure having uncertain material properties. Proc JSCE. 496:71-80.

[14] Madera AG (1996) Heat transfer from an extended surface at a stochastic heat-transfer coefficient and stochastic environmental temperature. Int J Eng Sci. 34:1093-9.

[15] Val'kovskaya VI, Lenyuk MP (1996) Stochastic Nonstationary Temperature Fields in a Solid Circular-Cylindrical Two-Layer Plate. J Math Sci. 79:1483-7.

[16] Gaikovich K (1996) Stochastic approach to the results of simultaneous solution of emission transfer and thermal conductivity equations. Radiophys Quantum Electronics. 39:273-81.

[17] Gaikovich KP (1996) Stochastic Theory of Temperature Distribution and Thermal Emission of Half-Space with Random Time-Dependent Surface Temperature. IEEE Trans Geosci Rem Sens. 34:582-7.

[18] Gaikovich K (1997) Correlation theory of the thermal regime and thermal radiation of a medium with random boundary conditions. Radiophys Quantum Electronics. 40:395-407.

[19] Gaikovich KP (1999) Spectral approach to correlation theory of thermal regime and thermal emission of the medium with random boundary conditions. IEEE Trans Geosci Rem Sens. 37:3-7.

[20] Nicola BM, Verlinden B, Beuselinck A, Jancsok P, Quenon V, Scheerlinck N, et al. (1999) Propagation of stochastic temperature fluctuations in refrigerated fruits. Int J Refrigeration. 22:81-90.

[21] Scheerlinck N, Verboven P, Stigter JD, Baerdemaeker JD, Impe JFV, Nicolai BM (2000) Stochastic finite-element analysis of coupled heat and mass transfer problems with random field parameters. Numer Heat Tran B. 37:309-30.

[22] Chantasiriwan S (2006) Error and variance of solution to the stochastic heat conduction problem by multiquadric collocation method. Int Comm Heat Mass Tran. 33:342-9.

[23] Nakamura T, Fujii K (2006) Probabilistic transient thermal analysis of an atmospheric reentry vehicle structure. Aero Sci Tech. 10:346-54.

[24] Zeman J (1965) Locally and temporally random distributed temperature and stress fields: part I. Acta Mech. 1:194-211.

[25] Zeman JL (1966) A method for the solution of stochastic problems in linear thermoelasticity and heat conduction. Int J Solids Struct. 2:581-9.

[26] Ibragimov MK, Merkulov VI, Subbotin VI (1966) Random thermoelastic stresses produced in a wall by temperature pulsations. Atomic Energy. 21:1223-5.

[27] Heller RA (1976) Thermal stress as a narrow-band random load. Proc ASCE J Eng Mech Div. 102:787-805.

[28] Lenyuk MP, Bukatar MI, Shelyag LK (1976) Random thermal stresses in an elastic half-space. Int Appl Mech. 12:452-8.

[29] Lenyuk MP, Shelyag LK (1978) Stress state of symmetric elastic body under random thermal perturbations. Int Appl Mech. 14:1137-42.

[30] Singh MP, Heller RA (1980) Random Thermal Stress in Concrete Containments. Proc ASCE J Struct Div. 106:1481-96.

[31] Miller AG (1980) Crack propagation due to random thermal fluctuations: Effect of temporal incoherence. Int J Pres Ves Pip. 8:15-24.

[32] Amada S (1982) Thermal stresses in bodies with random temperature distribution at their boundaries. J Soc Mater Sci. 31:251-7.

[33] Klevtsov I, Crane R (1994) Random Thermal Stress Oscillations and Fatigue Life Estimation for Steam Generator Tubes. J Pressure Vessel Technol. 116:110-4.

[34] Tanaka H, Toyoda M (1996) Random thermal fatigue in fast breeder reactor: a narrow-band spectrum. Nucl Eng Des. 160:333-45.

[35] Klevtsov I, Crane R (1998) Cyclic fatigue evaluation in metals under random oscillating stresses. Proceedings of the Estonian Academy of Sciences: Engineering. 4:286-306.

[36] Klevtsov I, Crane R (1998) Evaluation of Accumulated Fatigue in Metals Due to Random Oscillating Thermal Stresses. J Pressure Vessel Tech. 120:43-50.

[37] Nakamura H, Hamada S, Tanimoto T, Miyamoto A (1999) Estimation of Thermal Crack Resistance for Mass Concrete Structures with Uncertain Material Properties. ACI Struct J. 96:509-18.

[38] Sugano Y, kimoto J (1988) Thermal stresses in a nonhomogeneous plate due to random surface temperature. Trans Jpn Soc Mech Eng. 54A:1993-9.

[39] Sugano Y, Kimoto J (1991) A stochastic analysis of unaxisummetric thermal stresses in a nonhomogeneous hollow circular plate. Trans Jpn Soc Mech Eng. 57A:845-51.

[40] Sugano Y, Kanno T, Chiba R (2002) Stochastic thermal deformation and thermal stress in a laminated plate including FGM layer under random surface temperatures. Trans Jpn Soc Mech Eng. 68A:1588-93.

[41] Chiba R, Sugano Y (2007) Stochastic thermoelastic problem of a functionally graded plate under random temperature load. Arch Appl Mech. 77:215-27.

[42] Ahmadi G (1974) Heat conduction in solids with random initial conditions. J Heat Tran. 96:474-7.

[43] Becus GA (1977) Random generalized solutions to the heat equation. J. Math. Analysis Appl. 60:93-102.

[44] Grigorkiv VS, Okunenko VN, Timofeev YA (1982) Heat conduction in solids with finite rate of diffusion of heat and initial conditions in the form of random functions. J Eng Phys. 42:61-3.

[45] Tasaka S (1983) Convergence of statistical finite element solutions of the heat equation with a random initial condition. Comput Meth Appl Mech Eng. 39:131-6.

[46] Tasaka S (1985) Non-stationary statistical solutions of a class of random diffusion equations: analytical and numerical considerations. Int J Numer Meth Eng. 21:1097-113.

[47] Nicolai BM, Scheerlinck N, Verboven P, Baerdemaeker JD (2000) Stochastic perturbation analysis of thermal food processes with random field parameters. Transactions of the ASAE. 43:131-8.

[48] Chiba R, Sugano Y (2010) Thermal Stresses in Bodies with Random Initial Temperature Distribution. Math. Mech. Solids. 15:258-76.

[49] Sugano Y, Chiba R (2002) A stochastic analysis of temperature and thermal stress in functionally graded plates with randomly varying initial temperature. J Soc Mater Sci. 51:653-8.

[50] Chen YM, Tien CL (1967) Penetration of temperature waves in a random medium. J Math Phys. 46:188-94.

[51] Ahmadi G (1978) A perturbation method for studying heat conduction in solid with random conductivity. J Appl Mech. 45:933-4.

[52] Ahmadi G (1978) On functional methods for studying heat conduction in solids with random conductivity. Letter Heat Mass Transfer. 5:167-73.

[53] Fox RF, Barakat R (1978) Heat conduction in a random medium. J Stat Phys. 18:171-8.

[54] Keller JB, Papanicolaou GC, Weilenmann J (1978) Heat conduction in a one-dimensional random medium. Comm Pure Appl Math. 31:583-92.

[55] Tanaka K, Horie T, Sekiya T (1979) Mean temperature field variations in heterogeneous media. Acta Mech. 32:153-64.

[56] Drewniak J (1985) Boundary elements for random heat conduction problems. Eng Anal. 2:168-70.

[57] Nielsen SB (1987) Steady state heat flow in a random medium and the linear heat flow-heat production relationship. Geophys Res Letters. 14:318-21.

[58] Tzou DY (1988) Stochastic analysis of temperature distribution in a solid with random heat conductivity. J Heat Tran. 110:23-9.

[59] Sluzalec A (1991) Temperature field in random conditions. Int J Heat Mass Tran. 34:55-8.

[60] Nicolai BM, De Baerdemaeker J (1993) Computation of heat conduction in materials with random variable thermophysical properties. Int J Numer Meth Eng. 36:523-36.

[61] Emery AF, Fadale TD (1997) Handling temperature dependent properties and boundary conditions in stochastic finite element analysis. Numer Heat Tran A. 31:37-51.

[62] Hien TD, Kleiber M (1998) On solving nonlinear transient heat transfer problems with random parameters. Comput Meth Appl Mech Eng. 151:287-99.

[63] Srivastava K, Singh RN (1999) A stochastic model to quantify the steady-state crustal geotherms subject to uncertainties in thermal conductivity. Geophys J Int. 138:895-9.

[64] Kaminski M, Hien TD (1999) Stochastic finite element modeling of transient heat transfer in layered composites. Int Comm Heat Mass Tran. 26:801-10.

[65] Emery AF (2001) High order perturbation analysis of stochastic thermal systems with correlated uncertain properties. J Heat Tran. 123:390-8.

[66] Liu J (2001) Uncertainty analysis for temperature prediction of biological bodies subject to randomly spatial heating. J Biomech. 34:1637-42.

[67] Srivastava K (2005) Modelling the variability of heat flow due to the random thermal conductivity of the crust. Geophys J Int. 160:776-82.

[68] Saleh MM, El-Kalla IL, Ehab MM (2007) Stochastic Finite Element Technique for Stochastic One-Dimension Time-Dependent Differential Equations with Random Coefficients. Differential Equations and Nonlinear Mechanics. 2007:Article ID 48527 16 pages.

[69] Kotulski Z (1984) One-dimensional thermoelastic wave in solid with randomly fluctuating coefficient of linear expansion. Arch Mech. 36:499-514.

[70] Smith CO (1984) Probabilistic design criteria for cylinders and spheres under thermal stresses. J Vib Acoust Stress Reliab Des. 106:523-8.

[71] Tani S, Nakagiri S, Suzuki K, Higashino T (1987) Stochastic Finite Element Analysis of Thermal Deformation and Thermal Stresses of CFRP Laminated Plates : In Problems of Probabilistic Fiber Orientations. Trans Jpn Soc Mech Eng. 53A:1197-202.

[72] Aida Y, Yoshikawa H (1999) Thermal stress analysis in the stochastic field and identification of cracking probability for massive concrete structure. Proc JSCE. 620:43-54.

[73] Sluzalec A (2000) Thermoelastic analysis in random conditions. J Therm Stresses. 23:131-41.

[74] Hosseini SM, Shahabian F (2011) Transient analysis of thermo-elastic waves in thick hollow cylinders using a stochastic hybrid numerical method, considering Gaussian mechanical properties. Appl Math Model. 35:4697-714.

[75] Tzou DY (1989) Stochastic modelling for contact problems in heat conduction. Int J Heat Mass Tran. 32:913-21.

[76] Madera AG (1994) Simulation of stochastic heat conduction processes. Int J Heat Mass Tran. 37:2571-7.

[77] Manolis GD, Shaw RP (1996) Boundary integral formulation for 2D and 3D thermal problems exibiting a linearly varying stochastic conductivity. Comput Mech. 17:406-17.

[78] Barrett PR (1975) Stochastic thermal stress analysis of clad cylindrical fuel elements. Nucl Eng Des. 35:41-58.

[79] Chiba R, Sugano Y (2004) An Analysis of Stochastic Thermoelastic Problem in Functionally Graded Plates with Uncertain Material Properties Using Monte Carlo Simulation Method. J Soc Mater Sci. 53:967-73.

[80] Ferrante FJ, Graham-Brady LL (2005) Stochastic simulation of non-Gaussian/non-stationary properties in a functionally graded plate. Comput Meth Appl Mech Eng. 194:1675-92.

[81] Sugano Y, Chiba R, Kanno T (2006) Stochastic Analysis of Thermoelastic Problems in a Nonhomogeneous Plate with Random Material Properties. Trans Jpn Soc Mech Eng. 72A:247-54.

[82] Chiba R, Sugano Y (2008) Stochastic analysis of a thermoelastic problem in functionally graded plates with uncertain material properties. Arch Appl Mech. 78:749-64.

[83] Hosseini SM, Shahabian F (2011) Stochastic Assessment of Thermo-Elastic Wave Propagation in Functionally Graded Materials (FGMs) with Gaussian Uncertainty in Constitutive Mechanical Properties. J Therm Stresses. 34:1071-99.

[84] Hosseini SM, Shahabian Moghadam F (2011) Stochastic meshless local Petrov-Galerkin (MLPG) method for thermo-elastic wave propagation analysis in functionally graded thick hollow cylinders. Comput Model Eng Sci. 71:39-66.

[85] Kuznetsov AV (1996) Stochastic modeling of heating of a one-dimensional porous slab by a flow of incompressible fluid. Acta Mech. 114:39-50.

[86] Emery AF, Dillon H, Mescher AM (2010) The Effect of Spatially Correlated Roughness and Boundary Conditions on the Conduction of Heat Through a Slab. J Heat Trans. 132:051301-11.

[87] Mori M, Kondo M (1993) Temperature and thermal stress analysis in a structure with uncertain heat transfer boundary conditions. Trans Jpn Soc Mech Eng. 59A:1514-8.

[88] Ishikawa M (1999) On the mathematical modeling of thermoelastic systems with random emission under physical constraints and its simulation studies. Trans Inst Electro Inform Comm Eng. J82-A:972-9.

[89] Chiba R (2007) Stochastic thermal stresses in an annular disc with spatially random heat transfer coefficients on upper and lower surfaces. Acta Mech. 194:67-82.

[90] Chiba R (2009) Stochastic heat conduction analysis of a functionally graded annular disc with spatially random heat transfer coefficients. Appl Math Model. 33:507-23.

[91] Chiba R (2009) Stochastic thermal stresses in an FGM annular disc of variable thickness with spatially random heat transfer coefficients. Meccanica. 44:159-76.

[92] Vasseur G, Singh RN (1986) Effects of Random Horizontal Variations in Radiogenic Heat Source Distribution on Its Relationship With Heat Flow. J Geophys Res. 91:10,397-10,404.

[93] Ishikawa M (1997) On the strict heat conduction model with stochastic inputs. Trans Inst Systems Control Inform Eng. 10:556-62.

[94] Srivastava K, Singh RN (1998) A model for temperature variations in sedimentary basins due to random radiogenic heat sources. Geophys J Int. 135:727-30.

[95] Zeman JL (1965) Locally and temporally random distributed temperature and stress fields: part II. Acta Mech. 1:371-85.

[96] Clarke NS (1984) Heat diffusion in random laminates. Quarterly J. Mech. Appl. Math. 37:195-230.

[97] Shvets RN, Eleiko VI (1977) Stochastic temperature stresses in a cylinder with a rough surface. Sov Appl Mech. 13:1209-13.

[98] Mori M, Ukai O, Kondo M (1990) Applications of the stochastic finite-element method in a thin-walled pressure vessel containing uncertainty. Trans Jpn Soc Mech Eng. 56A:1455-60.

[99] Hutchinson DE, Norton MP (1986) Applicability of stochastic process theory to heat conduction in solids with random temperature fields. Appl Energ. 22:241-69.

[100] [100] Heller RA, Thangjitham S (1987) Probabilistic methods in thermal stress analysis. In: Hetnarski RB, editor. Thermal Stresses. New York: Elsevier Science Publishers. pp. 190-268.

[101] Emery AF (2004) Solving stochastic heat transfer problems. Eng Anal Bound Elem. 8:279-91.

[102] Nicolai BM, Scheerlinck N, Hertog MLATM (2006) Probabilistic Modeling. In: Sablani S, Datta AK, Rehman MS, Mujumdar AS, editors. Handbook of Food and Bioprocess Modeling Techniques. Boca Raton: CRC Press. pp. 265-89.

[103] Miyata H, Iijima S, Ooshima R, Abe T, Hisamatsu T, Hamamatsu T (1988) Application echnology on ceramics for structural components of high temperature machines. Trans pn Soc Mech Eng. 54A:1700-7.

[104] Ko WL (1988) Solution accuracies of finite element reentry heat transfer and thermal tress analyses of space shuttle orbiter. Int J Numer Meth Eng. 25:517-43.

[105] Stefanou G (2009) The stochastic finite element method: Past, present and future. Comput Meth Appl Mech Eng. 198:1031-51.

[106] FukutanI K, ToyoshimA S, Yutaka H, Yamamoto A (2003) Numerical Computation of Diameter Fluctuation in Optical Fiber Drawing from Silica Glass Preform by Perturbation Method. Trans Jpn Soc Mech Eng. 69C:2403-10.

[107] Prasad RC, Karmeshu, Bharadwaj KK (2002) Stochastic modeling of heat exchanger esponse to data uncertainties. Appl Math Model. 26:715-26.

[108] Becus GA, Cozzarelli FA (1976) The Random Steady State Diffusion Problem. I: Random Generalized Solutions to Laplace's Equation. SIAM J Appl Math. 31:134-47.

[109] Becus GA, Cozzarelli FA (1976) The Random Steady State Diffusion Problem. II: Random Solutions to Nonlinear, Inhomogeneous, Steady State Diffusion Problems. SIAM J Appl Math. 31:148-58.

[110] Becus GA, Cozzarelli FA (1976) The Random Steady State Diffusion Problem. III: Solutions to Random Diffusion Problems by the Method of Random Successive Approximations. SIAM J Appl Math. 31:159-78.

[111] Becus GA (1978) Solutions to the random heat equation by the method of successive approximations. J Math Anal Appl. 64:277-96.

[112] Willis JR, Hill JM (1988) On Heat Diffusion in Random Laminates. Quarterly J Mech Appl Math. 41:281-99.

[113] Case WR, Walston WH (1977) A finite element technique for non-deterministic thermal deformation analyses including temperature dependent material properties. Int J Numer Meth Eng. 11:915-32.

[114] Madera AG (1993) Method of analysis of stochastic nonstationary heat conduction equations. J Eng Phys Thermophys. 64:195-201.

[115] Nicolai BM, Baerdemaeker JD (1997) Finite element perturbation analysis of non-linear heat conduction problems with random field parameters. Int J Numer Meth Heat Fluid Flow. 7:525-44.

[116] Fadale TD, Emery AF (1994) Transient Effects of Uncertainties on the Sensitivities of Temperatures and Heat Fluxes Using Stochastic Finite Elements. J Heat Tran. 116:808-14.

[117] Nicolai BM, Baerdemaeker JD (1999) A variance propagation algorithm for the computation of heat conduction under stochastic conditions. Int J Heat Mass Tran. 42:1513-20.

[118] Nicolai BM, Verboven P, Scheerlinck N, De Baerdemaeker J (1998) Numerical analysis of the propagation of random parameter fluctuations in time and space during thermal ood processes. J Food Eng. 38:259-78.

[119] Nicolai BM, Scheerlinck N, Verboven P, Baerdemaeker JD (2001) Stochastic Finite-Element Analysis of Thermal Food Processes. In: Irudayaraj J, editor. Food Processing Operations Modeling. New York: M. Dekker.

[120] Scheerlinck N, Verboven P, Stigter JD, De Baerdemaeker J, Van Impe JF, Nicolai BM 2001) A variance propagation algorithm for stochastic heat and mass transfer problems n food processes. Int J Numer Meth Eng. 51:961-83.

[121] Liu N, Hu B, Yu Z-W (2001) Stochastic finite element method for random temperature n concrete structures. Int J Solids Struct. 38:6965-83.

[122] Le Maitre OP, Knio OM, Debusschere BJ, Najm HN, Ghanem RG (2003) A multigrid solver for two-dimensional stochastic diffusion equations. Comput Meth Appl Mech Eng. 192:4723-44.

[123] Xiu D, Karniadakis GE (2003) A new stochastic approach to transient heat conduction modeling with uncertainty. Int J Heat Mass Tran. 46:4681-93.

[124] Xiu D, Em Karniadakis G (2002) Modeling uncertainty in steady state diffusion problems via generalized polynomial chaos. Comput Meth Appl Mech Eng. 191:4927-48.

[125] Xiu D, Tartakovsky DM (2006) Numerical Methods for Differential Equations in Random Domains. SIAM J Sci Comput. 28:1167-85.

[126] Nicolai BM, Egea JA, Scheerlinck N, Banga JR, Datta AK (2011) Fuzzy finite element analysis of heat conduction problems with uncertain parameters. J Food Eng. 103:38-46.

[127] Gurvich MR (1993) Structural analysis of the random thermal expansion of laminated reinforced plastics. Mech Comp Mater. 29: 95-101.

Heat Transfer Applications

Entropy Generation Analysis of a Proton Exchange Membrane Fuel Cell (PEMFC) with a Fermat Spiral as a Flow Distributor

V.H. Rangel-Hernandez, C. Damian-Ascencio, D. Juarez-Robles, A. Gallegos-Muñoz, A. Zaleta-Aguilar and H. Plascencia-Mora

Additional information is available at the end of the chapter

1. Introduction

In these last decades, fuel cells technology have emerged as an alternative to conventional electricity generation systems [1, 2]. However, high operating and manufacturing costs, non-homogeneous current density production and low durability are some of the serious issues that have impeded their expansive application. Hence many research projects have been undertaken by academic and industrial sectors intended to improve the technology [3].

In this regard, some of these works have focused mainly on the optimization of the cell geometry [4–7]. An innovative design consisted of a bipolar plate with interdigitated flow field [8]. Such flow field conducts the reactants gases onto the active area and helps to eliminate the need for a separate cooling layer in the stack. Such improvement provides an increment in the performance of the PEMFC. Tuber et al. [9] proposes an optimization method based on fractal concepts to increase the performance of the bipolar plate fuel cell stacks. However, in terms of polarization curves, their performance was similar to that of a parallel channel configuration stack. The only advantage was a reduction of pressure drop. Others authors have used genetic algorithms [10, 11] or biomimetical models [12] to provide better designs, maximize the net power and minimize the pressure drop. In terms of the geometry of the channel, an Archimedes spiral has been used in [13]. Here the gas flows from the center to the external part of the spiral. A reduction of the pressure drop is observed. Besides, the model shows a uniform current density production along the cell. This model considerably improves the performance of the cell.

On the contrary, other research works have focused on highlighting the major causes of inefficiencies and their location in a fuel cell [3, 14]. The novelty of these works is that they provide crucial information such as heat and mass transfer as well as chemical and electrochemical reactions for improving the fuel cell design [15, 16].

Hence the goal of this study is to identify the main sources of irreversibility of a PEM Fuel Cell by application of the Entropy Generation Method. The main difference from other fuel cell is that it uses a Fermat spiral as flow distributor. In order to have a more complete analysis of the fuel cell, a new dimensional parameter is introduced. The parameter is helpful in unveiling the effect of the mass transfer on the overall entropy generation.

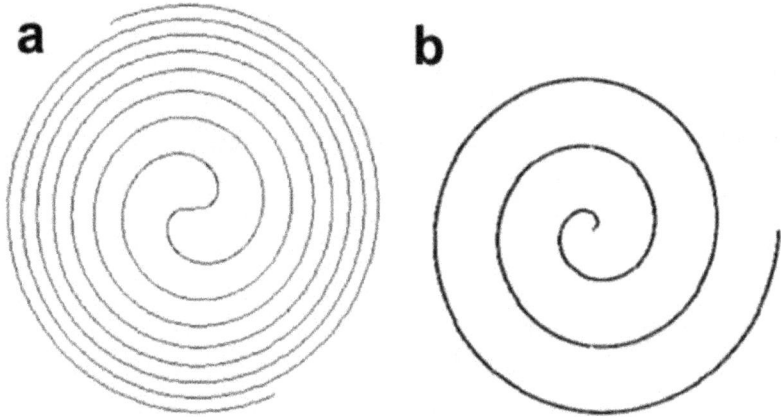

Figure 1. Spiral of (a) Fermat $r^2 = a\theta$ (b) Archimedes $r = b\theta$.

2. Model development

The particular geometry of the flow channels has been thought to have larger and more uniform current densities, to avoid gas stagnation and flooding of the channels as well as to reduce pressure drops.

Fig. 1 shows the Fermat and the Archimedes spiral configuration, respectively. It is observed that the parametrical equation characterizing the Fermat spiral does not increase proportional to the angle θ, which avoids generating gas flow channels of constant width. Thus four concentric Archimedes spirals have been used with a π radian phase angle between each couple of spirals, instead of using two Fermat spirals. This permits generating two channels of constant width.

Thus, in order to fulfill the requirements of the new geometry it has been necessary to modify two zones of the fuel cell, namely, the gas flow channels and the current collectors. While the gas diffusion layers, catalyst layers, anode and cathode and the membrane have been kept constant.

The geometrical parameters are shown in Table 1. Values are similar to those used by Shimpalee and Dutta [16]. The current collectors (electrodes), gas difussion layer (GDL), catalysts layers, and the membrane have the same width of the cell.

Fig. 2 shows the three-dimensional model developed in a co- flow pattern, where the gas reactant inlets are located at the center of the spirals. The outlets are located at the external part of the spirals. An alternative analysis was conducted by considering the gas inlets at the outer part of the spiral and the outlets at the center of it.

Since the gas inlet is under the level of the cell, a straight circular channel has been added to consider the height of the acrylic base where the fuel cell is mounted on.

During the analysis of the model, the following assumptions were considered:

- Steady state conditions
- Non-isothermal and single phase conditions
- Isotropic materials
- Three-dimensional mass diffusion model was considered

3. Mathematical model

3.1. Governing equations

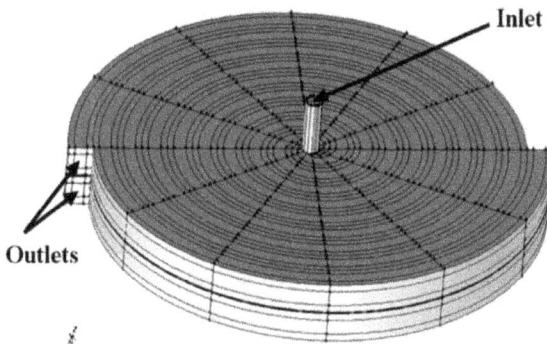

Figure 2. Computational model developed for the PEMFC with the shape of the Fermat spiral.

Side	Dimension (mm)
Current collector height	1.50
Channels height	1.00
Channels width	0.80
GDL height	0.50
Catalyst width	0.03
Cell width	1.60
Inlet radius	0.53
Inlet channel length	3.50
Number of turns	4.5

Table 1. Solutions to the motion equations.

The mathematical model consists of a series of partial differential equations which account for the continuity equation, the momentum equation, energy conservation and the conservation of species:

Conservation mass

$$\nabla \cdot (\rho \mathbf{u}) = S_i \tag{1}$$

Momentum Equations

$$\mathbf{u} \nabla \cdot (\rho \mathbf{u}) = -\nabla p + \nabla \cdot (\mu \nabla \mathbf{u}) + S_p \tag{2}$$

Hydrogen transport

$$\nabla \cdot (\rho \mathbf{u} m_{H_2}) = \nabla \cdot \mathbf{J} + S_{H_2} \tag{3}$$

Oxygen transport

$$\nabla \cdot (\rho \mathbf{u} m_{O_2}) = \nabla \cdot \mathbf{J} + S_{O_2} \tag{4}$$

Water transport

$$\nabla \cdot (\rho \mathbf{u} m_{H_2O}) = \nabla \cdot \mathbf{J} + S_{H_2O} \tag{5}$$

Solid phase

$$\nabla \cdot (\sigma_{sol} \nabla \phi_{sol}) + S_{sol} = 0 \tag{6}$$

Membrane phase

$$\nabla \cdot (\sigma_{mem} \nabla \phi_{mem}) + S_{mem} = 0 \tag{7}$$

Table 2 shows the source terms used in the calculations. From Table 2, the last equations use the following source terms:

$$S_{sol} = j_{an}^{ref} \left(\frac{[H_2]}{[H_2]_{ref}} \right)^{\gamma_{an}} \left(e^{a_{an} F \eta_{an}/RT} - e^{-a_{cat} F \eta_{an}/RT} \right) \tag{8}$$

$$S_{cat} = j_{cat}^{ref} \left(\frac{[O_2]}{[O_2]_{ref}} \right)^{\gamma_{cat}} \left(e^{a_{an} F \eta_{cat}/RT} - e^{-a_{cat} F \eta_{cat}/RT} \right) \tag{9}$$

For Eqs. (4) and (5), η is the electrode over-potential which results from the difference between the solid and the membrane potentials, ϕ_{sol} and ϕ_{mem}. At the anode side:

$$\eta_{an} = \phi_{sol} - \phi_{mem} \tag{10}$$

At the cathode side:

$$\eta_{cat} = \phi_{sol} - \phi_{mem} - V_{oc} \tag{11}$$

Where V_{oc} is the open circuit voltage of the cell and can be computed as follows:

$$V_{oc} = 0.0025T + 0.2329 \tag{12}$$

where, T must be provided in Kelvin degrees and V_{oc} in volts. The equation for the water transport coefficient, α, is given as:

$$\alpha(x,y) = n_d(x,y) - \frac{F}{I(x,y)} D_w(x,y) \frac{C_{wc} - C_{wa}}{t_a} \tag{13}$$

From Eq. (13), it can be noticed that this is a function of the water activity at the anode side. The electrosmotic drag coefficient, n_d, is computed as shown in Eq. (14) and (15):

Equation	Source terms	
(1)	$S_i = S_{H_2} + S_{H_2O}$	at the anode catalyst
	$S_i = S_{O_2} + S_{H_2O}$	at the cathode catalyst
	$S_i = 0$	otherwise
(2)	$S_p = -\frac{\mu}{\beta}\mathbf{u}$	in the porous region
	$S_p = 0$	otherwise
(3)	$S_{H_2} = -\frac{M_{H_2}}{2F}R_{an}$	at the anode catalyst
(4)	$S_{O_2} = --\frac{M_{O_2}}{4F}R_{cat}$	at the cathode catalyst
(5)	$S_{H_2O} = --\frac{M_{H_2=}}{2F}R_{cat}$	at the anode/cathode catalyst
(6)	$S_{sol} = -S_{an}$	at the anode catalyst
	$S_{sol} = +S_{cat}$	at the cathode catalyst
	$S_{sol} = 0$	otherwise
(7)	$S_{mem} = S_{an}$	at the anode catalyst
	$S_{mem} = -S_{cat}$	at the anode catalyst
	$S_{mem} = 0$	otherwise

Table 2. Solutions to the motion equations.

$$n_d(x,y) = 0.0049 + 2.02a_a - 4.53a_a^2 + 4.09a_a^3 \quad a \leq 1 \tag{14}$$

$$n_d(x,y) = 1.59 + 0.159(a_a - 1) \quad a > 1 \tag{15}$$

where, a_a is the water activity at the anode side. The water diffusion coefficient, D_w, is calculated by applying Eq. (16) [15]:

$$D_w = 5.5x10^{-11}n_d e^{2416\left(\frac{1}{303} - \frac{1}{T_s}\right)} \tag{16}$$

The concentration of water depends on both density and weight of the dry membrane as shown in Eq. (17) and (18):

$$C_{wk}(x,y) = \frac{\rho_{m,dry}}{M_{m,dry}}\left(0.043 + 17.8a_K - 39.8a_K^2 + 36.0a_K^3\right) \quad a_k \leq 1 \tag{17}$$

$$C_{wk}(x,y) = \frac{\rho_{m,dry}}{M_{m,dry}}(14.0 + 1.4(a_k - 1)) \quad a_k > 1 \tag{18}$$

Where the k subscript stands for either the anode or the cathode, and $\rho_{m,dry}$, $M_{m,dry}$ are the density and the weight of the dry membrane, respectively. The water activity is defined as:

$$a_k = \frac{X_{w,k}(x,y)P(x,y)}{P_{w,k}^{sat}} \tag{19}$$

Where $X_{w,k}$ is the molar fraction of the water, either at the anode or at the cathode side. The water saturation pressure is calculated as follows:

$$log_{10}P_{w,k}^{sat} = -2.1794 + 0.02953(T - 273.15) + 10^{-5}(T - 273.15)^2 + 1.4454x10^{-7}(T - 273.15)^3 \tag{20}$$

The membrane conductivity, σ_{mem} is computed by using Eq. [15]:

$$\sigma_{mem} = \epsilon(0.514\lambda - 0.326)e^{1268\left(\frac{1}{303} - \frac{1}{T}\right)} \tag{21}$$

Where λ, the water content, is calculated by applying Eq. [16]:

$$\lambda = 0.043 + 17.18a_a - 39.85a_a^2 + 36.0a_a^3 \quad a_a \leq 1 \tag{22}$$

$$\lambda = 14.0 + 1.4(a_a - 1) \quad a > 1 \tag{23}$$

3.2. Entropy generation equations

In order to determine the main sources of irreversibility in a fuel cell, it is necessary to use a homogeneous thermodynamic property. In this regard, the entropy generation method is an effective tool. However the analysis needs to be seen from a local point of view, so entropy generation equations have been derived for a local application.

An expression for the local rate of entropy generation can be provided in terms of fluxes and gradients in the system, as follows:

$$\sigma^S = \mathbf{q} \cdot \nabla\left(\frac{1}{T}\right) - \Sigma J_k \cdot \left(\frac{\mu_k}{T}\right) - \frac{1}{T}p^v \cdot \nabla v - P^v : \nabla v - \frac{1}{T}i \cdot \epsilon \tag{24}$$

From a local point of view, Eq. (21) provides important information about the phenomenological causes of the irreversibility. The first term is the entropy generation due to heat transfer, the second due to mass transfer, the third and fourth due to mechanical dissipation and the fifth is due to ohmic losses. Strictly, the entropy generation may be expressed in terms of physical phenomena:

$$\sigma^S = \sigma_{HT} + \sigma_{MF} + \sigma_{MD} + \sigma_{Ohm} \tag{25}$$

Thus the total entropy generation can be derived by integration of the local entropy generation, σ, in each computational domain, as follows

$$\sigma_{T,i} = \int_V \sigma_i dV \tag{26}$$

In many local rate entropy generation applications is usual to involve dimensionless parameters in order to indicate the dominant phenomenon in a system [17]. However such dimensionless parameters involve only heat transfer and fluid friction phenomena. For example, the most used is the Bejan number (Be) [17]. So when the entropy generation is due only to mechanical dissipation and heat transfer, i.e. $Be \gg 1/2$, this dimensionless parameter indicates that heat transfer irreversibilities are dominant. In particular, however, in fuel cells the entropy generation not only depends on heat transfer and fluid friction but only on mass diffusion. In view of this discussion, a new dimensionless parameter is proposed in Eq. (27) [13]:

$$\Pi = \frac{\sigma_{MF}}{\sigma_S} \tag{27}$$

Accordingly, Π expresses the ratio of the entropy generation due to mass diffusion to the total entropy production. As mass diffusion effects become insignificant, i.e. $\Pi << 1 = 3$, the entropy distribution is completely expressed.

4. Boundary conditions

4.1. Cell operating parameters

The analysis considers that the fuel cell is operating at a pressure of 101.325 kPa and a temperature of 343 K. The open circuit voltage at this temperature is assumed to be $V_{oc} = 1.0904\ V$. The gas inlet condition is defined as the cell velocity inlet. The gas outlet condition boundary was defined as the cell pressure outlet, with a gauge pressure of zero, i.e. gases are exhausted at atmospheric pressure. The inlet conditions of the gases, velocity and humidity are listed in Table 3.

	VLH	LH	HH	VHH
Anode				
Velocity (m/s)	1.7325	1.830	2.210	2.560
y_{H_2}	0.7286	0.6367	0.4073	0.2962
y_{H_2O}	0.2713	0.3632	0.5926	0.7037
Cathode				
Velocity m/s	7.330	7.910	9.050	12.90
y_{O_2}	0.3020	0.2880	0.2637	0.2085
y_{H_2O}	0.6489	0.6178	0.5694	0.4384
y_{N_2}	0.0489	0.0940	0.1667	0.3528

Table 3. Solutions to the motion equations.

As for the humidity conditions, these have been labeled as very high humidity (VHH), high humidity (HH), very low humidity (VLH) and low humidity (LH) according to Dutta et al. [16].

4.2. Boundary conditions at the current collector

The external faces of the current collectors are regarded as walls at constant electrode potential, i.e. potentiostatic cell. Thus the electric potential at the anode side is $V_{sol} = 0$ and at the cathode side is $V_{sol} = V_{cell}$.

4.3. Boundary conditions at the interface between channel and porous media

The interfaces between the gas flow channels and the GDL (and between the diffuser and the catalyst layer) are considered as porous media with a permeability of $\beta = 2x10^{-10} m^2$ and a porosity of $\epsilon = 0.7$.

The reference current density for the anode and cathode sides are, respectively,

$$j_a^{ref} = 2 \times 10^9\ A/m^3, \quad j_c^{ref} = 4 \times 10^6\ A/m^3$$

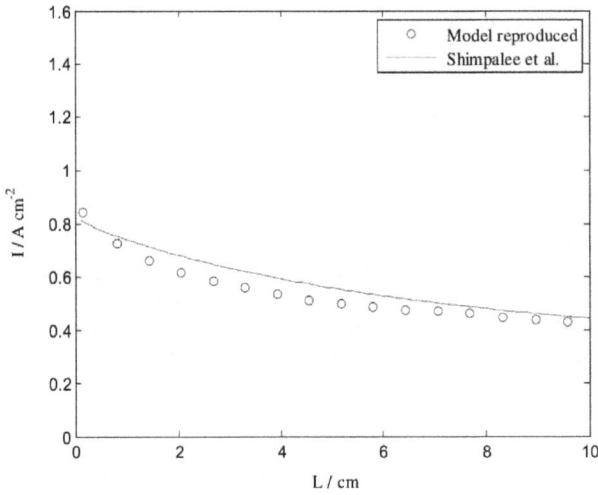

Figure 3. Computational model developed for the PEMFC with the shape of the Fermat spiral.

And the reference diffusivities are,

$$D_{H_2} = 6 \times 10^{-5} m^2/s, \quad D_{O_2} = 3.5 \times 10^{-5} m^2/s$$

$$D_{H_2O} = 6 \times 10^{-5} m^2/s, \quad D_{N_2} = 8 \times 10^{-5} m^2/s$$

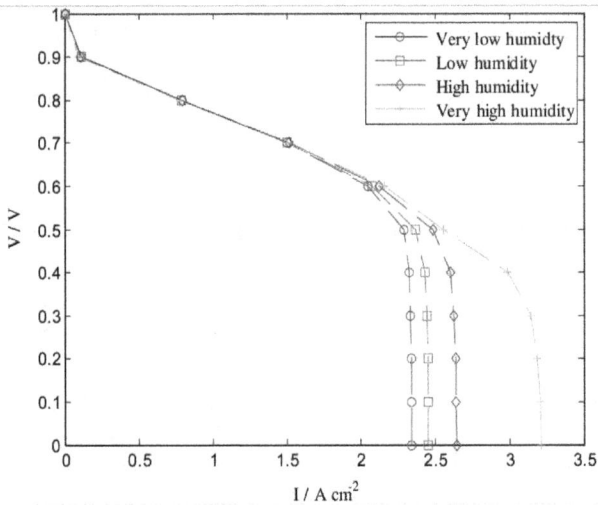

Figure 4. Computational model developed for the PEMFC with the shape of the Fermat spiral.

5. Validation of the model

The analysis of the model presented in the previous sections considers three cases. Case 1, analyzes the behavior of the cell with the velocities and mass fractions listed in Table 3. Besides, this case considers the gas inlets at the center of the spiral and the outlets at the external side of the spiral. In Case 2, the velocity inlets are increased twofold, i.e. mass flow is doubled, and the rest of the conditions remain unchanged. In Case 3, the inlets and outlets are reversed, i.e. the inlet gases are located on the external side of the spiral and the outlet gases at the center. The inlet conditions for Case 3 remain as in Table 3. Case 1 and 2 are analyzed considering four humidity conditions, while Case 3 is analyzed considering two conditions only, VHH and VLH.

The governing equations are solved with a specialized code that is based on the finite volume technique, and the application of an added module. The SIMPLE algorithm is used in the solution of the governing equations in a segregated form. This algorithm uses a relationship between velocity and pressure corrections to enforce mass conservation and to obtain the pressure path using the following steps: first the NaviereStokes equations are solved in the x, y and z direction; then, a pressure correction equation is used to enforce a mass balance; next, the species equations are solved with the data obtained in the previous steps; finally, the potential fields are solved.

The model validation has been performed by comparing the results with those provided by Dutta et al. [16] for the current distribution in a straight channel. Fig. 3 shows the comparison between both models. The results of Dutta and Shimpalee range from $0.8A/cm^2$ at the entrance of the channels to about $0.42A/cm^2$ at the exit of them, whereas the present model reports values ranging from $0.81A/cm^2$ at the inlet and to $0.41A/cm^2$ at the outlet of the channel.

6. Analysis of results

Fig. 4 shows the polarization curves for each one of the humidity conditions proposed in the present paper. It can be observed that the dependence of the cell voltage on the humidity concentrations and on the current density is strongly nonlinear. Noteworthy that as the humidity increases the maximum current density so does. The slight slope present at low values of the current density in the polarization curves indicates minor ohmic losses, however, at high values of the current density the polarization curves fall quickly as a consequence of higher concentration losses. For the case of very high humidity condition, the polarization curve extends significantly.

Fig. 5 shows the power density and entropy production curves at different humidity conditions. As observed, the left y-axis indicates the entropy generation, whereas the right y-axis the power density. It is evident that at low current density values, i.e. $< 2A/cm^2$, the entropy generation rate is insignificant. In fact, as current density goes down to values below 1.5, the entropy generation rate is practically constant. This results from the fact that activation losses are not predominant in this zone.

On the contrary, in the zone where the concentration losses are predominant, i.e. at higher current density values, the entropy production rate sharply increases. This demonstrates that the larger irreversibilities are mostly associated with the concentration losses.

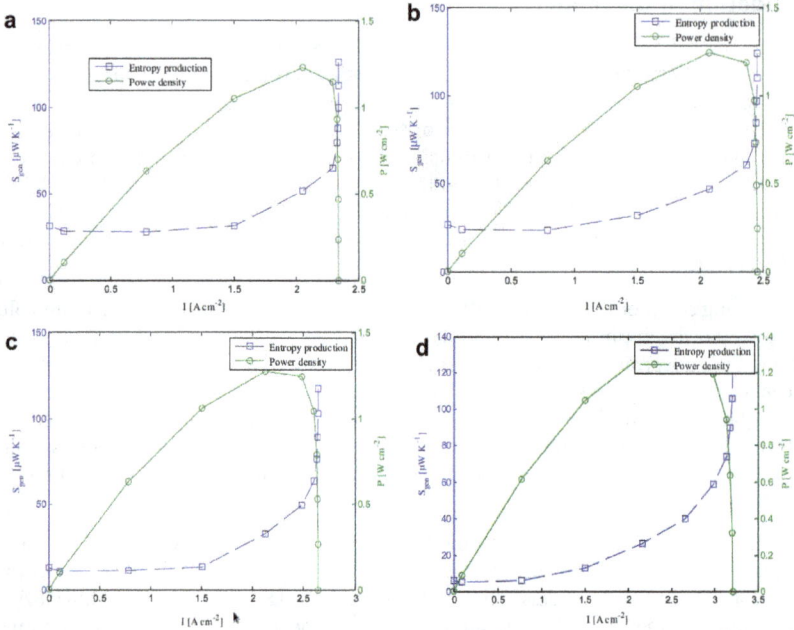

Figure 5. Power produced and entropy production for a) very low b) low c) high and d) very high humidity condition.

An important characteristic in the design of fuel cells has to do with the broadening of the power density zone as long as possible. This can only be achieved by increasing the humidity concentration as can be inferred from Fig. 5(a-d). In fact, it can be observed that the highest power density is roughly $1.25W/cm^2$ at current density of roughly $2.1A/cm^2$ for the cases when the humidity is not so high (i.e. very low, low and high humidity conditions). For the particular case of very high humidity condition, it can be observed that the domain for the highest power density values stretches consider- ably, which can be explained due to the low entropy production rate at very high humidity conditions.

The entropy generation technique points that the lesser the entropy production rate, the better the thermodynamic performance of a system. Accordingly, it is implied that the optimal flow conditions are high humidity conditions (Fig. 2). In this flow condition, the entropy generation rate is low at any current density. As aforementioned, different sources of losses in a fuel cell leads to different distributions of entropy generation, hence a more detailed analysis is required.

Graphs in Fig. 6(a-d) provide the effect of the humidity condition on the Bejan number and the entropy production ratio proposed in this paper (i.e. Π). It is important to recall that this dimensionless parameter was proposed in order to have a more complete representation of the irreversibilities present in a fuel cell. So it can be inferred that the main contribution to the entropy generation rate is obtained when the gases are mixed, since it is practically impossible for the reverse process of separation to occur spontaneously. The second term in Eq. (21) demonstrates that the entropy generation rate due to mass transfer depends proportionally on the diffusion coefficient, the concentration and the chemical potential, but inversely on the temperature. Thus it can be implied that the dependence of these irreversibilities on the temperature is high.

From Fig. 6(a-d) it can be observed that the concentration losses decrease as the current density increase. In fuel cells the reaction temperature remains nearly constant at high current densities but the total entropy generated increase asymptotically (as demonstrated by the Bejan number), therefore the entropy generated due to species diminishes quickly.

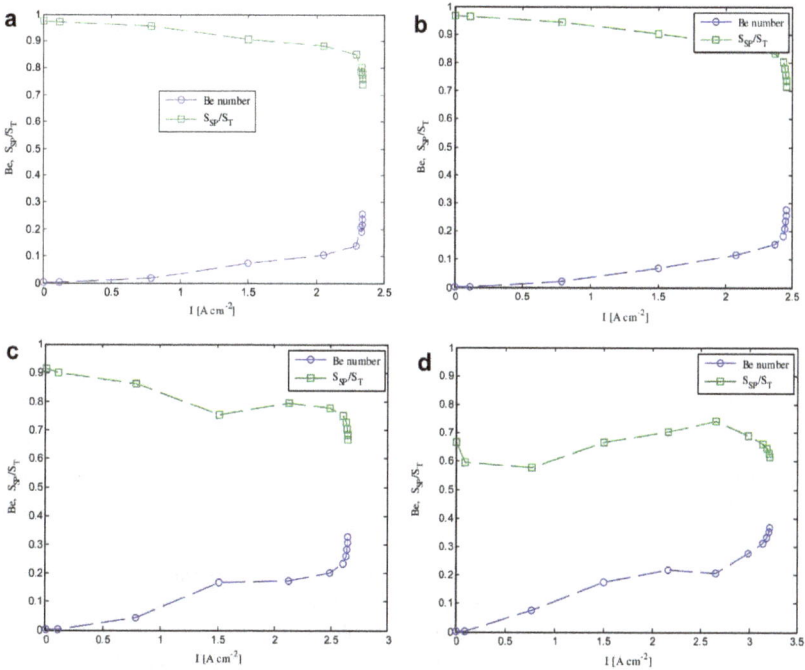

Figure 6. Entropy distribution for a) very low b) low c) high and d) very high humidity condition.

The entropy generation due to heat transfer depends on the temperature gradient (refer to Eq. (21)). Then because of a higher rate reaction in the anode, the current density and the temperature gradient increase, thereby conducting to a more generation entropy. Likewise at higher values of humidity the entropy generation due to heat transfer increases, Fig. 6. As mentioned above, this source of entropy generation depends directly on the temperature gradient, and an increment in humidity improves the conductivity in the cell. So, heat transfer losses become more relevant.

Finally, Fig. 7 shows the entropy distribution due to ohmic losses. As the current density increases, the entropy generation so does. This is the result of the reaction rate occurring in the membrane. This reaction diminishes the surface area where it occurs and an elec- trical resistance is produced. Then the ohmic losses are reduced as long as the humidity conditions are increased. Such behavior comes from the fact that at higher humidity conditions the amount of water on the electrodes improves the conductivity.

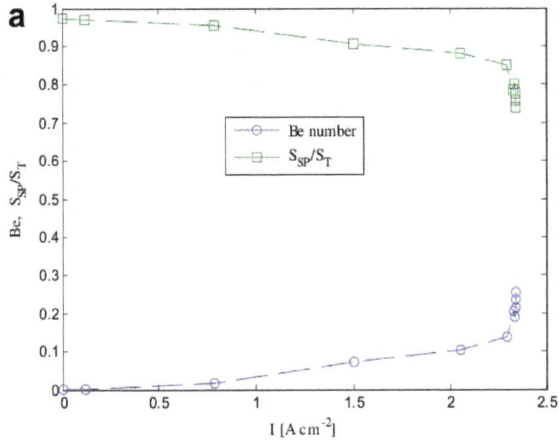

Figure 7. Entropy distribution for ohmic losses.

7. Conclusions

The approach of the entropy generation has been used in order to identify the main sources of irreversibility in a proton exchange membrane fuel cell. It has been shown that through a dimensionless parameters it is easier to identify the phenomenon that most influence on the irreversibility sources.

The analysis presented in this Chapter has turned out to be somewhat ambivalent since the lowest values of entropy generation are present at very low humidity conditions, whereas at high humidity conditions the entropy generation is high but, at this point, the highest density production is reached. In designing a fuel cell it comes to think what is better: a technical improvement or an economical minimization.

As for the activation losses, they remain steady for all of the cases studied because of the same operating conditions. The ohmic losses are similar too, since the properties of the material of the fuel cell do not change when the humidity is varied. In the case of the water distribution, its effect on the concentration losses is clearly noticeable.

The overall analysis demonstrates that the main sources of irreversibility in a fuel cell are the concentration losses for the most part of the operational domain. Contrarily, the irreversibilities due to the heat transfer phenomena are not dominant. As for the acti- vation losses, it is important to mention that they were not analyzed since they are already included in the concentration losses.

Finally, a new dimensionless parameter is proposed herein. The dimensionless parameter is defined as the ratio of entropy production due to mass transfer to the total entropy production as shown in Eq. (24). The order of magnitude of the Bejan number is lower than the ratio of entropy production. For this reason, it is recommended to use this ratio to characterize fuel cells along with the Bejan number.

The use of this new dimensionless parameter can be further investigated as well as the approach presented here to perform a more complete entropy generation analysis.

Acknowledgements

The authors thank to CONACYT for the partial support in the development of this work.

Author details

Rangel-Hernandez V. H.
Mechanical Engineering Department, Engineering Division, University of Guanajuato, Palo Blanco S/N, 36885, Salamanca, Mexico

Damian-Ascencio C. E.
Physics Department, Sciences Division, University of Guanajuato, C.P. 37150, León, Mexico

8. References

[1] Anonymous (2004) Fuel cell handbook, 7th ed. Morgantown: US Department of Energy, National Energy Technology Laboratory.
[2] Massardo AF, Lubelli F (2000) Internal reforming solid oxide fuel cell-gas turbine combined cycles (IRSOFC-GT): part A- cell model and cycle thermodynamic analysis. J Engg Gas Turbines Power 122: 27-35.
[3] Sciacovelli A, Verda V (2009) Entropy generation analysis in a monolithic-type solid oxide fuel cell (SOFC). Energy 34:850-865.
[4] Nguyen PT, Berning T, Djilali N (2004) Computational model of a PEM fuel cell with serpentine gas flow channels. J Power Sourc 130:149-157.
[5] Kazim A, Liu HT, Forges P. (1999) Modeling of performance of PEM fuel cells with conventional and interdigitated flow fields. J Appl Electrochem 29:1409-1416.
[6] Birgersson E, Vynnycky MA (2006) Quantitative study of the effect of flow- distributor geometry in the cathode of a PEM fuel cell. J Power Sourc 153:76-88.
[7] Li Xianguo, Sabir Imran (2005) Review of bipolar plates in PEM fuel cells: flow-field designs. Int J Hydrogen Energ 30:359-371.
[8] Chow CY, Wozniczka B, Chan JK (1999) Integrated reactant and collant fluid flow field layer for a fuel cell with membrane electrode assembly. Canadian Patent No. 2,274,974.
[9] T?ber K, Oedegaard A, Hermann M, Hebling C. (2004) Investigation of fractal flow-fields in portable proton exchange membrane and direct methanol fuel cells. J Power Sourc. 31:175-181.
[10] Senn SM, Poulikakos D (2004) Tree network channels as fluid distributors constructing double-staircase polymer electrolyte fuel cells. J A Phys. 96:842-852.

[11] Senn SM, Poulikakos D.(2006) Pyramidal direct methanol cells. Int J Heat Mass Tran. 49:1516-1528.

[12] Chapman A, Mellor I. (2003) Development of biomimetic flow field plates for PEM fuel cells. Eighth grove fuel cell symposium, September, London, UK.

[13] Damian-Ascencio C E, Hern?ndez-Guerrero A, Ascencio-Cendejas F, Juarez- Robles D (2010) Entropy generation analysis for a PEM fuel-cell with a biomimetic flow field, IMECE, Lake Buena Vista, Florida U.S.

[14] Alvarez T, Valero A, Montes JM (2006) Thermoeconomic analysis of a fuel cell hybrid power system from the fuel cell experimental data. Energy 31:1358-1370.

[15] Springer TE, Zawodzinski TA, Gottesfeld S.(1991) Polymer electrolyte fuel cell model. J Electrochem Soc. 138:2334-2341.

[16] Dutta S, Shimpalee S, Van Zee JW. (2000) Three-dimensional numerical simulation of straight channel PEM fuel cells. J Appl Electrochem. 30:135-146.

[17] Bejan (1996) A. Entropy generation minimization. New York: CRC Press.

Chapter 12

Thermal Interaction Between a Human Body and a Vehicle Cabin

Dragan Ružić and Ferenc Časnji

Additional information is available at the end of the chapter

1. Introduction

The main functions of a vehicle cabin are to provide a comfortable environment for its occupants and to protect them from vibrations, noise and other adverse influences. Since the conditions inside mobile machinery cabins (agricultural tractors and construction equipment) affect the health, performance and comfort of the operator, this research focuses on these types of cabins. Modern agricultural tractors are complex, highly efficient systems, and the aim of the development of these systems is the reduction of the negative impact they have on the environment. Main goals are achieving higher fuel economy, lower emission and less soil compaction caused by tractor wheels. There is also a tendency to improve the ergonomics of operator enclosures. An operator enclosure can be treated as workspace and the conditions inside a tractor cab have significant impact on the performance of the operator, and in that way on the total efficiency of the *operator-tractor-environment* system as well. From the operator's point of view, tractor cab ergonomics is a key factor in ensuring his optimum working performance, which could easily become the weakest link in the working process. While, on the one hand, the tractor cab offers mechanical protection and the protection from adverse ambient conditions, on the other hand, even under moderate outside conditions, a closed tractor cab acts like a greenhouse and its interior could become unpleasant, unbearable and even dangerous.

The project presented in the report by Bohm et al. (2002), deals with the thermal effects of glazing in cabs with large glass areas. Using the thermal manikin AIMAN, they studied and evaluated the effects of different kinds of glass and design of the windows, as well as the effects of sun protection and insulation glazing. The results showed that neither in severe winter conditions, nor in sunny summer conditions, could acceptable climate be obtained with standard glazing in cabs with large glass areas.

In a comprehensive project which was aimed at the reduction of vehicle auxiliary load, done by National Renewable Energy Laboratory, Golden, USA (Rugh and Farrington, 2008), a variety of research methods was used to research and develop innovative techniques and technologies that reduce the amount of fuel needed for the air-conditioning system, by lowering thermal loads. They concluded that the reflection of the solar radiation incident on the vehicle glass is the most important factor in making significant reductions in thermal loads. The use of solar-reflective glass reduced the average air temperature and the seat temperature. The use of reflective shades and electrochromic switchable glazing are also effective techniques for the reduction of the solar energy that enters the passenger compartment. They also found that solar-reflective coatings on exterior opaque surfaces and vehicle body insulation can reduce a vehicle's interior temperature, but they can do this to a lesser extent than solar-reflective glazing, shades, and parked-car ventilation.

Both projects aimed at the reduction of thermal loads in vehicle cabins, primarily considering the advanced techniques of heat rejection and insulation. The research described in the paper by Currle and Maue (2000) deals with the numerical study of the influence of air flow and vent geometrical parameters (area and position of the vents) on thermal comfort of passenger car occupants. The results showed significant influence of the vent area, not only on velocity levels in the cabin, but also on overall as well as local thermal comfort.

This chapter deals with the characteristics of the agricultural tractor cabs that have direct or indirect effect on thermal ergonomics of operator enclosures. For this purpose, thermal processes between hot (summer) environment and a tractor cab, as well as between the cab and an operator are analysed. The hot conditions are more complex and more demanding in terms of achieving comfortable thermal conditions in a tractor cab. The complexity of thermal interactions between a human, a tractor cab and the environment is shown in Fig. 1.

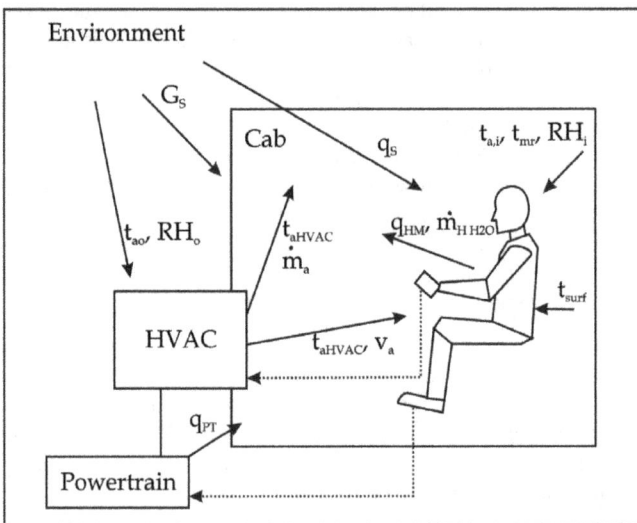

Figure 1. Thermal interaction between a human, a cabin and the environment. Dotted line arrows are control paths

In order to be able to improve thermal ergonomics, it is necessary to have a good knowledge of thermal properties of materials and other design parameters of importance for the cabin thermal processes, as well as the knowledge of the properties of the human body and its thermoregulation system. The main methods for both the improvement of the microclimatic conditions and the reduction of air-conditioning energy consumption in any type of cabin are:

- Passive:
  - prevention of cabin heat gain, using principles of heat rejection and insulation;
  - prevention of heat transfer to the operator's body by reducing the transferred solar radiation, as well as thermal radiation and heat transfer from the cabin interior surfaces.
- Active: cooling the operator's body by the effective use of airflow from the air-conditioning system.

The aims of the research are:

- to identify and evaluate the most important influences on heat transfer processes in the human-cabin-environment system,
- to explain the effects of the ventilation (air distribution) system design and settings on the operator's heat loss and thermal sensation in hot conditions.

2. Main design features of tractor cabs

An overview of the design features related to the thermal processes and HVAC characteristics was done using relevant technical documentation and by analysing the design of several agricultural tractors (Ružić and Časnji, 2011). Orchard and vineyard tractors (narrow track tractors) were not included in the analysis. All tractors from the sample were 4WD wheeled tractors, with power ranging from 40.5 to 155 kW, their weight ranging from 2750 to 8410 kg and their wheelbase ranging from 2.055 to 3.089 m. The outer lengths of their cabs were between 1.40 and 1.77 m, the cab widths were between 1.38 and 1.70 m, and the cab heights were between 1.45 and 1.80 m.

2.1. Materials

The basic materials for tractor cab frames are steel profiles, which primarily must meet the demands for mechanical protection of the operator. Cab frames usually have six pillars, although there are designs with four pillars.

The floor is made of a steel sheet supported by a steel frame. For the purposes of sound and heat insulation, the floor has an interior lining and, in some designs, an outer lining. The powertrain (transmission) of typical tractors is positioned beneath the cab. The powertrain of smaller tractors partly "protrudes" into the cab space, while larger tractors have a flat floor. In the interior, rubber or polymer flooring is used. Some cabs have a foam insulation material between the flooring and the floor. If there is additional heat and sound insulation beneath the floor, composite materials are used. The typical composite heat shield is made of reflective outer aluminium layer bonded on synthetic fibre core.

Cab roofs are made of polymers, and the cab can be equipped with a roof-window. The components of the ventilation system and the air-conditioning are generally placed in the roof.

Cab glazing in modern tractors takes approximately 60% of the total cab surface area. Glass is generally tempered and tinted. In the sample of analysed tractors, the windshield inclination is in the range of 8 – 20° and side windows inclination is in the range of 7 – 10°.

An overview of thermal properties of materials used in tractor cabs is given in Table 1.

	thermal conductivity, k (W/mK)	surface emissivity, ε	solar absorptivity, α_s	solar transmissivity normal to surface, τ
tempered single glass, clear	0.8	0.8 – 0.95	0.08	0.84 – 0.90
tempered single glass, green tinted	0.8	0.8 – 0.95	0.45	0.49
metal, painted white	40 – 45	0.85 – 0.96	0.21 – 0.25	-
polymer, white	0.12 – 0.20	0.90 – 0.97	0.23 – 0.49	-
metal, painted black	40 – 45	0.97	0.80	-
rubber, black	0.10 – 0.13	0.86 – 0.92	-	-
insulating foam	0.037 – 0.050	0.95	-	-
composite heat shield	0.040 – 0.060	0.10 – 0.20	-	-

Table 1. Thermal properties of tractor cab materials (McAdams, 1969; Incropera and DeWitt, 1981; Siegel & Howell, 1992; ASHRAE, 1997c; Saint-Gobain, 2003)

Cabs usually have an air distribution system with air vents (outlets of ventilation duct system) placed on the ceiling. Some of the cabs from the sample also have vents placed on the instrument panel, usually for heating. The air vents on the ceiling can have a symmetrical or an asymmetrical layout, while the vents on the instrument panel always have a symmetrical layout. The vents mostly have a circular cross-section and the operator can change the direction of the air jet.

Vapour compression systems are used for air-conditioning (AC), same as in automotive applications, with R134a as the refrigerant. Installed compressors can take around 4 – 8 kW from the tractor engine, which presents 2.5 – 15% of the rated power in this category of tractors.

3. Heat exchange between a tractor cab and the environment

Thermal processes in a tractor cab are in close relation to outside thermal conditions (air temperature, air velocity, intensity and direction of solar radiation). Thermal processes are more or less independent from tractor's working operations, whereas the noise and vibrations depend on them. Since the decrease in the cab heat gain means the reduction of the thermal load to the operator as well, the analysis was done for adverse environmental conditions that can be encountered over the summer period:

- maximum outdoor air temperature above 30°C,
- total solar irradiation on the outer surfaces up to 1000 W/m²,
- closed cab so that the operator is protected from noise and air pollution,
- low tractor velocity that does not promote natural tractor cab ventilation.

The sum of heat gain for a closed tractor cab in hot environment, under the steady-state conditions, consists of the following (Fig. 2.):

- heat transfer through the cab envelope due to the temperature difference Q_k,
- heat transfer through the cab roof caused by solar radiation Q_s,
- solar radiation transmission through glazing Q_{Gs},
- heat gain from the powertrain Q_{PT},
- sensible (Q_{HM}) and latent heat ($Q_{H\,H2O}$) released by the operator.

Heat load by infiltration is assumed to be zero due to the pressurisation of the tractor cab. In order to maintain the interior temperature constant, the heat removal by the air-conditioning (Q_{AC}) should be equal to the heat gain. According to the requirements for thermal comfort in summer conditions, interior air temperature should be in the range of 23 to 28°C (ASHRAE, 2003).

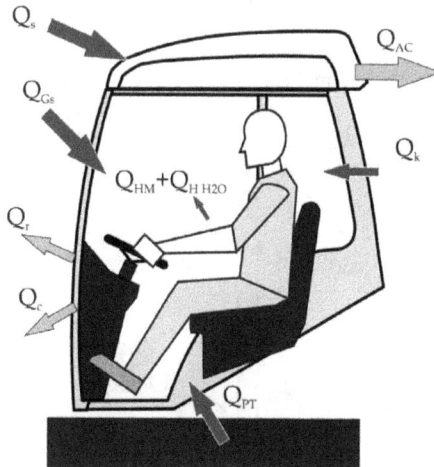

Figure 2. Thermal processes between a tractor cab and hot environment

For an analytical investigation of thermal processes, it was assumed that the heat flow is a one-dimensional steady flow through a plane wall with natural convection on both sides. A problem that comes up in the analytical determination of heat flux is the calculation of the convection heat transfer coefficients, which differ according to various sources (Incropera and DeWitt, 1981; ASHRAE, 1997a; Conceicao, 1999; Grossmann, 2010). Radiant heat exchanges among inner surfaces were neglected, because of the complexity of surface shapes and relatively small differences in temperature. The thermal radiation inside the cab is important for the thermal load of the operator, and this will be discussed later.

3.1. Heat transfer through cab walls due to temperature differences

Assuming that the air temperature in the vicinity of the walls is uniform, the one-dimensional heat load through this part of cab walls is (Fig. 3):

$$Q_k = A_k \cdot q_k = A_k \cdot U \cdot (t_{a,o} - t_{a,i}), \ W \tag{1}$$

Parameter U is the total heat transfer coefficient, which combines convections from each side of the wall with conduction. General equation for a wall of area A_j with s layers:

$$U_j = \frac{1}{\dfrac{1}{h_{c,i}} + \sum \dfrac{\delta_s}{k_s} + \dfrac{1}{h_{c,o}}}, \ W/m^2K \tag{2}$$

For all the surfaces with the heat transfer due to the difference in air temperature, the total value of U is:

$$U = \frac{1}{A_k} \sum U_j A_j, \ W/m^2K \tag{3}$$

This value is dependent on the thermal characteristics of the material that walls are made of (thermal conductivity k, W/mK), of the thickness of the layers and the surface area. The surfaces where this equation applies are the surfaces exposed neither to the solar radiation nor to the radiation from the powertrain. For example, these surfaces are windows on the shaded side of the cab.

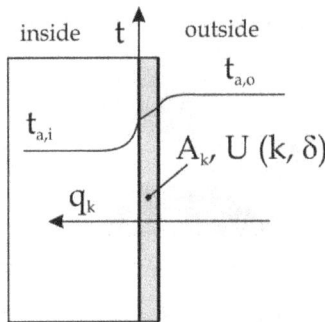

Figure 3. Heat transfer through the cab wall due to air temperature difference

Based on very few available sources, it is assumed here that the cab of a stationary vehicle has U value of around 10 W/m²K (Türler, 2003; Karlsson, 2007; Tavast, 2007; Grossmann,

2010). Consequently, the estimated heat flux through the cab wall (q_k) would be around 40 W/m²for the difference between the outside and the inside air temperature of $(30 - 26)°C = 4°$.

3.2. Effects of solar radiation

Cab glazing is a semitransparent medium, where solar irradiation can be partially reflected, absorbed and transmitted, Fig. 4.

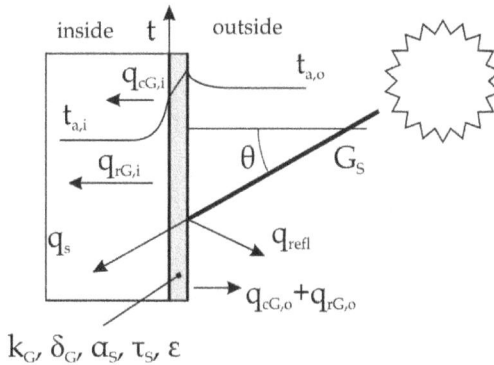

Figure 4. Solar energy transmission through glass

Solar irradiation on a tractor cab surface is variable depending on the position of the sun as well as the orientation of the cab surfaces. An example of maximum ("clear sky") intensity of normal irradiation and its variations in time and the influence of the surface inclination are shown in Fig. 5. The values are given for a vertical surface and for the surface inclined at 20°. The diagram shows that the vertically positioned glass receives around 15% less solar irradiation than the glass inclined at the angle of 20°.

The thermal balance equation for the outer glass surface can be written as:

$$G_s \cdot \alpha_s - \frac{k_G}{\delta_G}\left(t_{G,o} - t_{G,i}\right) - h_{G,o}\left(t_{G,o} - t_{a,o}\right) - \varepsilon \cdot \sigma\left(T_{G,o}^4 - T_{sky}^4\right) = 0 \tag{4}$$

The thermal balance on the inner glass surface is:

$$\frac{k_G}{\delta_G}\left(t_{G,o} - t_{G,i}\right) - h_{G,i}\left(t_{G,i} - t_{a,i}\right) - q_{rG,i} = 0 \tag{5}$$

The total heat flux through the glass exposed to the solar radiation consists of the transmitted part of the solar radiation, the convection on the inner side and longwave radiation (neglected here due to a small temperature difference between interior surfaces):

$$q_s = G_s \cdot \tau_s + h_c\left(t_{G,i} - t_{a,i}\right), \; W / m^2 \tag{6}$$

Using available values of the thermal properties of glass (Table 1) with the solar irradiation on inclined glass equal to 876 W/m^2, the estimated total heat flux through the glass will be around 740 W/m^2 for clear glass, and around 530 W/m^2 for tinted glass (green, with 75% transmittance of visible light; Saint-Gobain Sekurit, 2003). The air temperature difference was the same as above, i.e. 4°C. The drawback of use of the tinted glass is its solar radiation absorptivity, which causes the rise in the temperature of glass. In this example, the calculated glass temperatures will be around 30°C and 50°C, for clear and tinted glass respectively. In comparison to solar absorbing glass, a better but also a more expensive solution is the infra-red reflective glass that rejects almost a half of the solar radiation energy with less obstruction of visible light transmission (Bohm, 2002; Türler, 2003; Saint-Gobain, 2003). This kind of glass is used in automotive applications.

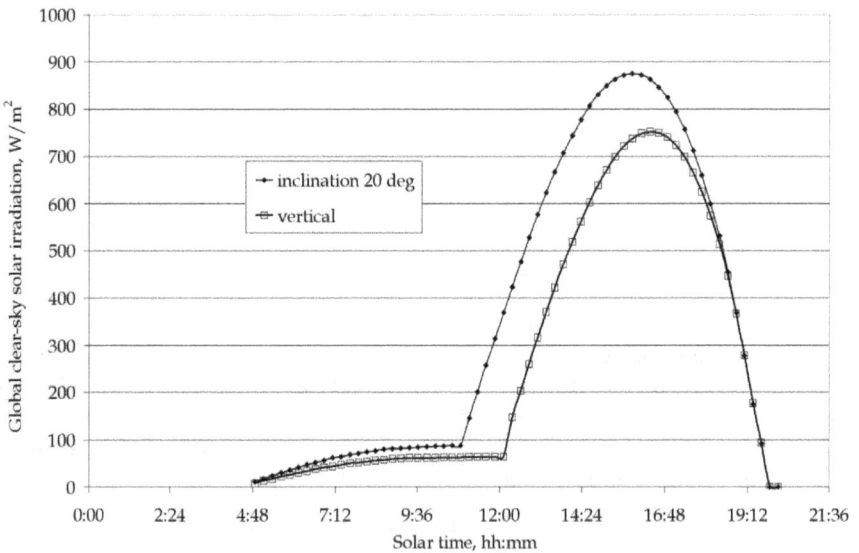

Figure 5. Global solar clear-sky irradiation G$_s$ on a west-faced vertical plane and on a plane inclined at the angle of 20° from vertical, in a central European region on a day in July (http://re.jrc.ec.europa.eu/pvgis/apps3/pvest.php)

The total amount of the heat transmitted through the glass caused by solar radiation is related to the normal projection of the tractor cab in the direction of radiation. The maximum solar transmissivity of glass is in the region of incident angles that are less than 30°, being the most inconvenient case both in terms of solar radiation transmittance to the cab interior as well as the direct effect on the operator.

The absorbed part of the solar irradiation heats the cab's outer opaque surfaces (the roof, for example) and their temperature rises. The maximum solar irradiation on a horizontal roof surface in the central Europe region on a summer day may exceed 900 W/m^2. After some exposure to solar radiation, the thermal balance between heat gain and heat release will be achieved. The heat from the outer surface is released into the environment (mostly to the sky) by longwave radiation, and into the surrounding air by convection. From the inner surface, the heat from the heated wall of the cabin is realeased into the air by convection and it is emitted by longwave radiation. The cab roof is an example of a surface where this mode of heat transfer takes place, Fig. 6.

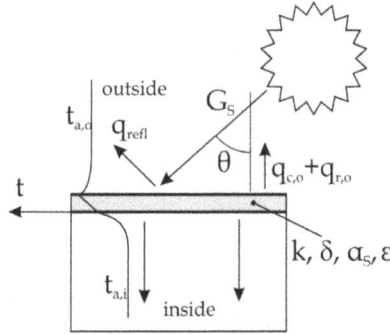

Figure 6. Heat transfer through the cab roof caused by solar radiation.

If there are any components of a ventilation system and an AC system in the roof, the heat transfer from the hot outer roof surface is not direct, but is transferred through the air space. The thermal balance equation for the outer surface of the roof can be written as:

$$G_s \cdot \alpha_s - \varepsilon \cdot \sigma \left(T_{roof,o}^4 - T_{sky}^4 \right) - h_{c,o}(t_{roof,o} - t_{a,o}) - \frac{k_{roof}}{\delta_{roof}}(t_{roof,o} - t_{roof,i}) = 0 \qquad (7)$$

The thermal balance on the inner roof surface is:

$$\frac{k_{roof}}{\delta_{roof}}(t_{roof,o} - t_{roof,i}) - h_{roof,i}\left(t_{roof,i} - t_{a,i} \right) - q_{r,roof,i} = 0 \qquad (8)$$

Using an example of the heat flux assessment for an opaque roof, treating the roof as a single layer wall, design parameters of the roof will be discussed. The radiation heat exchange between interior surfaces is neglected here, supposing that temperature differences among the surfaces are small. Hence, the equation of heat flux that flows through the roof, caused by solar radiation, is:

$$q_s = \alpha_s \cdot G_s - \varepsilon \cdot \sigma \left(T_{roof,o}^4 - T_{sky}^4 \right) - h_{c,o}(t_{roof,o} - t_{a,o}) =$$
$$= \frac{k_{roof}}{\delta_{roof}}(t_{roof,o} - t_{roof,i}) = h_{c,i}(t_{roof,i} - t_{a,i}) \qquad (9)$$

For example, the normalized data of heat flux and relative surface temperature, for 800 W/m² of solar irradiation on a horizontal surface, are given in Table 2. Four designs are analysed here, the cabs with roofs made of metal or polyester, both with and without a thermal insulating material on the inner side. Solar absorptivity of the outer surface is 0.3 and the AC keeps the interior air temperature at 26°C.

	Metal	Metal with insulation	Polyester	Polyester with insulation
Heat flux through the roof	100% (~30 W/m²)	70%	90%	60%
Inner surface temperature reduction	0 (surface temperature ~45°C)	6°	2°	7°
Inner surface radiation	100%	94%	99%	94%

Table 2. Evaluated relative values of the heat transfer parameters for different roof materials

As it was in the case with cab glazing, irradiation and solar absorptivity coefficient are dependent on the solar radiation incident angle, θ. The worst case is, of course, at noon, when the radiation is almost normal to the surface.

Solar absorptivity is also dependent on the colour of the surface. In terms of colour, light colours are preferable, and are widely used for tractor roofs. Under the same conditions, but with a dark coloured roof with solar absorptivity of 0.9, the heat transfer rate increases three times in comparison with the roof that has the solar absorptivity of 0.3. Also, the temperature of the roof interior will be 18 – 30°C higher, while the radiant heat flux emitted by the roof's inner surface would in this case rise more than 50%. Despite the high thermal conductivity, one of the most popular materials for the reflection of thermal radiation is aluminium ($\alpha_s = 0.09 - 0.15$; Incropera and DeWitt, 1981; Siegel & Howell, 1992), but it is not used for these purposes in tractor cabs.

Obviously, the solar absorptivity and longwave emissivity of the roof material, as well as thermal conductivity, are very important design factors. Therefore, the outer roof surface should have a low solar absorptivity coefficient and high emissivity (small α_s/ε ratio is preferable), with low thermal conductivity at the same time. However, in order to reduce thermal radiation towards the operator's body surface, the inner roof surface should have low thermal emissivity. Unfortunately, this is not the case with the kinds of materials usually used for the underside of the roof.

3.3. Heat gain from the powertrain

Heat transfer through the floor is modelled as the sum of the radiation from hot powertrain surfaces and the natural convection from the surrounding air, as in Fig. 7. It is assumed that the air and the outer surfaces of the powertrain have the temperature of 80°C (based on temperature of the transmission oil; data from tractor service manuals). In addition to the radiative properties and the temperature of surfaces, geometry also has some influence on the heat exchange. The geometry is described by the view factor. Because the distance between the powertrain and the floor is small, it is assumed here that the view factor is equal to unity (Incropera and DeWitt, 1981).

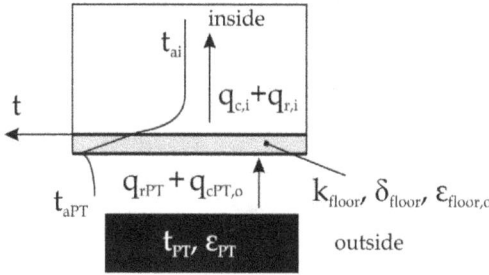

Figure 7. Heat gain from the powertrain

The heat flux that the floor absorbs from the hot powertrain surface and the surrounding air can be written as:

$$q_{PT} = \varepsilon_{PT:floor} \cdot \sigma\left(T_{PT}^4 - T_{floor,o}^4\right) + h_{c,o}\left(t_{a,o} - t_{floor,o}\right), \ W/m^2 \tag{10}$$

The factor $\varepsilon_{PT:floor}$ takes into account different emissivity of involved surfaces (Mc Adams, 1969):

$$\varepsilon_{PT:floor} = \cfrac{1}{\cfrac{1}{\varepsilon_{PT}} + \cfrac{1}{\varepsilon_{floor,o}} - 1} \tag{11}$$

Under the thermal equilibrium, the same heat flux is transferred through the floor and transmitted to the interior air:

$$q_{PT} = \frac{k_{floor}}{\delta_{floor}}\left(t_{floor,o} - t_{floor,i}\right) = h_{c,i}\left(t_{floor,i} - t_{a,i}\right), \ W/m^2 \tag{12}$$

For a simple floor without a heat shield on the outer side and with rubber flooring only, the estimated value of the heat flux through the cab floor caused by powertrain is around 110 W/m^2. If there is a heat shield beneath the floor and insulation between the flooring and the metal, the heat gain can be reduced to one third. Consequently, in the first case, the inner temperature of the floor will be above 50°C, while in the second case the temperature will be lower, around 34°C.

3.4. Total thermal load of the tractor cab

The total thermal load of the cab under the chosen summer conditions will depend on the size of the cab and on its orientation to the sun. The worst case would be that the largest side of the cab is facing the sun and the tractor cab is not equipped with the mentioned means for heat gain reduction. Therefore, for approximately 1/4 of the glazed area exposed to the sun (around 1.5 m²), the resulting solar heat gain would be around 1100 W (in tractors with clear glazing). The rest of the cab surface, (approximately more than 7 m²), would transfer up to 800 W. In total, adding the heat released by the operator, the estimated thermal load of the tractor cab would be around 2 kW. Fig. 8 shows the percentage of different modes of thermal loads and their reduction via the mentioned methods for heat rejection and insulation. The chart in Fig. 8 shows that glazing has the most important role in thermal processes that influence the microclimate conditions inside the tractor cab.

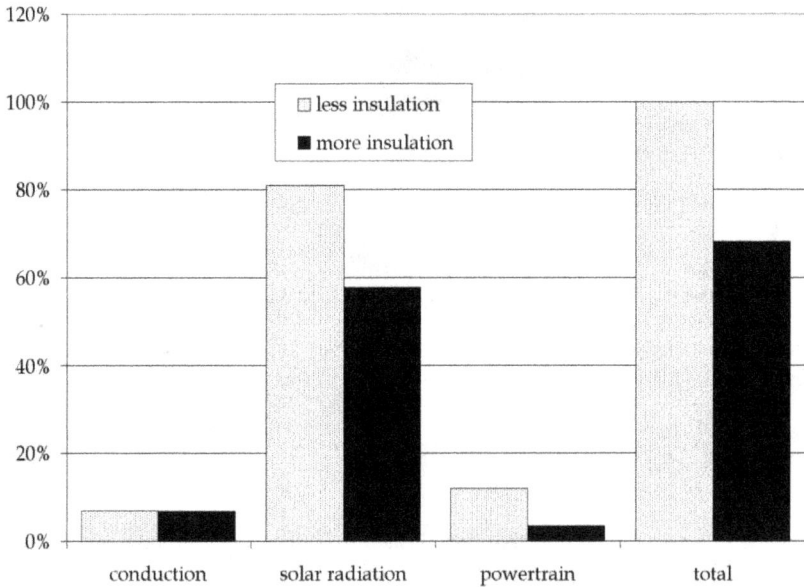

Figure 8. Percentage of heat fluxes of different modes of heat transfer to the tractor cab and the effects of application of the heat rejection and insulation

4. Thermal processes between an operator and a tractor cab interior

The conventional approach for the analysis of thermal processes is based on the balance between the sensible and latent thermal load of the cabin and the heat removal by supplied air. Therefore, quantity and conditions of the air supplied by the AC influences the overall thermal state in the cabin, as shown in Fig. 9. This approach, although technically correct, is not suitable for the assessment and prediction of the operator's thermal sensation. Due to the complex nature of the thermal environment in a tractor cab, it is not possible to accurately describe the thermal conditions inside a tractor cab by a single value, even when the value combines the microclimate factors (for example PMV index; Fanger, 1970). Because of this, other assessment parameters have been developed. One of them is the equivalent temperature (t_{eq}), defined as the temperature of a homogenous space, with mean radiant temperature equal to air temperature and with zero air velocity, in which a person exchanges the same heat loss by convection and radiation as in the actual conditions. The equivalent temperature can be evaluated for the whole body or for individual body parts.

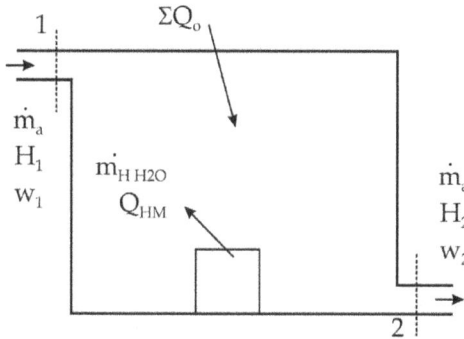

Figure 9. Schematic of the conventional approach to the analysis of air-conditioned space, considering only the air conditions at the inlet (1) and the outlet (2) and the incident rates of energy and moisture gains

In the conventional approach, an operator is treated only as a source of certain amount of sensible and latent heat. The human-based design takes into account the operator's shape, volume, and his thermal sensitivity.

4.1. Human body thermoregulation and thermal sensation

Microclimate conditions and hence human thermal sensation as well, are dependent on air temperature, air velocity, relative humidity and mean radiant temperature. However, individual differences regarding physiological and psychological response, clothing insulation, activity and preferences in terms of air temperature and air movement also have a strong impact on thermal sensation.

In an uncomfortable hot ambient, owing to the thermoregulation system, the human body sets off the process of vasodilatation and sweating, trying to prevent the rise of internal body temperature. Since these processes can cool down the body only to a certain degree, in order to prevent further rise in body temperature and to avoid the risk of hyperthermia, it is necessary to make the space comfortable or to enhance the process of releasing the heat from the body. Both of these methods are used in vehicles, the latter being more suitable for transient (cool-down) conditions.

The heat loss from the body surface in warm conditions relies on convective heat loss and the heat loss by sweat evaporation from the skin surface (Fanger, 1970; Parsons, 2003). The equation for the dry heat flux by convection can be written as:

$$C = f_{cl} \cdot h_c \cdot (t_{cl} - t_a), \ W/m^2 \tag{13}$$

Factor f_{cl} is the clothing area factor, which describes the ratio between the area of the clothed body and the area of the nude body. The temperature of the clothing surface (t_{cl}) will be equal to the skin temperature (t_{sk}) on body surfaces without clothing: face, forearms, hands and neck.

According to the equation (13), the higher the temperature difference between the air and the body surface, the higher the heat loss from the body. In addition, the coefficient of heat transfer by convection (h_c) is dependent on air velocity and the shape of a body part. The coefficient increases with the increase of air velocity (Fanger, 1970; De Dear et al., 1997; Parsons, 2003). Therefore, the ventilation system is able to influence this part of the heat transfer directly, by changing the airflow rate through the ventilation outlets (air vents). The general form of the equation for forced and natural convective heat transfer coefficient for the whole body is (Fanger, 1970; De Dear et al., 1997):

$$h_c = B_f \cdot v_a^n, \ \text{for forced convection, and} \tag{14}$$

$$h_c = B_n \cdot v_a^m, \ \text{for natural convection.} \tag{15}$$

The exponents n and m and the constants B_f and B_n are mostly dependent on body posture as found in various literary sources. The overview of the equations for the calculations of the heat transfer coefficient can be found in Fanger (1970), ASHRAE (1997b), De Dear et al. (1997) and Parsons (2003).

Evaporative heat loss from the skin is dependent on air humidity, skin wetness and evaporative heat transfer coefficient (h_e). The evaporative heat transfer coefficient is calculated using the Lewis relation (Parsons, 2003):

$$h_e = LR \cdot h_c, \ W/m^2 kPa \tag{16}$$

Lewis ratio (LR = 16.5 K/kPa) makes the evaporative heat transfer coefficient dependent on air velocity as well. In addition, air-conditioning lowers the relative humidity of the interior air and consequently improves the latent heat loss by the evaporation of sweat.

The body's overall thermal sensation is affected by the local thermal state of individual body parts, at the same level of cooling, due to different physiological properties (different sensibility for warm versus cool sensations, different sweating rate etc.; Zhang, 2003):

- Back, chest, and pelvis strongly influence overall thermal sensation, which closely follows the local sensation of these parts during local cooling;
- The head region, arms and legs have an intermediate influence on the body's overall thermal sensation;
- Hands and feet have much less impact on the overall sensation.

4.2. Local air velocities around a human body

The working principle of the vehicle air-conditioning and the ventilation system is based on driving the conditioned air through adjustable air vents in the vehicle cabin. This airflow causes changes in local and overall microclimate conditions, consequently also changing the heat loss from the operator's body. This airflow is characterized by spatially distributed local air velocities and air temperatures.

Local velocity of the airflow should have the ability to penetrate the operator's free convection flow. For example, Melikov (2004) recommends local air velocities around 0.3 m/s as the minimum values. ASHRAE 55-2009 standard suggests elevated air velocities up to 1.2 m/s, but only if the air speed is under the control of an exposed person and in the temperature range of 28 - 31°C. These values apply to a lightly clothed sedentary person (0.5 - 0.7 Clo), with the metabolic rate in the range between 1.0 and 1.3 Met. In the case of higher difference between the temperatures of ambient air and (colder) local airflow, the operator will prefer lower air velocities. On the other hand, higher metabolic activity, up to 3.2 Met for the operation of mobile machinery, could allow higher air velocities (ASHRAE, 2003; Melikov, 2004; Arens et al., 2009).

Other very important parameters for thermal sensation and the sensation of air quality are characteristics of the air in the operator's breathing zone. The breathing zone is a semi-spherical space around the mouth and the nose. In terms of perceived air quality, still air in the breathing zone is not acceptable in a warm ambient, even when overall comfort is achieved. The temperature of the localized air in the breathing zone should be equal or 3 - 4°C lower than the interior air temperature. In order to achieve the mentioned requirements, it is recommended to place air vents at distances from 0.4 to 0.6 m from the operator's breathing zone (Melikov, 2004). This should be done in agreement with the geometry of the interior space of the vehicle cabin.

4.3. Effects of thermal radiation and transmitted solar radiation

When a tractor cab is exposed to the sun, the operator's body receives the heat partly by solar radiation transmitted through the glass and partly by longwave thermal radiation from the surrounding surfaces. The amount of solar radiation energy that will be absorbed by the body will depend on the effective projected area and on the solar absorptivity of the body surface. The largest effective projected radiation area A_{eff} of a person in the sitting position is with the azimuth and altitude angles of 30° and 15° respectively, and the total surface is equal to (Fanger, 1970; Parsons, 2003):

$$A_{eff} = 0.72 \cdot 0.33 A_{Du} = 0.238 A_{Du}, \; m^2 \tag{17}$$

The total surface of the human body (A_{Du}) can be calculated from the body weight and height (after Dubois; Fanger, 1970; Parsons, 2003). The factor 0.72 in equation (17) is the effective radiation area factor, and the factor 0.33 is the projected area factor (Fanger, 1970; Parsons, 2003). Solar absorptivity of the human skin is around 0.62, while solar absorptivity for clothing depends on the colour (Incropera and DeWitt, 1981; Parsons, 2003). Tractor cab design and the operator's position on the seat do not offer good protection from solar radiation, although large cabs create better shading for the operator. As it can be seen in Fig. 10, the head and the chest are exposed and protected only by solar properties of glass, unless some kind of solar shading devices are used. Shading devices can be placed on the inner side, or as a better solution, on the outer side of the cab. In both cases, the shading devices must not restrict the operator's normal field of vision in the working area of the tractor. For that reason, sun visors or curtains must be easily adjustable. In addition to shading the operator, the outer shading devices prevent exposure of glass and other surfaces to solar radiation. A common solution for these purposes is the use of cab roof overhangs, since aerodynamics is not an issue in agricultural tractors.

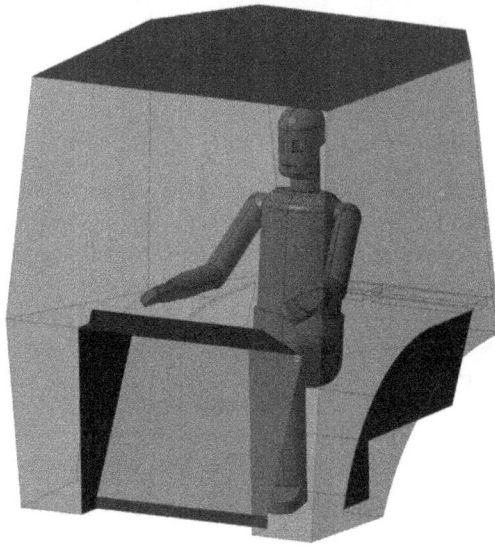

Figure 10. Projection of the operator inside the tractor cab from the direction of the azimuth and altitude angles of 30° and 15°, respectively

When surrounded by surfaces with higher temperatures than that of their skin and clothing, a person receives the heat by thermal radiation. Thermal radiation flux from hot surrounding surfaces depends on the wall surface temperature and the emissivity of the wall. The wall temperature and its emissivity are expressed by mean radiant temperature (t_{mr}) and linear radiative heat transfer coefficient (h_r):

$$q_r = h_r \cdot f_{cl} \cdot (t_{mr} - t_{cl}), \, W/m^2 \tag{18}$$

$$h_r = 2.88 \cdot \varepsilon \cdot \sigma \left(273.2 + \frac{t_{cl} + t_{mr}}{2} \right)^3, \, W/m^2 K \tag{19}$$

The emissivity of non-metallic surfaces is generally high, 0.9 and higher, just like the emissivity of the human skin (0.95; Incropera and DeWitt, 1981; ASHRAE 1997c) and clothing (0.77 for cotton; Siegel & Howell, 1992). For example, under the thermal conditions mentioned above, in the case of a single-layer metal sheet roof, the thermal radiation load to the operator's head would be around 70 W/m^2. In the case of having the polyester roof with internal thermal insulation, the thermal load would be less than 30 W/m^2.

The AC system is not able to directly change mean radiant temperature, except by decreasing the temperature of the inner surfaces (although very slowly).

5. Design of the air distribution system

The ventilation system drives and distributes the air inside the tractor cab using the air distribution system. The air distribution system consists of a blower, air ducts and several air vents usually positioned on the cab ceiling. The tasks of the ventilation system are:

- to remove excess heat from the operator's body,
- to supply the breathing zone with fresh air,
- to reduce the temperature inside the cab, and
- to pressurize the cab.

The air supply rate of tractor cab ventilation systems are generally in the range of several hundreds of cubic meters of air per hour, and are sufficient for up to 10 kW of cooling power.

According to the air jet theory, the air velocity along the airflow centreline is in direct proportion with the outlet velocity and in inverse proportion with the distance from the air vent. Consequently, the air jet direction, the distance from the body and the velocity profile on the outlet of the air vent are the main factors that influence the characteristics of the airflow over the body surface. These factors are determined by the interior geometry of the tractor cab, the operator's position in relation to the air vents and by the design of the ventilation ducts and vents (cross-section, number of vents), as well.

Control of the microenvironment is necessary due to individual differences among various operators, and airflow characteristics must be suited to individual human thermal sensation. For example, a powerful air-conditioner that easily decreases the temperature inside the tractor cab could produce an unpleasant stream of air, a draught. On the other hand, despite sufficient cooling power of the AC system in some cases, there is the possibility that a part of the operator's body would not be cooled satisfactorily, especially if exposed to solar radiation.

5.1. Evaluation of the air distribution system's efficiency and thermal sensation

The evaluation of thermal sensation in a tractor cab can be done by using human subjects, by directly measuring the microclimate physical quantities at discrete points in the cab or by using special human-shaped sensors. Complex human-shaped measuring instruments, the so-called thermal manikins, are the most suitable tool for a reliable, repetitive and objective evaluation of thermal conditions in a non-homogenous environment. Numerical methods for the research of vehicle cabin microclimate are also used. When CFD (Computational Fluid Dynamic) methods for the modelling of the *human-cabin-environment* system are to be used, it is necessary to have a model of the human body that is geometrically and thermally appropriate, the so-called computer simulated person (CSP). Fig. 11.

The analysis of different tractor cab ventilation layouts was described in Ružić et al. (2011) and Ružić (2012). In this research, the heat losses from body parts were observed on a virtual thermal manikin using the CFD simulation. All the simulations were performed without solar radiation. The airflow through the ventilation system was approximately 380 m³/h, with the air temperature of 20°C. In one group of simulations, the conditions were isothermal (Ružić et al., 2011), while in the other (Ružić, 2012) the initial conditions were the same as the hot ambient with the cab temperature of 30°C.

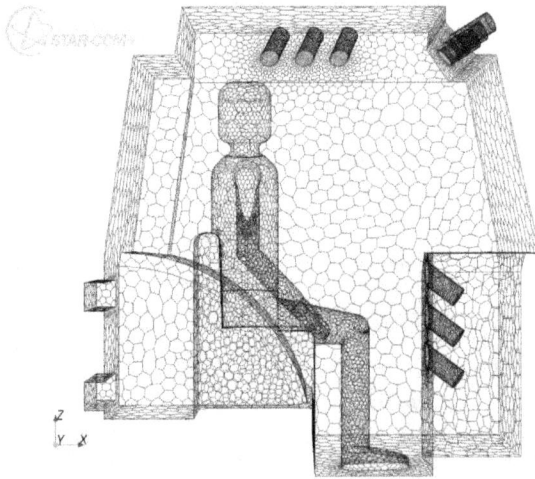

Figure 11. An example of a model of an agricultural tractor cab with a virtual thermal manikin, divided into the volume mesh for the CFD simulation

The results showed that the heat loss from the operator's body can vary significantly depending on the position and the direction of the vents, but all the simulations were done with the same cooling power. Consequently, the operator's thermal sensation will vary accordingly. As it can be seen in Fig. 12, under the same tractor cab thermal load and the same heat removal, but with two different air distribution designs, in the example *A* the total heat loss from the operator is 42% lower than in the example *B*. This indicates that human body characteristics and air distribution systems are two very important factors as well.

In terms of heat loss for the whole body and for thermally most sensitive body segments, the statistical analysis of results for various combinations of vents positioning and airflow direction shows that airflow direction is a more significant parameter than positioning of the vents. Therefore, it is important that the operator be able to change the characteristics of air jet, that is, the air flow direction and the velocity.

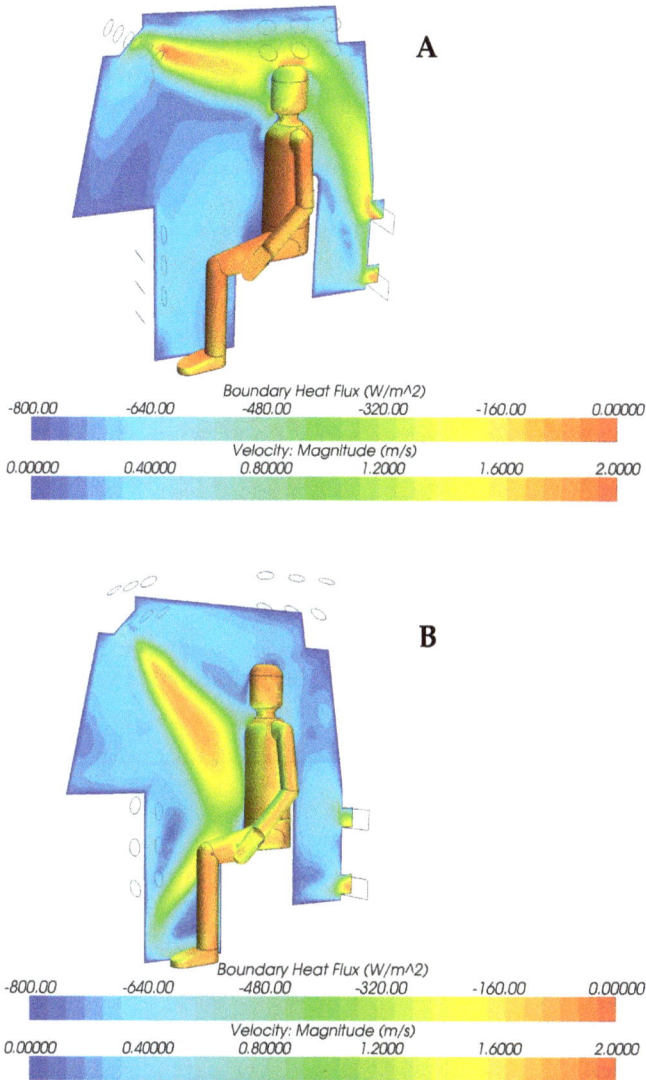

Figure 12. Comparison of air velocity distribution and heat loss from different parts of the operator's body using the CFD technique with virtual thermal manikin (Ružić, 2012)

In the non-isothermal cases (Ružić, 2012), the assessment criterion was the deviation from neutral values of equivalent temperatures. For example, under these conditions, the lowest deviation of the equivalent temperature for the whole body was achieved with the vents placed in the front part of the ceiling, directed at the operator's chest. The highest rate of heat loss from the operator's body was achieved in the layout with the vents placed in the instrument panel, directed at the operator's pelvis (H-point). This mode of operation would be of interest for cooling down of the operator under the conditions of high thermal load, for example caused by solar radiation. The difference between the best and the worst cases regarding equivalent temperature for the whole body was 10 degrees. For individual body segments, the difference was considerably larger. Consequently, if the operator sets the air distribution system in a wrong way, unwanted thermal conditions can be produced. For that reason, the interface for the AC and the ventilation system control must be designed in such a way that it is easy to use.

More information about the significance of certain factors could be obtained by expanding the range of factor variations. However, there is almost an unlimited number of different settings of just one air distribution design, as well as a variety of ambient conditions. Fortunately, the cabs of modern middle-range agricultural tractors are all similar in design and shape, and most of these cabs have the AC air distribution system with the vents on the ceiling. According to the results of those simulations, it is possible to improve and optimize thermal conditions in tractor cabs by implementing conventional air distribution systems.

6. Conclusions

Although the cabs of modern tractors have many common features, the results show that there are significant differences regarding thermal loads, caused by variations in tractor cab designs. Conclusions can be summarized as follows:

- The highest heat flux that enters the cab is caused by solar radiation through the glass (which is several times higher than the heat transferred by other modes). Paying attention to solar characteristics of glass is a direct way to reduce the operator's thermal load.
- Further improvements in the reduction of the heat load are the use of roof overhangs, adjustable sun visors and less inclined glass. A thermally reflective heat shield placed beneath the cab floor also improves thermal conditions. However, some of the existing technologies for thermal load reduction are not used to the full extent, especially in low-class tractors.
- The ventilation system and the AC system should be designed in such a way that the optimum heat loss from individual body parts can be obtained in most conditions. The direction of the air jet is one of the most crucial factors regarding heat losses from the body. Due to a wide range of boundary conditions and different individual preferences regarding thermal conditions, the air distribution system must be adequately adjustable and easy to use.
- For the analysis and optimization of the air conditioning system, experimentally verified numerical methods (CFD) with computer simulated person or virtual thermal manikin are an inevitable tool.

Author details

Dragan Ružić and Ferenc Časnji
University of Novi Sad, Faculty of Technical Sciences, Serbia

Acknowledgement

This research was done as a part of the project TR31046 - "Improvement of the Quality of Tractors and Mobile Systems with the Aim of Increasing Competitiveness and Preserving Soil and Environment", which was supported by the Serbian Ministry of Science and Technological Development.

Nomenclature

A – area, m^2
f_{cl} – clothing factor, (-)
G_s –solar irradiation, W/m^2
h – heat transfer coefficient, W/m^2K
H – enthalpy, J
k – thermal conductivity, W/mK
LR – Lewis ratio (16.5 K/kPa)
$\dot{m}$ – mass flow, kg/s
q – heat flux, W/m^2
Q – heat transfer, W
RH – relative humidity, %
t, T – temperature, °C, K, respectively
U – total heat transfer coefficient, W/m^2K
v – air velocity
w – humidity ratio, kg/kg

Greek symbols

α – surface absorptivity, (-)
δ – layer/wall/glass thickness, m
ε – surface emissivity, (-)
σ – Stefan-Boltzmann constant ($5.670 \cdot 10^{-8}$ W/m^2K^4)
θ – incident angle of solar radiation, degrees
τ – normal solar transmissivity, (-)

Subscripts

a – air
AC – air-conditioning
c – convection

cl – clothing
e – evaporation
eff – effective
G – cab glazing
H – human
HVAC – heating, ventilation, air-conditioning
i – inside
k – conduction
M – metabolic activity
mr – mean radiant
o – outside
PT – powertrain
r – thermal radiation
refl - reflection
s – solar
surf – surface

7. References

Arens, E.; Turner, S.; Zhang, H. & Paliaga, G. (2009). A Standard for Elevated Air Speed in Neutral and Warm Environments, *ASHRAE Journal*, Vol.51, No.25, (May 2009), pp. 8-18, ISSN 0001-2491

ASABE/ISO (1997). 14269-2 Tractors and Self-propelled Machines for Agriculture and Forestry - Operator Enclosure Environment - Part 2: Heating, Ventilation and Air-conditioning Test Method and Performance

ASHRAE. (1997a). *Fundamentals Handbook, Chapter 3: Heat Transfer*, American Society of Heating, Refrigerating and Air Conditioning Engineers, Atlanta, USA

ASHRAE. (1997b). *Fundamentals Handbook, Chapter 8: Thermal Comfort*, American Society of Heating, Refrigerating and Air Conditioning Engineers, Atlanta, USA

ASHRAE. (1997c). *Fundamentals Handbook, Chapter 36: Physical Properties of Materials*, American Society of Heating, Refrigerating and Air Conditioning Engineers, Atlanta, USA

ASHRAE. (1999a). *Book of Applications, Chapter 8: Surface Transportation*, American Society of Heating, Refrigerating and Air Conditioning Engineers, Atlanta, USA

ASHRAE. (1999b). *Book of Applications, Chapter 32: Solar Energy Use*, American Society of Heating, Refrigerating and Air Conditioning Engineers, Atlanta, USA

ASHRAE. (2003). Standard 55P: Thermal Environmental Conditions for Human Occupancy, Third Public Review, American Society of Heating, Refrigerating and Air Conditioning Engineers, Atlanta, USA

Bohm, M.; Holmer, I.; Nilsson, H. & Noren, O. (2002). *Thermal effects of glazing in driver's cabs, JTI-rapport 305*, JTI – Institutet for jordbruks, ISSN 1401-4963, Uppsala, Sweden

Conceicao, E.; Silva, M.; Andre, J. & Viegas, D. (1998). A Computational Model to Simulate the Thermal Behaviour of the Passengers Compartment of Vehicle, *SAE Paper* 1999-01-0778, Society of Automotive Engineers, USA

Currle, J. & Maue, J. (2000). Numerical Study of the Influence of Air Vent Area and Air Mass Flux on the Thermal Comfort of Car Occupants, *SAE Paper* 2000-01-0980, Society of Automotive Engineers, USA

De Dear, R.; Arens, E.; Hui, Z. & Oguro, M. (1997). Convective and Radiative Heat Transfer Coefficients for Individual Human Body Segments. *International Journal of Biometeorology*, Vol.40, No.4, (May 1997), pp. 141–156, ISSN 1432-1254

Fanger, P. O. (1970). *Thermal Comfort*, McGraw-Hill, ISBN 8757103410, Copenhagen, Denmark

Grossmann, H. (2010). *Pkw-klimatisierung*, Springer-Verlag, ISBN 9783642054945, Berlin, Germany

Incropera, F. P. & DeWitt, D. P. (1981). *Fundamentals of Heat and Mass Transfer*, John Wiley & Sons, ISBN 9780471427117, New York, USA

Karlsson, Anika (2007). *Simulation of Cooling Demand in a Truck Compartment*, Master thesis, ISSN 1650-8300, Uppsala Universitet, Sweden

Mc Adams, W. (1969). *Heat Transmission*, (in Serbian) Građevinska knjiga, Beograd, Serbia

Melikov, A. (2004). Personalized Ventilation, *Indoor Air*, Vol.14 (Suppl. 7), (December 2004), pp. 157–167, ISSN 0905-6947

Parsons, K. (2003). *Human Thermal Environments: The Effects of Hot, Moderate and Cold Environments on Human Health, Comfort and Performance*, 2nd ed. Taylor & Francis, ISBN 0-415-23792-0, London, UK

Rugh, J. & Farrington, R. (2008). Vehicle Ancillary Load Reduction Project Close-Out Report, *Technical Report* NREL/TP-540-42454, January 2008, National Renewable Energy Laboratory, Golden, USA

Ružić, D. & Časnji, F. (2011). Agricultural Tractor Cab Characteristics Relevant for Microclimatic Conditions. *Journal of Applied Engineering Science* Vol.2, No.9, (June 2011), pp. 323-330, ISSN 1451-4117

Ružić, D.; Časnji, F. & Poznić, A. (2011). Efficiency Assessment of Different Cab Air Distribution System Layouts, *Proceedings of 15th Symposium on Thermal Science and Engineering of Serbia*, pp. 819-828, ISBN 978-86-6055-018-9, Sokobanja, Serbia, October 18 – 21, 2012

Ružić, D. (2012). Analysis of Airflow Direction on Heat Loss From Operator's Body in an Agricultural Tractor Cab, *Proceedings of 40th International Symposium on agricultural engineering "Actual Tasks on Agricultural Engineering"*, pp. 161-169, ISBN 1333-2651, Opatija, Croatia, February 21 – 24, 2012

Saint-Gobain Sekurit. (2003). *Glazing Manual*, available at http://www.sekurit.com,

Siegel, R. & Howell, J.R. (1992). *Thermal Radiation Heat Transfer, 3rd Edition*, Hemisphere Publishing Corporation, ISBN 0-89116-271-2, Washington, USA

Sorensen, D. N. & Voigt, L. K. (2003). Modelling Flow and Heat Transfer Around a Seated
    Human Body by Computational Fluid Dynamics, *Building and Environment* Vol.38,
    No.6, (June 2003), pp. 753–762, ISSN 0360-1323
Tavast, J. (2007). *Solar Control Glazing for Trucks*, Master thesis, ISSN 1650-8300, Uppsala
    Universitet, Sweden
Türler, D.; Hopkins, D. & Goudey, H. (2003). Reducing Vehicle Auxiliary Loads Using
    Advanced Thermal Insulation and Window Technologies, *SAE Paper* 03HX-36, Society
    of Automotive Engineers, USA
Zhang, H. (2003). *Human Thermal Sensation and Comfort in Transient and Non-Uniform Thermal
    Environments*, PhD thesis, University of California, Berkeley, USA

Non-Isothermal Spontaneous Imbibition Process Including Condensation Effects and Variable Surface Tension

Bautista Oscar, Sánchez Salvador, Méndez Federico, Bautista Eric and Arcos Carlos

Additional information is available at the end of the chapter

1. Introduction

Due to the continued miniaturization of semiconductor devices, power electronics, biosensors and aerospace equipment, problems associated with overheating of these components have increased. Accordingly, the innovative cooling techniques are required to meet the demands of heat load removal from highly integrated electronic circuits and the electronic components of spacecraft designed for advanced long-term spacecraft missions. Those demands result in the rapid development of improved heat rejection techniques such as the interfacial vaporization and condensation heat transfer of thin liquid films. Closed two-phase devices such as heat pipes and thermosyphons have been and are being used successfully for the above application. Whatever configuration is used, the heat energy removed at the chip is transported away and rejected from the system by condensation at a remote location. Therefore, a fundamental understanding of the condensation process in minichannels and capillaries is important to optimize design considerations. In this direction, for possible use in electronic cooling applications, Begg et al. [1] developed a mathematical model of annular film condensation in a miniature tube. In this model, the liquid flow has been coupled with the vapor flow along the liquid-vapor interface through the interfacial temperature, heat flux, shear stress and pressure jump conditions. The model predicts the position of the liquid-vapor interface. The numerical results show that complete condensation of the incoming vapor is possible at comparatively low heat loads. L. P Yarin et al. [2] present a quasi-one dimensional model of laminar flow in a heated capillary. In the frame of this model, the effects of channel size, initial temperature of the working fluid, wall heat flux and gravity on two phase capillary flow are studied. It is shown that hydrodynamical and thermal characteristics of laminar flow in a heated capillary are determined by the physical properties of the liquid and its vapor, as well as the heat flux at the wall. The effect of dimensionless parameters such as the Peclet, Jakob numbers, and dimensionless heat flux or Nusselt number on the velocity, temperature and pressure within the liquid and vapor domains has been studied. In addition, the above authors conducted an experimental analysis for showing that the flow in micro-channels appear to have distinct phase domains. On the other hand, the theory of two-phase laminar flow in a heated microchannels was presented by Yarin et al. [3]. They studied the thermohydrodynamic characteristics of a two-phase capillary flow with phase change at the meniscus by using a quasi-one-dimensional model for the flow. It takes into account the principal characteristics of the phenomenon, namely, the effects of the inertia, pressure, gravity, friction forces and capillary pressure due to the curvature of the interface surface, as well as the thermal and dynamical interactions of the liquid and vapor phases. To describe the flow outside of the meniscus in the domains of the pure liquid or vapor, the one-dimensional mass, momentum and energy equations are used. The possible states of the flow are considered, and the domains of steady and unsteady states are outlined. Meanwhile, Qu et al. [4] established the physical and mathematical models to account for the formation of evaporating thin liquid film and meniscus in capillary tubes. Their results show that in regard to the capillary tubes of micron scale, the calculation results show that, the bigger the inner radius or the smaller the heat flow, the longer the evaporating interfacial region will be. There only exists meniscus near the wall, and nearby the axial center is flat interface. While as to the capillary tubes of scale about 100 μm, the evaporating interfacial region will increase with heat flux. Compared to the capillaries of micron scale, the meniscus region will extend to the center of capillary axis. Recently, the Lucas-Washburn equation [5, 7], was extended by Ramon and Oron [8], describing the motion of a liquid body in a capillary tube so as to account for the effect of interfacial mass transfer due to phase change, either evaporation or condensation. They showed that the phase change affects the equilibrium height of the meniscus, the transition threshold from monotonic to oscillatory dynamics, and the frequency of oscillations. At higher mass transfer rates and/or large capillary radii, vapor recoil is found to be the dominant factor. In general, evaporation decreases the equilibrium height, increases the oscillation frequency and diminishes the transition threshold to oscillations. For condensation, two regimes are identified: at high mass transfer rates similar trends to those of evaporation are observed, whereas the opposite is found for low mass transfer rates, resulting in an increased equilibrium height, lower oscillation frequencies and a shift of the transition threshold toward monotonic dynamics. It must be noted that in the last mentioned work, a difference of temperature or interfacial temperature resistance at the interface is imposed for causing the phase change process. However, the temperature of the liquid bulk was assumed constant. We consider that a more realistic case occurs when the temperature field in the liquid is taken into account, causing a local variation of the interfacial temperature. In fact, this point was commented by Ramon and Oron [8]. Therefore, the main purpose of this paper is to solve the Lucas-Washburn equation in conjunction with the energy equation for the liquid penetrating a capillary tube, considering that the imbibition front is also controlled by a direct condensation process and considering a linear dependence on temperature of the surface tension.

2. Formulation

A long vertical capillary tube of radius R, as shown in Figure 1, is filled with a saturated vapor at temperature T_s whose density is ρ_v. At time $t = 0$, the bottom of the capillary tube comes into contact with a large reservoir of a liquid whose density is ρ, viscosity μ and thermal diffusivity α, assumed constant. The liquid reservoir is kept at temperature T_0, which is below the vapor temperature T_s; for $t > 0$, the contact originates a spontaneous non-isothermal imbibition process of the liquid into the capillary tube. It is assumed that the liquid wets the capillary inner wall completely, for which case the contact angle, θ, is set equal to zero [9]. For instance, a typical substance that satisfies this condition for the equilibrium contact angle, is the silicone oil [10]. Due the temperature difference, $T_s - T_0$, condensation occurs at the moving vapor-liquid interface. In addition, we assume that the temperature interface

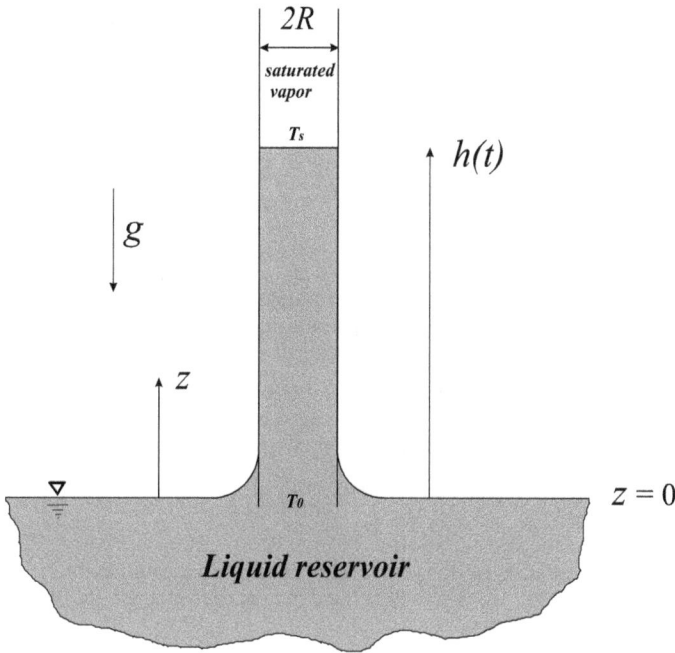

Figure 1. Schematic diagram of the studied physical model.

is maintained at T_s. Using a one-dimensional formulation of the average conservation laws, we derive the corresponding dimensionless momentum and energy equations for the liquid penetrating the capillary. In addition, the imbibition front is characterized by a uniform capillary pressure. In this manner, the fluid flow is governed by the continuity, momentum and energy equations, given by

$$\nabla \cdot \mathbf{v} = 0 \tag{1}$$

$$\rho \frac{D\mathbf{v}}{Dt} = \rho \mathbf{g} - \nabla p + \mu \nabla^2 \mathbf{v} \tag{2}$$

$$\rho c_p \frac{DT}{Dt} = k\nabla^2 T + \Phi \tag{3}$$

where $\mathbf{v}\,(\mathbf{r},t)$ and $p\,(\mathbf{r},t)$ represent the velocity and pressure fields, respectively. In cylindrical coordinates $\mathbf{r} = (r,\theta,z)$, with the origin of the coordinates at the tube inlet and the z axis along its axis, the radial distance $(0 \leq r \leq R)$ is measured from the z axis (Figure 1). The velocity components $\mathbf{v} = (v_r, v_\theta, v_z)$ are along the $\hat{\mathbf{r}}$, $\hat{\theta}$ and $\hat{\mathbf{z}}$, respectively. In this work, unidirectional flow is assumed and admits the azimuthally symmetrical solution $\mathbf{v} = v_z\hat{\mathbf{z}}$, with $v_\theta = v_r = 0$. Therefore, Eq. (1) yields $\partial v_z/\partial z = 0$, which means that the v_z is a function of the radial coordinate and time, i.e., $v_z = v_z\,(r,t)$. However, at the capillary inlet and in a small zone beneath the meniscus interface, the velocity deviates from its value $v_z(r,t)$ inside the column and depends on z. Therefore, with Eq. (2), the z component acquires the form

$$\frac{\partial v_z}{\partial t} + v_z\frac{\partial v_z}{\partial z} = -\frac{1}{\rho}\frac{\partial p}{\partial z} - g + \frac{\mu}{\rho r}\frac{\partial}{\partial r}\left(r\frac{\partial v_z}{\partial r}\right) \tag{4}$$

We assume further that the fully developed velocity profile in the capillary tube that gives rise to the viscous force is given by the equation

$$v_z(r,t) = 2\langle v_z(t)\rangle \left[1 - \left(\frac{r}{R}\right)^2\right] \tag{5}$$

where

$$\langle v_z(t)\rangle = \frac{2}{R^2}\int_0^R v_z(r,t)\,r\,dr \tag{6}$$

is the flow velocity averaged over the capillary cross section. In the specialized literature, it is accepted that the parabolic velocity profile in a pipe is well established at a distance Z from the fluid entry such that $Z/d \approx Re/30$, where d is the pipe diameter and Re is the Reynolds number, defined as $Re = \rho \langle v_z(t)\rangle d/\mu$. For typical values of capillary rise, where $Re < 1$, the profile of Eq. (5) is well established at a distance smaller than the capillary radius.

Because it is required to determine the temporal change of the meniscus height $h(t)$, the azimuthally symmetric terms of Eq. (4) are averaged over the volume of the moving liquid column [11]. The time derivative of the velocity on the left-hand side of Eq. (4) is:

$$\left\langle\frac{\partial v_z}{\partial t}\right\rangle = \frac{2}{h(t)R^2}\int_0^{h(t)} dz \int_0^R v_z r\,dr = \frac{d\langle v_z\rangle}{dt}. \tag{7}$$

The averaged pressure gradient term is obtained by taking into account that $p(z=0) = 0$ and $p(z = h(t)) = -2\sigma\cos(\theta_a)/R$, where θ_a is the dynamic contact angle (in the present work $\theta_a \approx 0$):

$$-\frac{2}{h(t)R^2}\int_0^R r\,dr \int_0^{h(t)} \frac{\partial p}{\partial z}\,dz = \frac{2\sigma}{h(t)R}. \tag{8}$$

Supposing that the flow rate at the tube inlet $v_z|_{z\to 0}\longrightarrow 0$ and at the meniscus interface $v_z(r,t)|_{z\to h(t)}\longrightarrow v_z(r,t)$, the averaged inertial convective term per Eq. (5) is

$$\frac{2}{h(t)R^2}\int_0^R r\,dr \int_0^{h(t)} \frac{\partial(v_z^2/2)}{\partial z}\,dz = \frac{2}{3}\frac{\langle v_z(t)\rangle^2}{h(t)} \tag{9}$$

For the averaged viscosity term,

$$\frac{2\mu}{h(t)R^2}\int_0^R \frac{1}{r}\frac{\partial}{\partial r}\left(r\frac{\partial u}{\partial r}\right)r\,dr \int_0^{h(t)} dz = -\frac{8\mu}{R^2}. \tag{10}$$

With the mass balance equation, and considering the possibility of phase change

$$\langle v_z(t)\rangle \equiv u = dh/dt + j/\rho, \tag{11}$$

where j is the interfacial mass flux due to the condensation process, given by [6].

$$j = \frac{k_l}{h_{lv}}\left(\frac{\partial T}{\partial z}\right)_{z=h(t)}. \tag{12}$$

In the above equation (12), k_l and h_{lv} are the thermal conductivity and the latent heat of vaporization of the liquid, respectively. Additionally, we need the liquid temperature field, T. Therefore, substituting Eqs. (7)-(10) into Eq. (4), we obtain the Lucas-Washburn equation that describes the rise of the liquid within the capillary [8]

$$\rho g h + \frac{8\mu}{R^2} h u + \frac{j^2}{\rho_v} + \rho \frac{d}{dt}(hu) = \frac{2\sigma}{R}, \tag{13}$$

with the following initial conditions

$$h(t = 0) = 0, \tag{14}$$

$$\left.\frac{dh}{dt}\right|_{t=0} = \left(\frac{2\sigma}{\rho R}\right)^{1/2}, \tag{15}$$

where the surface tension is assumed to be dependent on temperature through the relationship: $\sigma = \sigma_0 - \sigma_T(T - T_0)$, with σ_0, σ_T =constants> 0 [12]. It must be noted that Eq. (15) is obtained by a balance between inertia and capillary forces, ignoring the influence of viscosity and gravity close to the contact moment between the capillary and the liquid reservoir. Quéré [13] deduced this expression, analytically and experimentally, establishing that for a short time the height of the liquid column increases linearly.

In order to solve the Eq. (13), additionally we need the energy equation of the liquid phase:

$$\frac{\partial T}{\partial t} + u \frac{\partial T}{\partial z} = \alpha \frac{\partial^2 T}{\partial z^2}, \tag{16}$$

with the following initial and boundary conditions

$$T(z, t = 0) = T_0, \tag{17}$$

$$T(z = 0, t) = T_0, \tag{18}$$

$$T(z = h(t), t) = T_s. \tag{19}$$

In the above formulation we have assumed thermodynamic equilibrium condition at the interface, meaning that the phase change does not alter the interface temperature. In order to obtain the suitable characteristic scales for the problem, we conduct an order of magnitude analysis from the momentum and energy equations for the liquid penetrating the capillary tube. In this context, the competition between thermal and dynamics penetrations generates a non-isothermal capillary flow, which is developed inside the capillary tube. After an elapsed time t, the non-isothermal imbibition front reaches an average distance, $h(t)$. Therefore, the thermal and imbibition effects introduce two time scales: the thermal penetration scale, t_{th}, and the imbibition scale t_i, which will be determined in the Order of magnitude analysis section.

3. Order of magnitude analysis and theoretical analysis

In this section we carry out an order of magnitude analysis, similar to that used by O. Bautista et al. [14] and J. P. Escandon et al. [15], in order to determine the characteristic scales of this problem. From Eq. (13) we can readily identify, by using an order of magnitude analysis, one characteristic time scale, t_e, associated with the characteristic equilibrium height, h_e. Similarly, from Eq. (16), we recognize two characteristic time scales associated with the thermal energy transported by the imbibed fluid: a convective scale t_{conv} and a conductive scale t_{cond}. For estimating the characteristic scales mentioned before, we conduct the following order of magnitude analysis. From Eq. (13), the order of magnitude for the characteristic equilibrium time t_e of the imbibition process is determined by a balance between the surface tension and viscosity forces, as follows:

$$2R\sigma_0 \sim \frac{8\mu h_e}{t_e}, \tag{20}$$

where h_e can be easily evaluated from a balance between the surface tension and viscosity forces, that is,

$$2R\sigma_0 \sim 8\mu \left(\frac{h_e^2}{t_e} + \frac{h_e j}{\rho} \right). \tag{21}$$

Combining relationships (20) and (21), we obtain

$$h_e \sim \frac{2\sigma_0}{\rho g R} \left(\frac{1 - \frac{j^2 R}{2\sigma_0 \rho_v}}{1 + \frac{8\mu j}{R^2 \rho^2 g}} \right). \tag{22}$$

From this last relationship, the terms $j^2 R/2\sigma_0 \rho_v$ and $8\mu j/R^2 \rho^2 g$ are very small compared to the unity; in such case, the order of magnitude for the equilibrium height is $h_e \sim 2\sigma_0/\rho g R$, and the corresponding equilibrium time, from Eq. (20) is

$$t_e \sim 4\mu h_e^2/\sigma_0 R. \tag{23}$$

On the other hand, an energy balance between the transported thermal energy by the motion of the liquid and the accumulation energy term dictates that

$$\frac{\Delta T}{t_{conv}} \sim u_c \frac{\Delta T}{h_{th}}, \tag{24}$$

where, $\Delta T = T_s - T_0$ and u_c , h_{th} and t_{conv} represent the characteristic average velocity associated with the velocity of the imbibition front, the characteristic thermal penetration and the characteristic convective time scale, respectively. Therefore, from Eq. (24), the characteristic convective time scale is given as,

$$t_{conv} \sim \frac{h_{th}}{u_c}. \tag{25}$$

In a similar way, from Eq. (16), a balance between diffusive and accumulation terms, dictates that,

$$\frac{\Delta T}{t_{cond}} \sim \alpha \frac{\Delta T}{h_{th}^2}, \tag{26}$$

and using the above relationship, the characteristic diffusive time scale, t_{cond}, is given by

$$t_{cond} \sim \frac{h_{th}^2}{\alpha}. \tag{27}$$

To obtain the order of magnitude of the characteristic thermal penetration, we compare the convective and diffusive terms of the energy equation,

$$u_c \frac{\Delta T}{h_{th}} \sim \alpha \frac{\Delta T}{h_{th}^2}, \tag{28}$$

obtaining that

$$h_{th} \sim \frac{\alpha}{u_c}. \tag{29}$$

The characteristic average imbibition velocity, in a first approximation, is easily derived from Eq. (11), assuming that the mass flux j at the interface is very small, therefore

$$u_c \sim \frac{h_e}{t_e} = \frac{\rho g R^2}{8\mu\sigma_0}. \tag{30}$$

By considering Eqs. (23), (25) and (27), we obtain

$$t_e \ll \frac{t_{conv}}{t_{cond}} \sim 1. \tag{31}$$

The analysis of this transient heat transfer process can be characterized by adopting as characteristic time the imbibition time scale t_e. The characteristic scales determined previously will be used to nondimensionalize the governing equations properly in the following section.

4. Model formulation and numerical solution

Introducing the following dimensionless variables

$$\tau = \frac{t}{t_e}, \ \eta = \frac{z}{h(t)}, \ Y = \frac{h(t)}{h_e}, \ \theta = \frac{T_s - T}{T_s - T_0}, \tag{32}$$

the system of Eqs. (13)-(15) and (16)-(19), is transformed to

$$\varepsilon \frac{\partial}{\partial \tau} \left(Y \frac{dY}{d\tau} + \frac{Ja}{\beta} \frac{\partial \theta}{\partial \eta} \Big|_{\eta=1} \right) + Y + Y \frac{dY}{d\tau} + \frac{Ja}{\beta} \frac{\partial \theta}{\partial \eta} \Big|_{\eta=1} = 1 - \Gamma Ja, \tag{33}$$

where $\Gamma = h_{lv}\sigma_T/\sigma_0 c_p$

$$Y(\tau = 0) = 0, \tag{34}$$

$$\frac{dY}{d\tau} = \frac{1}{\varepsilon^{1/2}}, \tag{35}$$

and

$$\beta \frac{\partial \theta}{\partial \tau} + \left(\frac{\beta}{Y} \frac{dY}{d\tau} + \frac{Ja}{Y^2} \frac{\partial \theta}{\partial \eta} \Big|_{\eta=1} \right) \frac{\partial \theta}{\partial \eta} = \frac{1}{Y^2} \frac{\partial^2 \theta}{\partial \eta^2}, \tag{36}$$

together with the following initial and boundary conditions,

$$\theta(\tau = 0) = 1, \tag{37}$$

$$\theta(\eta = 0) = 1, \tag{38}$$

$$\theta(\eta = 1) = 0. \tag{39}$$

In the above equations, the dimensionless parameters ε, β and the Jakob number Ja are defined as

$$\varepsilon = \frac{\rho g R^2}{128\sigma_0} \frac{g R^3 \rho^2}{\mu^2}, \tag{40}$$

$$\beta = \frac{h_e}{h_{th}} = \frac{\sigma_0 R}{4\alpha\mu} \gg 1, \tag{41}$$

and

$$Ja = \frac{C_p(T_s - T_0)}{h_{lv}}. \tag{42}$$

The solution of the problem (33)-(39) shall provide $Y = Y(\tau; \varepsilon, \beta, Ja, \Gamma)$ and $\theta = \theta(\eta, \tau; \varepsilon, \beta, Ja, \Gamma)$. To solve the system of Eqs. (33)-(39), we use a regular perturbation method [16], based on the fact that in direct condensation, the characteristic values of the Jakob number are usually small. This permits to consider the Jakob number, Ja, as the parameter of perturbation. We assume the following perturbation expansions, in power of Ja, for the dimensionless imbibition front and dimensionless temperature:

$$Y = Y_0 + Ja\, Y_1 + \cdots \tag{43}$$

and

$$\theta = \theta_0 + Ja\, \theta_1 + \cdots, \tag{44}$$

where Y_0 and θ_0 are the leading-order solutions, i. e., the case where no phase change occurs at the interface; Y_1 and θ_1 are higher order corrections to the leading order that include the phase change. Substituting the expansions (43) and (44) in the system of Eqs. (33)-(39), and collecting terms of the same powers of Ja, we obtain, for $O\left(Ja^0\right)$:

$$\varepsilon \left(\frac{dY_0}{d\tau}\right)^2 + \varepsilon Y_0 \frac{d^2 Y_0}{d\tau^2} + Y_0 + Y_0 \frac{dY_0}{d\tau} = 1, \tag{45}$$

and

$$\beta Y_0^2 \frac{\partial \theta_0}{\partial \tau} + \beta Y_0 \frac{dY_0}{d\tau} \frac{\partial \theta_0}{\partial \eta} = \frac{\partial^2 \theta_0}{\partial \eta^2}. \tag{46}$$

The initial and boundary conditions associated with Eqs. (45) and (46) are, respectively

$$Y_0\,(\tau = 0) = 0, \tag{47}$$

$$\frac{dY_0\,(\tau = 0)}{d\tau} = \frac{1}{\varepsilon^{1/2}}, \tag{48}$$

$$\theta_0\,(\tau = 0) = 1, \tag{49}$$

$$\theta_0\,(\eta = 0) = 1, \tag{50}$$

$$\theta_0\,(\eta = 1) = 0. \tag{51}$$

For the first order solution, $O\left(Ja^1\right)$,

$$2\varepsilon \frac{dY_0}{d\tau}\frac{dY_1}{d\tau} + \varepsilon Y_1 \frac{d^2 Y_0}{d\tau^2} + \varepsilon Y_0 \frac{d^2 Y_1}{d\tau^2} + \frac{\varepsilon}{\beta}\frac{\partial}{\partial \tau}\left.\frac{\partial \theta_0}{\partial \eta}\right|_{\eta=1} \tag{52}$$

$$+ Y_1 + Y_1 \frac{dY_0}{d\tau} + Y_0 \frac{dY_1}{d\tau} + \frac{1}{\beta}\left.\frac{\partial \theta_0}{\partial \eta}\right|_{\eta=1} = -\Gamma$$

and

$$\beta^2 Y_0^2 \frac{\partial \theta_1}{\partial \tau} + 2\beta^2 Y_0 Y_1 \frac{\partial \theta_0}{\partial \tau} + \beta^2 Y_0 \frac{dY_0}{d\tau}\left[\frac{\partial \theta_1}{\partial \eta} + \left(1 + \beta^2 Y_1\right)\frac{\partial \theta_0}{\partial \eta}\right] = \frac{\partial^2 \theta_1}{\partial \eta^2}. \tag{53}$$

The initial and boundary conditions for Eqs. (52) and (53) are, respectively

$$Y_1(\tau = 0) = 0 \tag{54}$$

$$\left.\frac{dY_1}{d\tau}\right|_{\tau=0} = 0 \tag{55}$$

$$\theta_1(\tau = 0) = 0, \tag{56}$$

$$\theta_1(\eta = 0) = 0, \tag{57}$$

$$\theta_1(\eta = 1) = 0. \tag{58}$$

As a particular case and neglecting the inertial term, from Eq. (45), we can derive the well known case of the imbibition process without phase change, whose solution is given by,

$$Y_0 = 1 + W(x) \tag{59}$$

where, $W(x)$ represents the Lambert function [17], being x the argument of this function, given by $x = -\exp(-1 - \tau)$. Eq. (59) represents the zeroth-order solution for the momentum equation, without phase change and constant surface tension.

In order to obtain the zeroth-order solution for the dimensionless temperature in the liquid bulk, we apply the following finite difference scheme to the Eq. (46):

$$\theta_{0(i)}^{n+1} = \frac{1}{\beta\left(Y_{0(i)}^2\right)^2} \frac{\Delta\tau}{(\Delta\eta)^2} \left(\theta_{0(i+1)}^n - 2\theta_{0(i)}^n + \theta_{0(i-1)}^n\right) \tag{60}$$

$$-\frac{\left(Y_{0(i)}^{n+1} - Y_{0(i)}^n\right)}{Y_{0(i)}^n} + \left(\frac{\theta_{0(i+1)}^n - \theta_{0(i-1)}^n}{2\Delta\eta}\right) + \theta_{0(i)}^n$$

where, Y_0 is obtained by integrating numerically the Eq. (47) with the classical fourth order Runge-Kutta method. In Eq. (60), i is the spatial node in the liquid bulk and n denotes the time step. Introducing the numerical solution of Eqs. (47) and (60) in Eq. (52), we obtain the next higher order equation for the imbibition front, Y_1, which was solved by the conventional fourth order Runge-Kutta method. For simplicity, the details are omitted.

5. Results and discussion

In this chapter we conduct a semi-analytical approach to describe the imbibition process of a fluid into a capillary tube including the direct condensation at the imbibition front, considering that the surface tension at the liquid-vapor interface can be a function of the temperature. The present analysis serves to emphasize that the inertial terms in the momentum equation, the phase change at the imbibition front together with temperature-dependent surface tension effect, are very important for determining the temporal evolution of the imbibition front. This is reflected as a consequence of the main involved dimensionless parameters in the analysis: ε, the Jakob number, Ja and the parameter Γ. In all numerical calculations presented in this section, we have selected representative values for aforementioned parameters. Because the range of these parameters can be very large, the assumed values in our numerical results are representative of typical cases. The numerical results are plotted in Figs. 2-6. In these, we have emphasized the importance of the aforementioned dimensionless parameters on the imbibition process, and should be noted that the zeroth-order solution for the momentum equation is independent of the temperature field (see Eq. (46)). In Fig. 2 we evidence the importance of the parameter Γ on the imbibition front as a function of the dimensionless time. In this figure, the dimensionless height is plotted as a function of the dimensionless time for $\varepsilon = 0.5$, $\beta = 600$, $Ja = 0, 0.1, 0.3$. Here, it is very clear that the sensitivity of the imbibition front to the assumed values of the parameter Γ; for increasing values of above parameter, for a given dimensionless time, the dimensionless height of the imbibition front tends to decrease.

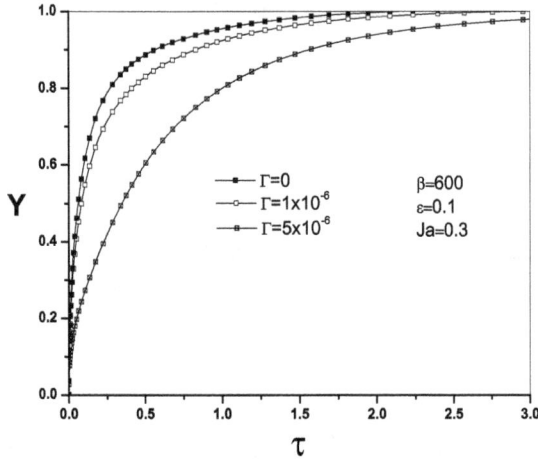

Figure 2. Dimensionless height of the imbibition front as a function of the dimensionless time for the case of $\varepsilon = 0.1$, $\beta = 600$, $Ja = 0.3$ and three values of the parameter $\Gamma (= 5 \times 10^{-6}, 1 \times 10^{-6}, 0.)$

In Fig. 3, the initial occurrence of oscillations is observed for the case of $\varepsilon = 0.5$. The oscillations are damped and tend to the Jurin's height ($h_e \sim 2\sigma / \rho g R$). The above result occurs for a dimensionless time $\tau \sim 4$, for the case of the ε used in this figure. From this figure it is clear that increasing the Jakob number increases the dimensionless height of the imbibition front. On the other hand, a similar behavior in comparison with Fig. 2 is appreciated by considering the effect of the parameter Γ; for increasing values of this parameter, the imbibition front decreases.

The influence of the parameter ε, is shown in Fig. 4. Here we plotted the imbibition height as a function of the dimensionless time, for $\beta = 600$, three values of ε and two values of the parameter Γ. Black and hollows symbols represent the cases of $\Gamma = 1 \times 10^{-6}$ and 1×10^{-5}, respectively. As can be seen, the parameter ε results in the occurrence of oscillations, which appear for a value of $\varepsilon = 0.5$. For higher values of this parameter, greater imbibition heights are reached. As can be appreciated, for values of the parameter ε, it takes a larger dimensionless time for reaching the equilibrium height, given by the Jurin's law. Therefore, the first term (inertial term) at the right hand of Eq. (33), has a great influence on the hydrodynamic solution of the present problem. From the physical point of view, this behavior occurs for liquids of very low viscosity or for increasing values of the capillary radius (see relationship (40)) for a given liquid.

In Fig. 5, the dimensionless interfacial velocity of the liquid as a function of the dimensionless time is plotted, for $\beta = 600$, $Ja = 0.1$, $\varepsilon = 0.1$ and four vales of the parameter Γ. This figure

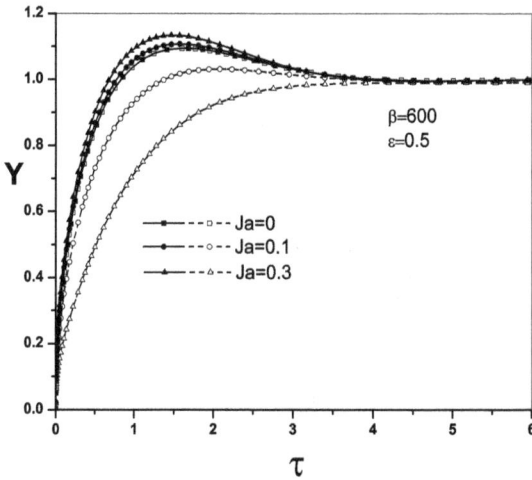

Figure 3. Dimensionless height of the imbibition front as a function of the dimensionless time for the case of $\varepsilon = 0.5, \beta = 600$, Ja(=0, 0.1, 0.3). Black and hollow symbols correspond to $\Gamma = 1 \times 10^{-6}$ and 1×10^{-5}, respectively.

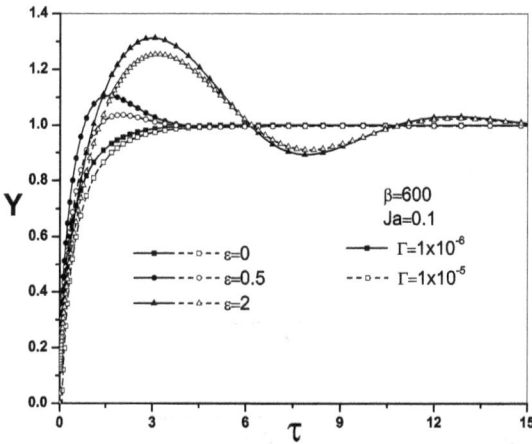

Figure 4. Dimensionless height of the imbibition front as a function of the dimensionless time for three values of ε (= 0,0.5,2), $\beta = 600$, Ja=0.1 and two values of Γ(=1 × 10^{-6}, 1 × 10^{-5}).

shows the case where no oscillations are present, and for the value of $\tau \sim 3$, the motion of the imbibition front is practically decelerated. From this figure, and according to the Fig. 2, increasing values of the parameter Γ tends to decrease the velocity of the imbibition front.

On the other hand, in Fig. 6, negative values for $dY/d\tau$ are obtained. In this figure, we show that for the case where oscillations are present, the equilibrium of the imbibition front occurs for larger dimensionless times in comparison with the case of $\varepsilon = 0$.

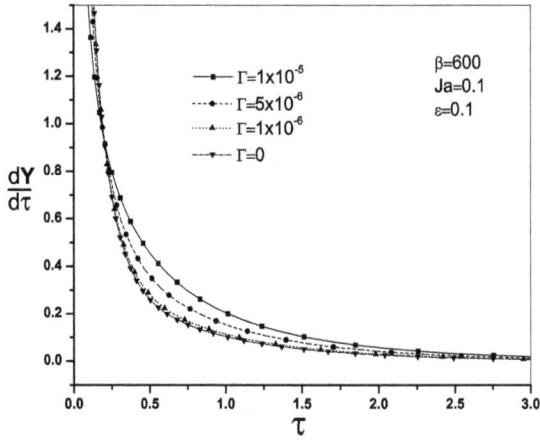

Figure 5. Dimensionless interfacial velocity as function of the dimensionless time, for $\beta = 600$, $Ja=0.1$, $\varepsilon = 0.1$, and four values of the parameter $\Gamma(=1\times10^{-5}, 5\times10^{-6}, 1\times10^{-6}, 0)$.

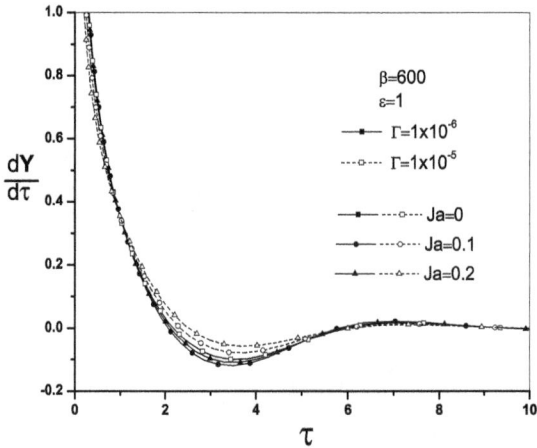

Figure 6. Dimensionless interfacial velocity as function of the dimensionless time, for $\beta = 600$, $\varepsilon = 1$, three values of the Jakob number ($Ja = 0, 0.1, 0.2$) and two values of the parameter $\Gamma(=1\times10^{-6}, 1\times10^{-5})$.

The effects of the Jakob number, the parameter ε and the dimensionless parameter Γ on the maximum dimensionless imbibition height are shown in Fig. 7. For the case of $Ja = 0$, the effect of the dimensionless parameter Γ has no influence on the imbibition front; however, for a fixed value of the Jakob number ($Ja \neq 0$), for increasing values of the parameter Γ, the maximum imbibition height decreases. This behavior is according to those shown in Figs. 2-4. From the same figure, $Ja = 0$ corresponds to that of isothermal case, i. e., there is no phase change process at the imbibition front.

The dimensionless temperature profiles in the liquid are shown in Fig. 8, as a function of the dimensionless coordinate η, for different values of the dimensionless time, two values of

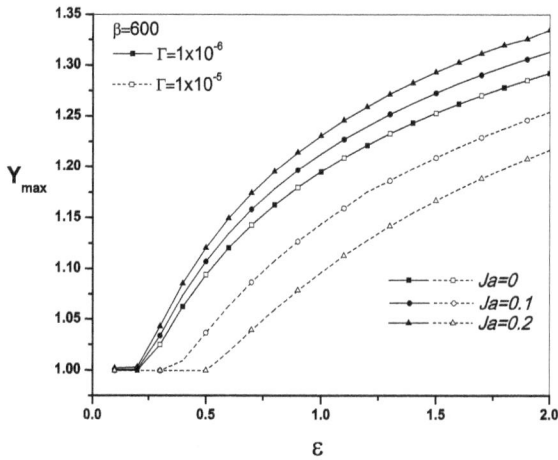

Figure 7. Maximum imbibition height as a function of the dimensionless parameter ε, for $\beta = 600$, three values of the Jakob number, $Ja(= 0, 0.1, 0.2)$ and two values of $\Gamma(= 1 \times 10^{-6}, 1 \times 10^{-5})$.

the parameter $\beta(= 600, 1000)$, with $\varepsilon = 0$. In this figure it is evident that the dimensionless temperature profile in the liquid shows a linear behavior for large values of the dimensionless time, whereas for short dimensionless time, large temperature gradients exist near the imbibition front.

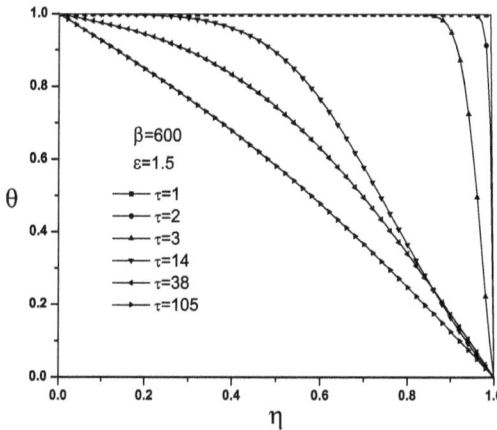

Figure 8. Dimensionless temperature of the liquid as a function of the dimensionless variable η, evaluated at different dimensionless times, $\beta = 600$ and $\varepsilon = 1.5$

Acknowledgements

This work has been supported by a research grant no. 20120706 of SIP-IPN at Mexico.

Author details

Bautista Oscar, Bautista Eric and Arcos Carlos
National Polytechnic Institute of Mexico, College of Mechanical and Electrical Engineering (ESIME Azcapotzalco), Deparment of thermofluids, México D. F., Mexico

Méndez Federico and Sánchez Salvador
National University of Mexico (UNAM), Division of Mechanical and Electrical Engineering, Department of Thermofluids, México D. F., Mexico

6. References

[1] E. Begg, D. Khrustalev, A. Faghri (1999) Complete Condensation of Forced Convection Two-Phase Flow in a Miniature Tube. J. Heat Transfer 121: 904-915.

[2] L. P. Yarin, A. Mosyak, G. Hestroni (2009) Fluid Flow, Heat Transfer and Boiling in Micro Channels. Springer.

[3] L.P. Yarin, L.A. Ekelchik, G. Hetsroni (2002) Two-phase laminar flow in a heated microchannels. Int. J. Multiphase Flow. 28: 1589-1616.

[4] W. Qu, T. Ma, J. Miao, J. Wang (2002) Effects of radius and heat transfer on the profile of evaporating thin liquid film and meniscus in capillary tubes. Int. J. Heat and Mass Transfer. 45: 1879-1887.

[5] R. Lucas (1918) Ueber das Zeitgesetz des Kapillaren Aufstiegs von Flussigkeiten. Kolloidn Zh. 23: 15-22.

[6] A. Faghri, Y. Zhang (2006) Transport Phenomena in Multiphase Systems. Elsevier Academic Press. pp. 209-210.

[7] E. W. Washburn (1921) The Dynamics of Capillary Flow. Phys. Rev. 17: 273-283.

[8] Guy Ramon, Alexander Oron (2008) Capillary rise of a meniscus with phase change. J. Colloid Interface Sci. 327: 145-151.

[9] Shong-Leih Lee, Hong-Draw Lee (2007) Evolution of liquid meniscus shape in a capillary tube. J. Fluids Eng. 129: 957-965.

[10] Stanley Middleman (1995) Modeling Axisymmetric Flows. Dynamics of Films, Jets, and Drops. Academic Press. numero ppppp.

[11] A. A. Duarte, D. E. Strier, and D. H. Zanette (1995), The rise of a liquid in a capillary tube revisited: A hydrodynamical approach. Am. J. Phys. 64 (4): 413-418.

[12] Yu. O., Kabova, V.V. Kuznetsov, O.A. Kabov (2012), Temperature dependent viscosity and surface tension effects on deformations of non-isothermal falling liquid film. International Journal of Heat and Mass Transfer. 55:1271-1278.

[13] D. Quéré (1997) Inertial capillarity. Europhys. Lett. 39: 3679-3682.

[14] O. Bautista, F. Méndez, E. G. Bautista (2010) Thermal dispersion driven by the spontaneous imbibition front. Applied Mathematical Modelling. 34: 4184-4195.

[15] J. P. Escandón, O. Bautista, F. Méndez, E. G. Bautista (2009) Influence of the wicking process on the heat transfer in a homogeneous porous medium. Journal of Petroleum Science and Engineering. 67: 91-96.

[16] C. M. Bender and S. A. Orzag (1999) Advanced Mathematical Methods for Scientists and Engineers: Asymptotic Methods and Perturbation Theory. Springer. pp. 319-367.

[17] N. Fries, M. Dreyer (2008) An analytic solution of capillary rise restrained by gravity. J. Colloid Interface Sci. 320: 259-263.

Chapter 14

The Effects of Power Characteristics on the Heat Transfer Process in Various Types of Motionless Mixing Devices

Rafał Rakoczy, Marian Kordas and Stanisław Masiuk

Additional information is available at the end of the chapter

1. Introduction

Mixing is probably the most frequently used chemical engineering operation in industry. It should be noticed that a good quality of mixing is generally very difficult to achieve. Static mixers, also known as motionless mixers, are usually applied in various types of chemical engineering fields. In recent decades, static mixers have been found in a wide range of industries. These mixers are designed to mix fluids without any recourse to a the mechanical motion and static mixers may be used as an alternative to mechanical agitators. A motionless mixer is a specially designed geometrical structure of different shapes and forms of the elements inserted in a long pipe. The number of elements and the type of a mixing device both depend on the mixing degree in any applications and a wide range of tasks. A static mixer consists of a contacting device with a series of internal stationary mixing elements of a specific geometry, inserted in a pipe. By using a specific geometrical combination of these elements one-after-another, the additional turbulence of the flowed fluid contributes to the improvement of mixing efficiency.

The benefits of a static mixer include low operating and maintenance costs. Morever, it needs little place, no mechanical seals, it is self-cleaning and low shear forces prevent excessive stressing of the product especially in bio-processes. By using a static mixer it is easy to control the output product quality. Furthermore, static mixers are characterized by the low energy consumption and an easy usage and installation.

The use of static mixers as alternative mixing devices for various fluids has been analysed in many scientific researches (Ruivo et al., 2006; Regner et al., 2005). Static mixers are nowadays used in many chemical processes as in-line contacting devices for gases and liquids. Typical applications range from the reactor in polymer systems (Fourcade et al., 2001), heat or mass transfer alternative devices (Simões et al., 2008), mixing liquid-liquid systems in the laminar or turbulent flow (Rama Rao et al., 2007) to intensifying gas-liquid mass transfer processes (Al

Taweel et al., 2005). Moreover, static mixers have a wide range of applications, such as continuous mixing, chemical reactors or mass and heat transfer processes or a thermal homogenization (Li et al., 1996; Visser et al., 1999; Lemenand et al., 2010). Several theoretical and experimental studies of the pressure drop are presented in (Liu et al., 2006; Engler et al., 2004; Hobbs et al., 1998; Kumar et al., 2008; Li et al., 1997; Joshi et al., 1995). The applications of static mixers to improve the heat transfer can be divided into the thermal homogenization (in many cases coupled with the compositional homogenization), a pure heat transfer in heat exchangers and a combined heat transfer with chemical reactions (Thakur et al., 2003).

In the case of the mixing operation, the pressure drop is used as a pertinent criterion to highlight the differences between various types of static mixers. This parameter is very common to all applications of static mixers. As follows form the literature (Thakur et al., 2003) the key parameter for the heat transfer process in static mixers is the Nusselt number. Azer and Lin defined the parameter which takes the heat transfer enhancement and the pressure drop increase into consideration (Thakur et al., 2003).

The main objective of the present work is to propose a new criterion for the selection of a static mixer in heat transfer experiments. This parameter is taking into account the hydrodynamic behaviour of the flow in a static mixer by using the fluid mechanic equations. Additionally, the proposed criterion is applied to the various types of static mixers (EMI, Ross, Komax and the patented construction of the static mixer (Masiuk & Szymański, 1997).

2. Theoretical background

The mathematical description of heat transfer operations in tubular static mixers is given by the energy equation. For a steady flow of an incompressible fluid with constant or averaged transport properties or physical parameters it gives

$$\rho c_p \left(\frac{\partial T}{\partial t} + \overline{w} \, grad T \right) = -T \left(\frac{\partial p}{\partial T} \right)_V div \overline{w} + div \left(k \, grad T \right) + \frac{dp}{dt} + \Phi_V \tag{1}$$

In the cylindrical coordinate system (r, θ, z), Eq.(1) can be rewritten in the equivalent form

$$\rho c_p \left[\begin{array}{l} \dfrac{\partial T(t;r,\theta,z)}{\partial t} + \\ + \left(w_r \overline{e_r} + w_\theta \overline{e_\theta} + w_z \overline{e_z} \right) \left(\dfrac{\partial T(t;r,\theta,z)}{\partial r} \overline{e_r} + \dfrac{1}{r} \dfrac{\partial T(t;r,\theta,z)}{\partial \theta} \overline{e_\theta} + \dfrac{\partial T(t;r,\theta,z)}{\partial z} \overline{e_z} \right) \end{array} \right] =$$

$$= -T(t;r,\theta,z) \left(\frac{\partial p(t;r,\theta,z)}{\partial T} \right)_V \left(\frac{1}{r} \frac{\partial(rw_r)}{\partial r} + \frac{1}{r} \frac{\partial w_\theta}{\partial \theta} + \frac{\partial w_z}{\partial z} \right) +$$

$$+ k \left[\frac{1}{r} \frac{\partial}{\partial r} \left(r \frac{\partial T(t;r,\theta,z)}{\partial r} \right) + \frac{1}{r^2} \frac{\partial^2 T(t;r,\theta,z)}{\partial \theta^2} + \frac{\partial^2 T(t;r,\theta,z)}{\partial z^2} \right] + \frac{\partial p(t;r,\theta,z)}{\partial t} +$$

$$+ \left(w_r \overline{e_r} + w_\theta \overline{e_\theta} + w_z \overline{e_z} \right) \left(\frac{\partial p(t;r,\theta,z)}{\partial r} \overline{e_r} + \frac{1}{r} \frac{\partial p(t;r,\theta,z)}{\partial \theta} \overline{e_\theta} + \frac{\partial p(t;r,\theta,z)}{\partial z} \overline{e_z} \right) + \Phi_V \tag{2}$$

In the above Eq.(2), the viscous dissipation function , Φ_V , may be expressed as follows

$$
\Phi_V = 2\eta \left[\left(\frac{\partial w_r}{\partial r} \right)^2 + \left(\frac{1}{r}\frac{\partial w_\theta}{\partial \theta} + \frac{\partial w_r}{\partial r} \right)^2 + \left(\frac{\partial w_z}{\partial z} \right)^2 \right] +
$$

$$
+\eta \left\{ \left[r\frac{\partial}{\partial r}\left(\frac{w_\theta}{r} \right) + \frac{1}{r}\frac{\partial w_r}{\partial \theta} \right]^2 + \left[\frac{1}{r}\frac{\partial w_z}{\partial \theta} + \frac{\partial w_\theta}{\partial z} \right]^2 + \left[\frac{\partial w_r}{\partial z} + \frac{\partial w_z}{\partial r} \right]^2 \right\} +
$$

$$
+ \left(\eta_V - \frac{2}{3}\eta \right)\left[\frac{1}{r}\frac{\partial(rw_r)}{\partial r} + \frac{1}{r}\frac{\partial w_\theta}{\partial \theta} + \frac{\partial w_z}{\partial z} \right]^2 \qquad (3)
$$

The practical application of the proposed Eq.(2) is limited by a complicated structure of this relation. Therefore, this relationship can be simplified to a more efficient form. It should be noticed that in this case the following assumptions are taken into consideration:

- for cylindrical coordinates

$$
r = [0, r_0]; \quad \theta = [0, 2\pi]; \quad z = [0, L] \qquad (4a)
$$

- for the steady-state of heat transfer process

$$
\frac{\partial T(t;r,\theta,z)}{\partial t} = 0; \quad \frac{\partial p(t;r,\theta,z)}{\partial t} = 0 \qquad (4b)
$$

- for the incompressible fluid

$$
div\,\overline{w}(r,\theta,z) = 0 \qquad (4c)
$$

- for physical parameters of the fluid are calculated for the averaged temperature of the mixed liquid

$$
\left(\rho, c_p, \eta, k \right) = f\left(\left[T_{in} \right]_{avg} \right) \qquad (4d)
$$

- for the axial symmetry of the fluid flow

$$
\left(T, \overline{w}, p \right) \neq f(\theta); \quad w_r = 0 \quad if \quad r = 0 \wedge r = r_0 \qquad (4e)
$$

- for the pressure drop in the radial direction

$$
\frac{\partial p}{\partial r} \cong 0 \qquad (4f)
$$

- for the temperature of steam (heating medium)

$$T_{steam} = const \tag{4g}$$

- for the velocity (the z coordinate represents the main flow direction)

$$w_z = f\left(\overset{\bullet}{m}\right) \tag{4h}$$

- for the temperature

$$T(r,z) = f\left(\overset{\bullet}{m},r,z\right) \tag{4i}$$

- for the pressure

$$p(z) = f\left(\overset{\bullet}{m},z\right) \tag{4j}$$

- for the temperature driving force

$$\Delta\widetilde{T}_{in}\left(\overset{\bullet}{m},r,z\right) = \left[\widetilde{T}_{w,in}\left(\overset{\bullet}{m},r,z\right) - \widetilde{T}_{in}\left(\overset{\bullet}{m},r,z\right)\right] \tag{4k}$$

- for the viscous dissipation function Φ_V

$$\Phi_V = 0 \tag{4l}$$

The proposed Eq.(2) may be simplified satisfying the proposed assumptions and conditions (see Eqs (4a-4l)). Thus, the rewritten form of Eq.(2) in the cylindrical coordinate system may be expressed as follows

$$\rho c_p w_r \frac{\partial \widetilde{T}_{in}\left(\overset{\bullet}{m},r,z\right)}{\partial r} + \rho c_p w_z\left(\overset{\bullet}{m}\right)\frac{\partial \widetilde{T}_{in}\left(\overset{\bullet}{m},r,z\right)}{\partial z} = k\left(\frac{\partial^2 \widetilde{T}_{in}\left(\overset{\bullet}{m},r,z\right)}{\partial r^2} + \frac{1}{r}\frac{\partial \widetilde{T}_{in}\left(\overset{\bullet}{m},r,z\right)}{\partial r}\right) +$$

$$+k\frac{\partial^2 \widetilde{T}_{in}\left(\overset{\bullet}{m},r,z\right)}{\partial z^2} + w_z\left(\overset{\bullet}{m}\right)\frac{\partial p\left(\overset{\bullet}{m},z\right)}{\partial z} \tag{5}$$

taking into account the mass flow rate $\overset{\bullet}{m}$.

The graphical presentation of the obtained Eq.(5) is presented in Fig.1.

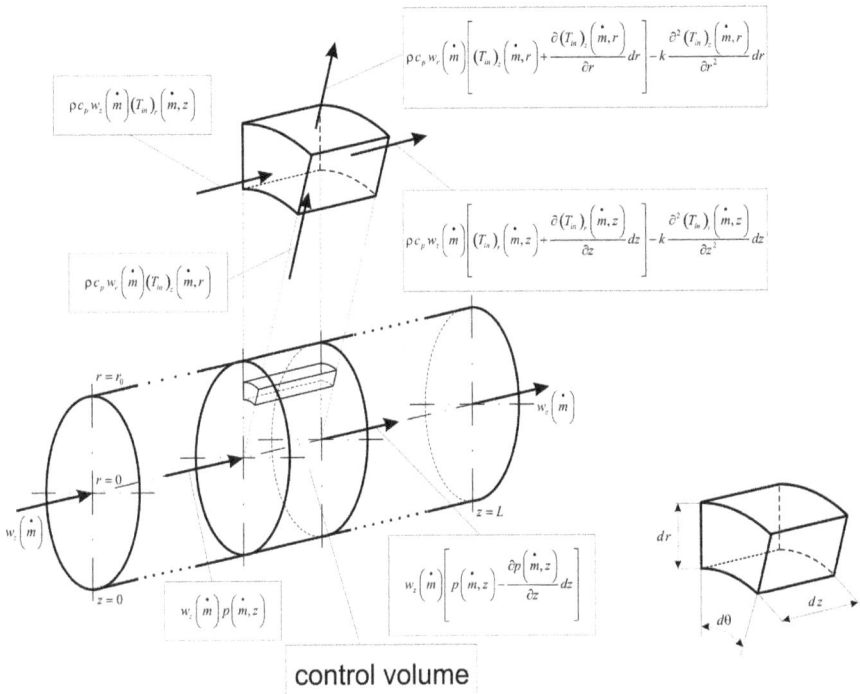

control volume

Figure 1. Graphical presentation of Eq.(5)

The term $k\left(\dfrac{\partial^2 \widetilde{T_{in}}\left(\overset{\bullet}{m},r,z\right)}{\partial r^2} + \dfrac{1}{r}\dfrac{\partial \widetilde{T_{in}}\left(\overset{\bullet}{m},r,z\right)}{\partial r}\right)$ in the above Eq.(5) represents the heat

conduction in the radial direction. This term may be defined by means of the heat transfer coefficient as follows

$$\left\{ \int_0^{r_0} k\left[\frac{\partial^2 \widetilde{T_{in}}\left(\overset{\bullet}{m},r,z\right)}{\partial r^2} + \frac{1}{r}\frac{\partial \widetilde{T_{in}}\left(\overset{\bullet}{m},r,z\right)}{\partial r}\right] dr = \int_0^{r_0} \partial\left(kr\frac{\partial \widetilde{T_{in}}\left(\overset{\bullet}{m},r,z\right)}{\partial r}\right) dr = kr\frac{\partial \widetilde{T_{in}}\left(\overset{\bullet}{m},r,z\right)}{\partial r}\Bigg|_{r=0}^{r=r_0}\right\} =$$

$$= r_0 h\left(\overset{\bullet}{m}\right)\left[\widetilde{T_{in}}\left(\overset{\bullet}{m},r,z\right)\right]_{avg} \tag{6}$$

Taking into consideration the above relation (Eq.(6)) and the definition of the averaged temperature driving force

$$\left[\widetilde{T}_{in}\left(\overset{\bullet}{m},r,z\right)\right]_{avg} = \frac{\int_0^{r_0} w_r \widetilde{T}_{in}\left(\overset{\bullet}{m},r,z\right)rdr}{\int_0^{r_0} w_r\, rdr} \tag{7}$$

we obtain the following form of Eq.(6)

$$r_0 h\left(\overset{\bullet}{m}\right)\frac{\int_0^{r_0} w_r \widetilde{T}_{in}\left(\overset{\bullet}{m},r,z\right)rdr}{\int_0^{r_0} w_r\, rdr} = -\int_0^{r_0} \rho c_p w_r \frac{\partial \widetilde{T}_{in}\left(\overset{\bullet}{m},r,z\right)}{\partial r}rdr -$$

$$+\int_0^{r_0}\left[\rho c_p w_z\left(\overset{\bullet}{m}\right)\frac{\partial \widetilde{T}_{in}\left(\overset{\bullet}{m},r,z\right)}{\partial z}-k\frac{\partial^2 \widetilde{T}_{in}\left(\overset{\bullet}{m},r,z\right)}{\partial z^2}\right]rdr + \int_0^{r_0} w_z\left(\overset{\bullet}{m}\right)\frac{\partial p\left(\overset{\bullet}{m},z\right)}{\partial z}rdr \tag{8}$$

According to the established conditions (Eqs (4a-4l)) the first term of the right side of Eq.(8) is defined as follows

$$\int_0^{r_0} \rho c_p w_r \frac{\partial \widetilde{T}_{in}\left(\overset{\bullet}{m},r,z\right)}{\partial r}rdr = 0 \tag{9}$$

Moreover, the integral $\int_0^{r_0}\widetilde{T}_{in}\left(\overset{\bullet}{m},r,z\right)rdr$ in Eq.(8) may be expressed in the following form (for $r\in\left(0,r_0\right)$)

$$\int_0^{r_0}\widetilde{T}_{in}\left(\overset{\bullet}{m},r,z\right)rdr = T_{w,in}\left(\overset{\bullet}{m},r=r_0,z\right)-T_{in}\left(\overset{\bullet}{m},r=r_0,z\right)\Rightarrow \widetilde{T}_{in}\left(\overset{\bullet}{m},z\right) \tag{10}$$

Introducing relations (9) and (10) in the Eq.(8) gives the differential form of the energy balance equation

$$\frac{2}{r_0}h\left(\overset{\bullet}{m}\right)\widetilde{T}_{in}\left(\overset{\bullet}{m},z\right) = -\rho c_p w_z\left(\overset{\bullet}{m}\right)\frac{\partial \widetilde{T}_{in}\left(\overset{\bullet}{m},z\right)}{\partial z}+k\frac{\partial^2 \widetilde{T}_{in}\left(\overset{\bullet}{m},z\right)}{\partial z^2}+w_z\left(\overset{\bullet}{m}\right)\frac{\partial p\left(\overset{\bullet}{m},z\right)}{\partial z} \tag{11}$$

The dependence of the liquid density on the temperature may be taken into consideration by applying the following expression

$$\rho\left(\widetilde{T_{in}}\left(\overset{\bullet}{m},z\right)\right)=\rho\left(1+\beta\widetilde{T_{in}}\left(\overset{\bullet}{m},z\right)\right) \tag{12}$$

Then, Eq.(11) may be written as follows

$$\frac{2}{r_0}h\left(\overset{\bullet}{m}\right)\widetilde{T_{in}}\left(\overset{\bullet}{m},z\right)=$$

$$=-\rho\left(1+\beta\widetilde{T_{in}}\left(\overset{\bullet}{m},z\right)\right)c_p w_z\left(\overset{\bullet}{m}\right)\frac{\partial\widetilde{T_{in}}\left(\overset{\bullet}{m},z\right)}{\partial z}+k\frac{\partial^2\widetilde{T_{in}}\left(\overset{\bullet}{m},z\right)}{\partial z^2}+w_z\left(\overset{\bullet}{m}\right)\frac{\partial p\left(\overset{\bullet}{m},z\right)}{\partial z} \tag{13}$$

The governing Eq.(13) may be rewritten in a symbolic shape which is useful for a dimensionless analysis. The introduction of non-dimensional quantities denoted by sign $\left({}^{*}\right)$ into this relationship yields

$$\left(2\frac{h_0\widetilde{T_{in_0}}}{r_0}\right)\left[h^*\left(\overset{\bullet}{m}\right)\widetilde{T_{in}}^*\left(\overset{\bullet}{m},z\right)\right]=$$

$$=-\left(\frac{\rho_0 c_{p_0}w_{z_0}\widetilde{T_{in_0}}}{l_0}+\frac{\rho_0\beta_0\widetilde{T_{in_0}}c_{p_0}w_{z_0}\widetilde{T_{in_0}}}{l_0}\right)\left[\rho^*\left(1+\beta^*\widetilde{T_{in}}^*\left(\overset{\bullet}{m},z\right)\right)c_p^*w_z^*\left(\overset{\bullet}{m}\right)\frac{\partial\widetilde{T_{in}}^*\left(\overset{\bullet}{m},z\right)}{\partial z^*}\right]+ \tag{14}$$

$$+\left(\frac{k_0\widetilde{T_{in_0}}}{l_0^2}\right)\left[k^*\frac{\partial^2\widetilde{T_{in}}^*\left(\overset{\bullet}{m},z\right)}{\partial z^*\partial z^*}\right]+\left(\frac{w_{z_0}\Delta p_0}{l_0}\right)\left[w_z^*\left(\overset{\bullet}{m}\right)\frac{\partial p^*\left(\overset{\bullet}{m},z\right)}{\partial z^*}\right]$$

The temperature $\widetilde{T_{in_0}}$ in the above Eq.(14) is defined by using the following relationship

$$\frac{1}{l_0}\int_0^{l_0}\widetilde{T_{in}}\left(\overset{\bullet}{m},z\right)dz=\widetilde{T_{in_0}}\left(\overset{\bullet}{m}\right)\Rightarrow\widetilde{T_{in_0}} \tag{15}$$

The non-dimensional form of Eq.(14) may be scaled against the convective term $\left(\frac{k_0\widetilde{T_{in_0}}}{l_0^2}\right)$.

The dimensionless form of this equation includes the following dimensionless groups characterizing the heat transfer process in a static mixer

$$\left(2\frac{h_0\widetilde{T_{in_0}}}{r_0}\cdot\frac{l_0^2}{k_0\widetilde{T_{in_0}}}\right)\Rightarrow\left(4\frac{h_0 d_0}{k_0}\left(\frac{l_0}{d_0}\right)^2\right)\Rightarrow 4\text{Nu}\left(\frac{l_0}{d_0}\right)^2 \tag{16a}$$

$$\left(\frac{\rho_0 c_{p_0} w_{z_0} \widetilde{T_{in_0}}}{l_0} \cdot \frac{l_0^2}{k_0 \widetilde{T_{in_0}}} + \frac{\rho_0 \beta_0 \widetilde{T_{in_0}} c_{p_0} w_{z_0} \widetilde{T_{in_0}}}{l_0} \cdot \frac{l_0^2}{k_0 \widetilde{T_{in_0}}}\right) \Rightarrow \left(\frac{\rho_0 c_{p_0} w_{z_0} l_0}{k_0} + \frac{\rho_0 \beta_0 \widetilde{T_{in_0}} c_{p_0} w_{z_0} l_0}{k_0}\right) \Rightarrow$$

$$\Rightarrow \left(\frac{\rho_0 w_{z_0} d_0}{\eta_0} \cdot \frac{c_{p_0} \eta_0}{k_0} \cdot \frac{l_0}{d_0} + \frac{\rho_0^2 \beta_0 g_0 \widetilde{T_{in_0}} l_0^3}{\eta_0^2} \cdot \frac{\eta_0}{\rho_0 w_{z_0} d_0} \cdot \frac{w_{z_0}^2}{g_0 l_0} \cdot \frac{c_{p_0} \eta_0}{k_0} \cdot \frac{d_0}{l_0}\right) \Rightarrow \qquad (16b)$$

$$\Rightarrow \left(\operatorname{Re}\operatorname{Pr}\frac{l_0}{d_0}\left(1 + \frac{\operatorname{Gr}\operatorname{Fr}}{\operatorname{Re}^2}\left(\frac{d_0}{l_0}\right)^2\right)\right)$$

$$\left(\frac{w_{z_0}\Delta p_0}{l_0} \cdot \frac{l_0^2}{k_0 \widetilde{T_{in_0}}}\right) \Rightarrow \left(\frac{\Delta p_0}{\rho_0 w_{z_0}^2} \cdot \frac{\eta_0 w_{z_0}^2}{k_0 \widetilde{T_{in_0}}} \cdot \frac{\rho_0 w_{z_0} d_0}{\eta_0} \cdot \frac{l_0}{d_0}\right) \Rightarrow \left(\operatorname{Eu}\operatorname{Br}\operatorname{Re}\left(\frac{l_0}{d_0}\right)\right) \qquad (16c)$$

Taking into account the proposed relations (16a)-(16c), we find the following dimensionless governing equation

$$\left(4\operatorname{Nu}\left(\frac{l_0}{d_0}\right)^2\right)\left[h^*\left(\overset{\bullet}{m}\right)\widetilde{T}_{in}^{\,*}\left(\overset{\bullet}{m},z\right)\right] =$$

$$= -\left(\operatorname{Re}\operatorname{Pr}\frac{l_0}{d_0}\left(1 + \frac{\operatorname{Gr}\operatorname{Fr}}{\operatorname{Re}^2}\left(\frac{d_0}{l_0}\right)^2\right)\right)\left[\rho^*\left(1 + \beta^*\widetilde{T}_{in}^{\,*}\left(\overset{\bullet}{m},z\right)\right)c_p^* w_z^*\left(\overset{\bullet}{m}\right)\frac{\partial \widetilde{T}_{in}^{\,*}\left(\overset{\bullet}{m},z\right)}{\partial z^*}\right] + \qquad (17)$$

$$+ \left[k^*\frac{\partial^2 \widetilde{T}_{in}^{\,*}\left(\overset{\bullet}{m},z\right)}{\partial z^*\partial z^*}\right] + \left(\operatorname{Eu}\operatorname{Br}\operatorname{Re}\left(\frac{l_0}{d_0}\right)\right)\left[w_z^*\left(\overset{\bullet}{m}\right)\frac{\partial p^*\left(\overset{\bullet}{m},z\right)}{\partial z^*}\right]$$

The above non-dimensional relationship (see Eq.(17)) is obtained by means of a scaling analysis. Scaling means non-dimensionalization of an equation describing the system under study. In this case this technically simple procedure is applied in order to obtain a dimensionless equation and relations between relevant dimensionless numbers, which are used to describe the effects of power characteristics on the heat transfer process in a motionless mixing device. It should be noticed that the proposed dimensionless equation (see Eq.(17)) has certain advantages, such as: the obtained dimensionless numbers are relevant for the analysed problem, the proportion between the individual terms may be established and the relationships that indicate the relevant proportions between these terms may be worked out.

It should be noticed that the proposed Eq.(17) consists of the combinations of the dimensionless numbers containing the heat transfer and the mixing process. The complex structure of this relation results from the combination of momentum and heat transports in a static mixer. Moreover, the establishment of the new criterion for the selection of a static mixer in heat transfer investigations is available by using the elaborated form of Eq.(17). The dimensionless numbers or their combinations may be used for making correlations, which are commonly used for design, scale-up and optimization purposes. In these correlations, some quantities or effects may be neglected. Moreover, the proposed correlations are presented as a linear relation between complex dimensionless terms.

From the dimensionless form of Eq.(17) it follows that the dimensionless Nusselt number (this number is commonly applied to the correlations for the heat transfer problems) may be separately compared with the dimensionless numbers terms. It should be noticed that the Nusselt number estimates the heat passing through the interface without computing the temperature and velocity fields. Therefore, the relationships obtained for the analysed problem may be expressed in the following form taking into account that the term $\left(\mathrm{Eu\,Br\,Re}\left(\dfrac{l_0}{d_0}\right) \right)$ is neglected

$$\left(\mathrm{Nu}\left(\frac{l_0}{d_0}\right)^2\right) \propto \left(\mathrm{Re\,Pr}\frac{l_0}{d_0}\left(1+\frac{\mathrm{Gr\,Fr}}{\mathrm{Re}^2}\left(\frac{d_0}{l_0}\right)^2\right)\right) \Rightarrow \mathrm{Nu} \propto \mathrm{Re\,Pr}\frac{d_0}{l_0}\left(1+\frac{\mathrm{Gr\,Fr}}{\mathrm{Re}^2}\left(\frac{d_0}{l_0}\right)^2\right) \Rightarrow$$

$$\Rightarrow \mathrm{Nu} \propto \mathrm{Pe}\frac{d_0}{l_0}\left(1+\frac{\mathrm{Gr\,Fr}}{\mathrm{Re}^2}\left(\frac{d_0}{l_0}\right)^2\right)$$

(18)

or when the term $\left| \mathrm{Re\,Pr}\dfrac{l_0}{d_0}\left(1+\dfrac{\mathrm{Gr\,Fr}}{\mathrm{Re}^2}\left(\dfrac{d_0}{l_0}\right)^2\right) \right|$ is neglected

$$\left(\mathrm{Eu\,Br\,Re}\left(\frac{l_0}{d_0}\right)\right) \propto \left(\mathrm{Re\,Pr}\frac{l_0}{d_0}\left(1+\frac{\mathrm{Gr\,Fr}}{\mathrm{Re}^2}\left(\frac{d_0}{l_0}\right)^2\right)\right) \Rightarrow \mathrm{Eu} \propto \frac{\mathrm{Re\,Pr}}{\mathrm{Br\,Re}}\left(1+\frac{\mathrm{Gr\,Fr}}{\mathrm{Re}^2}\left(\frac{d_0}{l_0}\right)^2\right) \Rightarrow$$

$$\Rightarrow \mathrm{Eu} \propto \frac{\mathrm{Pe}}{\mathrm{Br}}\left(\frac{1}{\mathrm{Re}}+\frac{\mathrm{Gr\,Fr}}{\mathrm{Re}^3}\left(\frac{d_0}{l_0}\right)^2\right)$$

(19)

When the operational objective is the heat transfer process the key parameter may be defined as follows

$$\varepsilon = \frac{\dot{Q}}{N} \Rightarrow \varepsilon = \frac{hF\widetilde{T_{in}}}{V\Delta p} \Rightarrow \varepsilon = \frac{\pi dl h\widetilde{T_{in}}}{\frac{\pi d^2}{4}w\Delta p} \Rightarrow \varepsilon = 4\left(\frac{hd}{k}\right)\cdot\left(\frac{\rho w^2}{\Delta p}\right)\cdot\left(\frac{k\widetilde{T_{in}}}{\eta w^2}\right)\cdot\left(\frac{\eta}{\rho w d}\right)\cdot\left(\frac{l}{d}\right) \Rightarrow$$

$$\Rightarrow \varepsilon = 4\frac{\mathrm{Nu}}{\mathrm{Eu\,Br\,Re}}\left(\frac{l}{d}\right)$$

(20)

The proposed criterion (see Eq.(20)) takes both the heat transfer enhancement and the pressure drop into account. According to the obtained Eq.(17), we find the following relationship

$$\left(4\mathrm{Nu}\left(\frac{l_0}{d_0}\right)^2\right) \propto \left(\mathrm{Eu}\,\mathrm{Br}\,\mathrm{Re}\left(\frac{l_0}{d_0}\right)\right) \Rightarrow 4\mathrm{Nu}\left(\frac{l_0}{d_0}\right)^2 = \varepsilon\mathrm{Eu}\,\mathrm{Br}\,\mathrm{Re}\left(\frac{l_0}{d_0}\right) \Rightarrow$$

$$\Rightarrow \varepsilon = 4\frac{\mathrm{Nu}}{\mathrm{Eu}\,\mathrm{Br}\,\mathrm{Re}}\left(\frac{l_0}{d_0}\right)$$

(21)

It should be noticed that the proposed criterion ε (see Eq.(21)) is worked out by applying the theoretical analysis of the non-dimensional form of Eq.(17). The proposed parameter ε may be used as a criterion suitable for the applications of static mixers as the heat exchangers. Form the theoretical point of view, this criterion is dependent on hydrodynamic conditions (the dimensionless Reynolds number) and the real effect of viscous dissipation on the heat transfer process (the dimensionless Brinkman number). Table 1 summarizes all the essential and independent dimensionless numbers given in Eqs (17-21).

3. Experimental details

3.1. Experimental investigations of pressure drop in static mixers

Experimental studies of the pressure drop device were carried out using the static mixer experimental set-up shown in Fig.2.

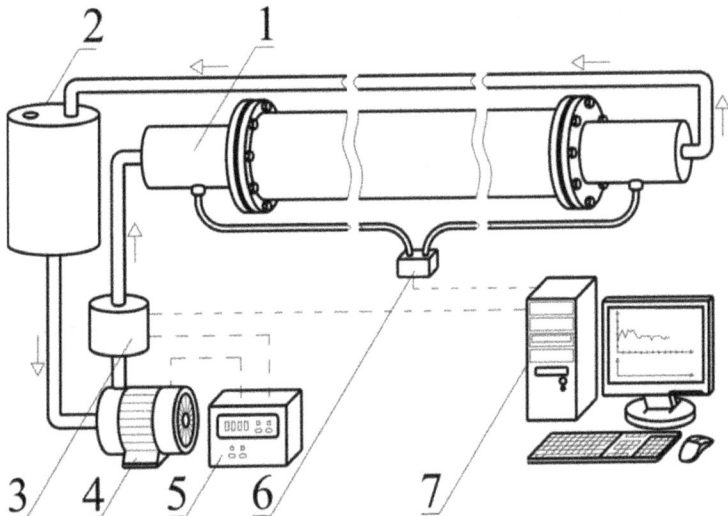

Figure 2. Sketch of experimental set-up: 1 – static mixer, 2 – storage vessel, 3 – electromagnetic flow meter, 4 – circulating pump, 5 – electronic control box, 6 – pressure drop converter FCX-C, 7 – personal computer

Name	Symbol	Definition	Significance	Interpretation and Remarks
Brinkman	Br	$\dfrac{\eta_0 w_{z_0}^2}{k_0 \widetilde{T_{in_0}}}$	$\dfrac{\text{momentum transfer}}{\text{conductive heat transfer}}$	Dimensionless number related from a wall to a flowing fluid. This number is a measure of the importance of the viscous heating relative the conductive heat transfer.
Euler	Eu	$\dfrac{\Delta p_0}{\rho_0 w_{z_0}^2}$	$\dfrac{\text{pressure force}}{\text{inertial force}}$	It express the relationship between a local pressure drop and the kinetic energy.
Froude	Fr	$\dfrac{w_{z_0}^2}{g_0 l_0}$	$\dfrac{\text{inertial force}}{\text{gravitational force}}$	Dimensionless number defined as a ratio of characteristic velocity to a gravitational wave velocity.
Grashof	Gr	$\dfrac{\rho_0^2 \beta_0 g_0 \widetilde{T_{in_0}} l_0^3}{\eta_0^2}$	$\dfrac{\text{buoyancy force}}{\text{viscous force}}$	It approximates the ratio of the buoyancy to viscous force acting on a fluid.
Nusselt	Nu	$\dfrac{h_0 d_0}{k_0}$	$\dfrac{\text{convective heat transfer}}{\text{conductive heat transfer}}$	This number is the ration of convective to conductive heat transfer across the boundary.
Péclet	Pe	$\dfrac{\rho_0 w_{z_0} d_0 c_{p_0}}{k_0}$	$\dfrac{\text{rate of advection}}{\text{rate of diffusion}}$	This number is defined to be the ratio of the rate of advection to rate of diffusion of the same quantity driven by appropriate gradient. The presented number is applied in the case of the diffusion of heat (thermal diffusion).
Prandtl	Pr	$\dfrac{c_{p_0} \eta_0}{k_0}$	$\dfrac{\text{momentum diffusivity}}{\text{thermal diffusivity}}$	It defines the ratio of momentum diffusivity and thermal diffusivity.
Reynolds	Re	$\dfrac{\rho_0 w_{z_0} d_0}{\eta_0}$	$\dfrac{\text{inertial force / convection}}{\text{viscous force}}$	This number gives a measure of the ratio of inertial forces to viscous forces.

Table 1. Dimensionless numbers in Eqs (15-19) and their physical role

All the experimental work was performed using the static mixer in a horizontal tube closed on both ends by cylindrical caps. Inside the tube the mixing elements were placed to mix the flowing fluid. The cylindrical cups had central inlet and outlet valves. On the lower part of these cups the outlet valve and valves for the digital pressure drop converter FCX-CII (FHCV11V2-AKABY-AA, Fuji Electric France S.A., Clermont-Ferrand, France) were attached. This converter was used to measure the pressure drop of the fluid between the ends of the horizontal tube. On the upper part of the cylindrical cups venting valves was mounted. The experimental installation worked in a closed circuit. The mixed fluid (water) was sucked by a circulation pump from a buffer container and pumped through a flow meter to the tube of the static mixer. The fluid flow in the mixer's tube was measured by the electromagnetic flow meter (MPP-04, ENCO S.A., Warszawa, Poland) installed at the mixer's inlet.

3.2. Experimental investigations of the heat transfer process in the static mixer

The static mixer experimental set-up for the investigations of the heat transfer process is presented in Fig.3. This apparatus consists of an external heat exchanger, a circulating pump, an electrical steam generator and temperature sensors connected to a digital measuring equipment.

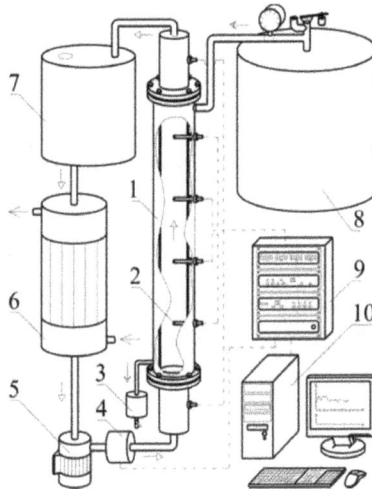

Figure 3. Static mixer of experimental set-up: 1 – static mixer with heating jacket, 2 – temperature sensors, 3 – condenser pot, 4 – electromagnetic flow meter, 5 – circulating pump, 6 – heat exchanger, 7 – storage vessel, 8 – steam generator, 9 – temperature sampling system, 10 – personal computer

The heat transfer enhancement was determined by measuring the profiles on each side of the heating pipe as well as the temperature field inside the static mixer. The heat from the condensing steam in the heating jacket was introduced to the flowing and mixed liquid by the novel mixing device. One system of the temperature sensors was tightened closely to both sides of the inner tube and other bunches of small sensors were placed inside the bulk of the static mixer. These sensors were used to measure the distribution of the mixer bulk temperature across the radius and the length of the inner tube. According to the theoretical consideration, the experimental investigations were realized for the steady-state condition of the heat transfer process. The measured temperature distribution was undertaken to determine the augmentation of the heat transfer by using the tested mixing device. Water was used as the liquid flowing through the inner pipe of the static mixer. The mass flow rate was measured by means of the electromagnetic flow-meter. The temperature within the mixed liquid varied significantly along the tube axis and increased much with the variation of the mass flow rate. The variations of the temperature in the cross section of the inner tube were so small, that the recorded distribution of the temperature might be accepted as no existing. The experimental set-up was equipped with a measuring instrument which controlled the temperature of the liquid and supervised the real-time acquisition of all the experimental data coming from the sensors. This device also measured the temperature fluctuations inside the static mixer during the process. Electric signals were sampled by using the special thermal sensors (LM-61B, National Semiconductor Corporation, Santa Clara, USA) and were passed through the converter (PCI-1710HG, Advantech, Milpitas, USA) to a personal computer for further processing.

3.3. The applied type of motionless inserts (elements)

Figure 4 shows schematic diagrams of motionless inserts used in the present work. The main purpose of these elements was to redistribute fluid in the radial and tangential directions transverse to the main flow. In the case of this experimental work, the enhancement of the heat transfer process was realized by applying the industrial mixing devices (EMI, Ross, Komax).

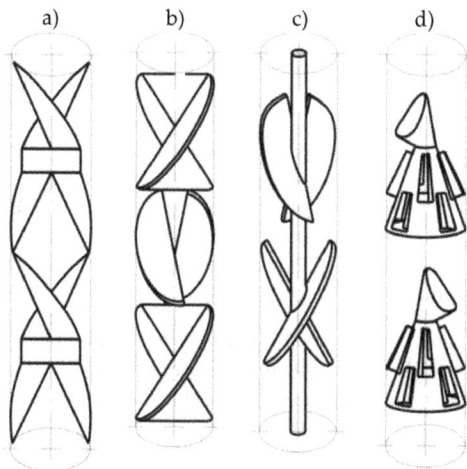

Figure 4. Applied motionless inserts: a) Komax, b) Ross, c) EMI, d) new type of mixing element

Moreover, the measurements were also carried out with the static mixer equipped with a new mixing insert, which was a modified version of the mixer described in the patent registration (Masiuk & Szymański, 1997) (see Fig.4d). The single mixing element had two truncated cones connected with each other. The diameter of the bigger cone was equal to the inner diameter of the apparatus. On the side surface of the inlet cone the longitudinal slots were placed symmetrically along the circumference. The slots were equipped with steering blades placed on their longitudinal edges. The side surface of the inlet cone had an oval shape near its lower base.

A stream of fluid flowing into the inlet cone was redistributed - the slots on the surface of the cone were steering the flow outside, while the steering blades bestowed the arising streams with the rotational movement. A part of the stream leaving the inside of the insert through the outlet cone became turbulent due to the cone's oval shape. The inner and outer streams were mixed before entering the next mixing element where the streams were redistributed once again, rotated in the opposite direction, and reconnected.

A novel mixing device was built up from five separate segments (see Fig.4d) in the form of two top opened cones. The large cone had longitudinal holes on the conical face with motionless outside baffles. The experiments were carried out with the cascade including 5 mixing elements. They were arranged in the tube of the static mixer along its full length. Neighbouring mixing elements differed in the spin direction of the blades and in the magnitude of the bulge of the outlet cone and were alternatively directed against each other. A good mixing and an increase of fluid homogeneity were achieved for the fluid that passed through the system of the mixing elements.

Table 2 gives the ranges of the operating conditions and the calculated dimensionless numbers for an empty duct and for static mixing elements.

4. Results and discussion

From the practical point of view the heat transfer problems may be modelled by means of the well-known Nusselt type equation

$$Nu = f(Re, Pr) \tag{22}$$

In the present report we consider the heat transfer in static mixers may be described by a similar but somewhat modified relationship between the dimensionless Nusselt number and the combination of numbers which were defined by the hydrodynamic effects in the tested experimental set-up containing different types of motionless devices.

The generalization of the results of the convective heat transfer measurements was correlated by using the worked out Nusselt-type equation (see Eq.(18)). It should be noticed that the proposed relation was obtained from the theoretical considerations. Therefore, the

dependency $Nu \propto Pe \dfrac{d_0}{l_0}\left(1 + \dfrac{Gr\,Fr}{Re^2}\left(\dfrac{d_0}{l_0}\right)^2\right)$ was gained as the linear relation. Evidently,

type of static mixer (or empty duct)	$d[m]$	$l[m]$	$T_{w,in}[°C]$	$[T_{m,avg}][°C]$	$\Delta p[Nm^{-2}]$	$m[kgs^{-1}]$	$h[Wm^{-2}K^{-1}]$	$\rho[kgm^{-3}]$	$c_p[Jkg^{-1}K^{-1}]$
empty duct			66.7-98.4	31-67.1	0.36-4.08	0.0125-0.2611	129-286	980-996	4173-4186
EMI			75.1-98.5	56-88.5	10.55-111.74	0.0104-0.2611	358-1009	969-987	4178-4204
Ross	0.1	1	72.5-99.8	46-94.5	9.37-98.84	0.0039-0.2611	210-824	966-991	4174-4208
Komax			75.5-98.5	56-88.7	7.89-86.79	0.0104-0.2611	357-1143	969-986	4178-4203
new type of mixing element (Masiuk & Szymański, 1997)			58.8-93.7	44-65.6	0.14-195.2	0.04-0.25	857-2078	980-991	4176-4187

type of static mixer (or empty duct)	$k[Wm^{-1}K^{-1}]$	$\eta[kgm^{-1}s^{-1}]$	$\beta[K^{-1}]$	$u_z[ms^{-1}]$	Re	Pr	Pe	Gr	Br
empty duct	0.6165-0.6638	0.000379-0.000736	0.000299-0.000586	0.0016-0.0151	419-2044	2.39-4.98	1004-10181	$1.94 \cdot 10^8 - 11.89 \cdot 10^8$	$4.87 \cdot 10^{-11} - 7.51 \cdot 10^{-9}$
EMI	0.6528-0.6759	0.000284-0.000437	0.000513-0.000692	0.0014-0.0168	466-3923	1.77-2.79	824-10553	$4.69 \cdot 10^8 - 8.52 \cdot 10^8$	$7.84 \cdot 10^{-11} - 1.11 \cdot 10^{-8}$
Ross	0.6402-0.6768	0.000244-0.000518	0.000437-0.000714	0.000523-0.0152	207-2901	1.51-3.38	314-9800	$4.15 \cdot 10^8 - 7.58 \cdot 10^8$	$1.86 \cdot 10^{-11} - 7.03 \cdot 10^{-9}$
Komax	0.6528-0.6759	0.00028-0.000437	0.000513-0.000693	0.0014-0.0168	468-2786	1.76-2.79	823-10594	$4.99 \cdot 10^8 - 8.59 \cdot 10^8$	$7.97 \cdot 10^{-11} - 9.67 \cdot 10^{-9}$
new type of mixing element (Masiuk & Szymański, 1997)	0.63-0.66	0.0004-0.0006	0.00042-0.00058	0.0052-0.0232	1273-5308	2.53-3.97	3227-21078	$1.68 \cdot 10^{11} - 9.56 \cdot 10^{11}$	$5.82 \cdot 10^{-10} - 6.55 \cdot 10^{-8}$

type of static mixer (or empty duct)	Fr	Eu	Nu	$Pe\left(\dfrac{1}{Br}+\dfrac{GrFr}{Re^3}\left(\dfrac{d_0}{l_0}\right)^2\right)$	$Pe\dfrac{d_0}{l_0}\left(1+\dfrac{GrFr}{Re^2}\left(\dfrac{d_0}{l_0}\right)^2\right)$
empty duct	$1.65 \cdot 10^{-5} - 0.00015$	1.23-10.74	19.52-46.35	$0.066 \cdot 10^{10} - 4.91 \cdot 10^{10}$	101-1018
EMI	$1.39 \cdot 10^{-5} - 0.00017$	33.06-313.14	53.09-153.84	$0.024 \cdot 10^{10} - 2.25 \cdot 10^{10}$	82-1056
Ross	$5.33 \cdot 10^{-6} - 0.00015$	29.83-278.21	31.09-128.84	$0.048 \cdot 10^{10} - 8.15 \cdot 10^{10}$	31-981
Komax	$1.39 \cdot 10^{-5} - 0.00017$	26.14-234	52-176	$0.029 \cdot 10^{10} - 2.21 \cdot 10^{10}$	82-1060
new type of mixing element (Masiuk & Szymański, 1997)	$5.29 \cdot 10^{-5} - 0.00033$	1.1-8	129-330	$0.006 \cdot 10^{10} - 0.44 \cdot 10^{10}$	423-2149

Table 2. The tabulated ranges of the operating conditions and the calculated dimensionless numbers

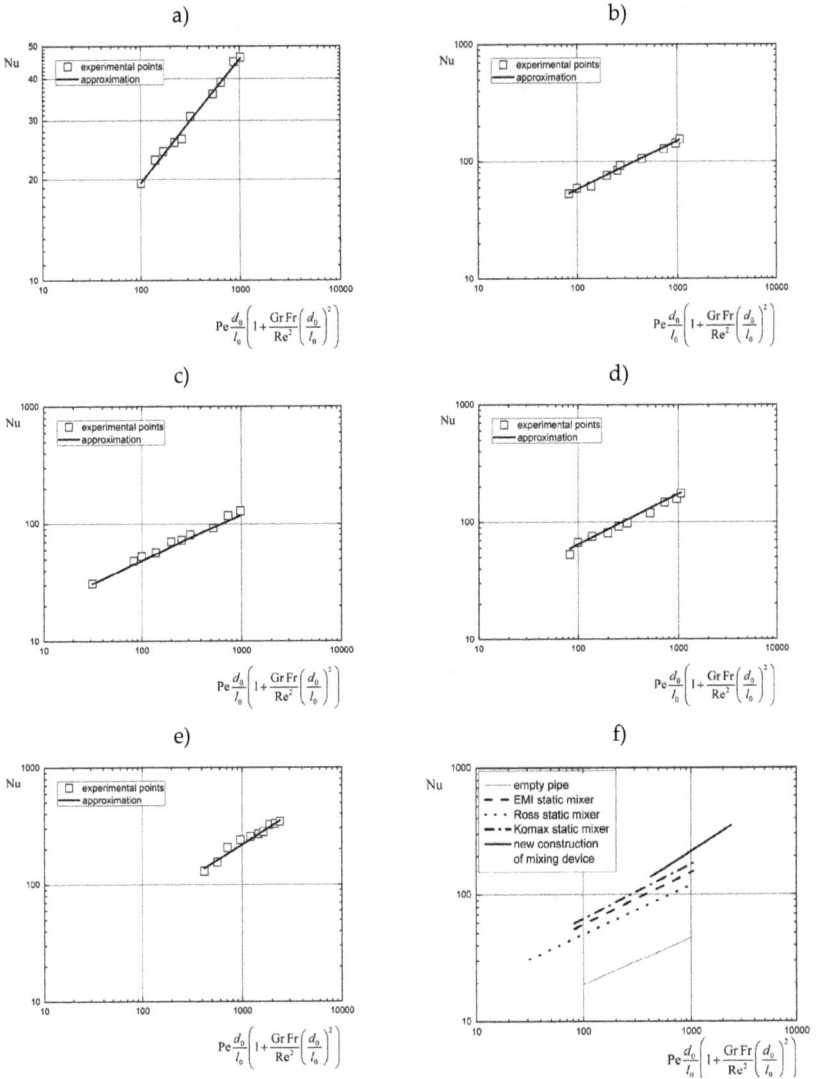

Figure 5. The generalization of the results of the convective heat transfer measurements:
a) empty pipe, b) EMI static mixer, c) Ross static mixer, d) Komax static mixer, e) new construction of motionless mixing device, f) comparison of the heat transfer correlations for the different type of motionless mixing device

Eq.(18) must be rendered quantitatively to be practically used. This may be done by assuming that the functional relation is in the following form

$$\mathrm{Nu} = f\left[\mathrm{Pe}\frac{d_0}{l_0}\left(1+\frac{\mathrm{Gr\,Fr}}{\mathrm{Re}^2}\left(\frac{d_0}{l_0}\right)^2\right)\right] \Rightarrow \mathrm{Nu} = p_1\left[\mathrm{Pe}\frac{d_0}{l_0}\left(1+\frac{\mathrm{Gr\,Fr}}{\mathrm{Re}^2}\left(\frac{d_0}{l_0}\right)^2\right)\right]^{p_2} \tag{23}$$

In order to establish the effect of the term $\left[\mathrm{Pe}\frac{d_0}{l_0}\left(1+\frac{\mathrm{Gr\,Fr}}{\mathrm{Re}^2}\left(\frac{d_0}{l_0}\right)^2\right)\right]$ on the dimensionless

Nusselt number the experimental data obtained in this work are graphically illustrated in the log-log system in Fig. 5.

The experimental results shown in Fig.5a-5e suggest that the heat transfer process in various types of motionless inserts may be analytically described by a unique monotonic function (see Eq.(23)). The constants and exponents were computed by employing the Matlab software and the principle of least squares and the obtained values are collected in Table 3. The calculated values of the approximation errors $(MAE, MRE, MPE, \sigma, R)$ are also presented in this table.

type of static mixer (or empty duct	Eq.(23)		approximation errors				
	p_1	p_2	MAE	MRE	MPE	σ	R
empty duct	3.55	0.37	0.62	0.02	2.05%	9.14	0.99
EMI	8.79	0.41	2.52	0.03	2.74%	35.07	0.99
Ross	8.05	0.39	4.58	0.06	5.57%	28.71	0.99
Komax	8.89	0.43	6.01	0.06	6.04%	41.48	0.99
new type of mixing element (Masiuk & Szymański, 1997)	5.24	0.54	10.34	0.05	4.58%	70.79	0.98

Table 3. The values of the parameters Eq.(23) with the list of approximation errors

It can be seen from Fig.5a-5e that the scatter of experimental points for various types of a motionless mixer may be described by the same type of relation (Eq.(23)) by using various values of the coefficients p_1 and p_2. It is clear that the heat transfer process is dependent on the geometrical configuration of the applied motionless insert. Therefore, the difference between coefficients in the proposed relation (23) (see Table 3) is due to different geometries of static mixers.

The enhancement of the heat transfer process can be attributed to the breakdown of the thermal boundary layer especially near the tube wall. It should be noticed that the enhancement is provided by the motionless mixer in which the higher heat flux or the heat transfer rate is achieved without the deterioration of the flowing fluid.

The comparison of the results of our own investigations is presented in Fig.5f. This figure shows that the significant enhancement in the heat transfer performance for the applied static mixers is obtained with respect to the empty tube. It can be seen that the values of the

Nusselt number for the new type of the static mixer are greater than the values of this dimensionless number obtained in this work for different types of the motionless devices (EMI, Ross and Komax). It should be noticed that the incensement of the heat transfer ratio is connected with the flow regime. The observed discrepancies between the obtained heat transfer characteristics increase with the intensification of the hydrodynamic conditions expressed by the term $\left[Pe\dfrac{d_0}{l_0}\left(1+\dfrac{Gr\,Fr}{Re^2}\left(\dfrac{d_0}{l_0}\right)^2\right)\right]$. The new construction of the mixing device

is given a modest improvement in the heat transfer rate but it is achieved by higher values of the mentioned term. Thus, standard motionless mixers (EMI, Ross or Komax) may be more useful to enhnace the heat transfer process for a lower value of the term. As follows from the analysis of Fig.5f, the new construction of the motionless insert is more effective for the turbulent region of the fluid flow. This leads to the conclusion that the heat transfer rate is mainly dependent on the geometrical configuration of the motionless insert and the generated hydrodynamic conditions in the flowing fluid through the static mixer.

It should be noticed that the hydrodynamic conditions in motionless mixers are strongly dependent on the pressure drop. The pressure drop database may be directly correlated by using the friction factor or the ratio of friction factors. From the practical point of view, this database may be elaborated by using the dimensionless Euler number. It can avoid confusion when comparing the static mixer to the empty tube or different types of motionless devices.

In the present report, the pressure drop for different types of static mixers is described by the relationship between the dimensionless Euler number and the obtained complex of the dimensionless numbers (see Eq.(19)). Accordingly to the obtained relationship and the above considerations the experimental data may be worked out by means of the following relationship

$$Eu = f\left[\frac{Pe}{Br}\left(\frac{1}{Re}+\frac{Gr\,Fr}{Re^3}\left(\frac{d_0}{l_0}\right)^2\right)\right] \Rightarrow Eu = p_3\left[\frac{Pe}{Br}\left(\frac{1}{Re}+\frac{Gr\,Fr}{Re^3}\left(\frac{d_0}{l_0}\right)^2\right)\right]^{p_4} \qquad (24)$$

Figure 6 illustrates the manner in which the Euler number varies in the operational range describing the term

$\left[\dfrac{Pe}{Br}\left(\dfrac{1}{Re}+\dfrac{Gr\,Fr}{Re^3}\left(\dfrac{d_0}{l_0}\right)^2\right)\right]$. The dashed lines in Fig.6a-6e are calculated from the general

Eq.(24), where the relation between the dimensionless numbers is obtained from the theoretical considerations. In the case of these experimental investigations, the established relation (24) was also proposed as a power function. This form is universally applied in the description of the heat and momentum transfer problem.

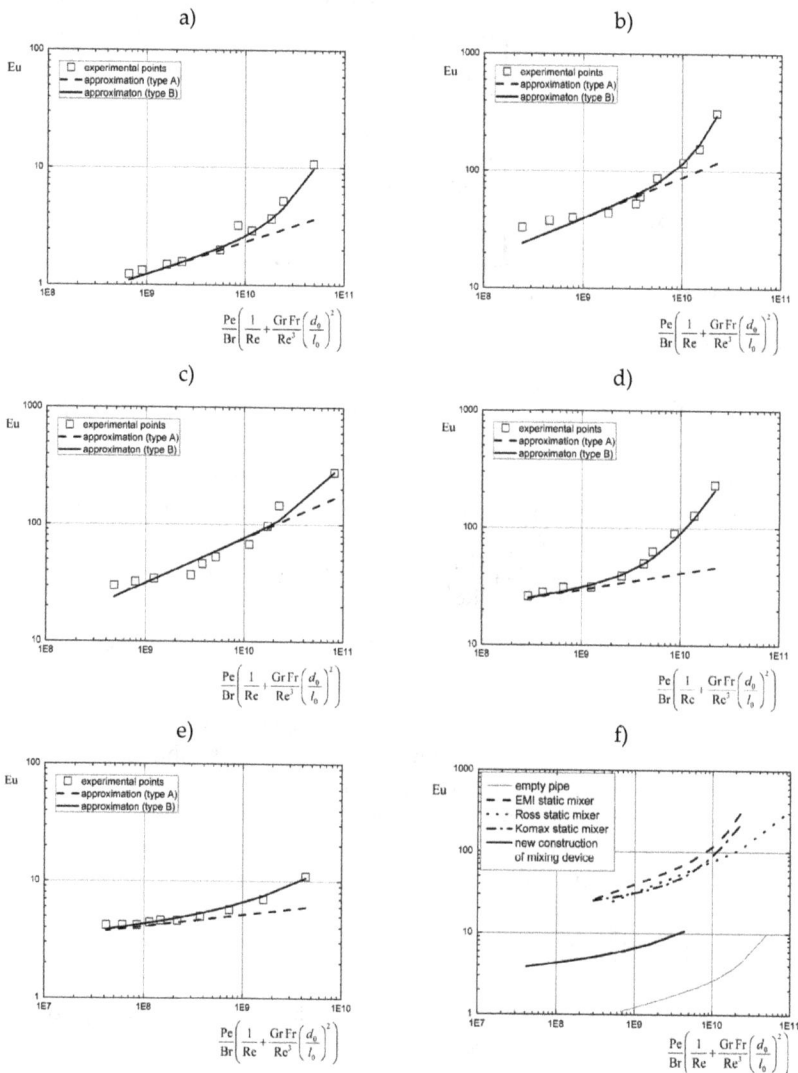

Figure 6. The generalization of the results of the power characteristic: a) empty pipe, b) EMI static mixer, c) Ross static mixer, d) Komax static mixer, e) new construction of motionless mixing device, f) comparison o the heat transfer correlations for the different type of motionless mixing device

The constants and exponents were computed by employing the Matlab software and the principle of least squares. The list of the parameter values (p_3, p_4) and the calculated values of the approximated errors are tabulated in the table 4 (approximation type A in Fig.6).

type of static mixer (or empty duct	Eq.(24)		approximation errors				
	p_3	p_4	MAE	MRE	MPE	σ	R
empty duct	0.004	0.28	1.21	0.21	21%	2.16	0,88
EMI	0.03	0.35	31.05	0.13	13%	65.62	0.89
Ross	0.012	0.38	19.99	0.04	3.57%	61.91	0.92
Komax	1.63	0.14	37.59	0.31	30.59%	49.29	0.85
new type of mixing element Masiuk & Szymański, 1997)	0.65	0.10	0.93	0.13	12.62%	1.61	0.91

Table 4. The values of the parameters Eq.(24) with the list of approximation errors

As follows from the graphical presentation of the obtained results (see Fig.6a-6e) the power characteristic as the $Eu \propto \left[\dfrac{Pe}{Br} \left(\dfrac{1}{Re} + \dfrac{Gr\,Fr}{Re^3} \left(\dfrac{d_0}{l_0} \right)^2 \right) \right]$ curve is a very curvilinear dependence.

Initially, the results given in Fig.6a-6e indicate the enhancement in the dimensionless Euler number with an increase of the complex dimensionless numbers. Up to a certain critical value of this complex $\left(\left[\dfrac{Pe}{Br} \left(\dfrac{1}{Re} + \dfrac{Gr\,Fr}{Re^3} \left(\dfrac{d_0}{l_0} \right)^2 \right) \right] \in \left(10^9, 10^{10} \right) \right)$, the experimental points follow a straight line and can be correlated by using the analytical description given by the proposed Eq.(24) (approximation type A in Fig.6).

It should be noticed that the dimensionless Euler number increases sharply with the increase of the complex dimensionless numbers and considerably higher increase takes place in the region of large values of this complex. The experimental results shown in Fig.6 suggest that the Euler number versus the mentioned term of the dimensionless relationship can be analytically described by a unique monotonic function including a special correction term. It should be noticed that the dimensionless Euler number versus $\left[\dfrac{Pe}{Br} \left(\dfrac{1}{Re} + \dfrac{Gr\,Fr}{Re^3} \left(\dfrac{d_0}{l_0} \right)^2 \right) \right]$ term is described by a similar relationship that is proposed in the relevant literature (Masiuk, 2000; Masiuk, 1999; Masiuk & Rakoczy, 2007). Then, the previously proposed relation may be adapted in the following general dimensionless correlation

$$Eu = p_3 \left[\frac{Pe}{Br} \left(\frac{1}{Re} + \frac{Gr\,Fr}{Re^3} \left(\frac{d_0}{l_0} \right)^2 \right) \right]^{p_4} \left\{ 1 + p_5 \left[\frac{Pe}{Br} \left(\frac{1}{Re} + \frac{Gr\,Fr}{Re^3} \left(\frac{d_0}{l_0} \right)^2 \right) \right]^{p_6} \right\} \qquad (25)$$

The constants and exponents in the proposed relation (25) depend on the geometrical configuration of the motionless mixer and their values are collected in Table 5.

type of static mixer (or empty duct	Eq.(25)				approximation errors				
	p_3	p_4	p_5	p_6	MAE	MRE	MPE	σ	R
empty duct	0.004	0.28	$3 \cdot 10^{-18}$	1.66	0.29	0.073	7.23%	2.68	0.99
EMI	0.03	0.35	$3 \cdot 10^{-21}$	2	6.41	0.008	0.78%	83.57	0.99
Ross	0.012	0.38	$1 \cdot 10^{-22}$	2	8.42	0.012	1.29%	74.57	0.98
Komax	1.63	0.14	$1.5 \cdot 10^{-14}$	1.39	5.56	0.049	4.91%	60.51	0.99
new type of mixing element (Masiuk & Szymański, 1997)	0.65	0.10	$0.9 \cdot 10^{-7}$	0.72	0.23	0.008	0.82%	2.061	0.99

Table 5. The values of the parameters Eq.(25) with the list of approximation errors

Figure 6 also illustrates the graphical forms of the above Eq.(25) for the tested motionless mixers as the full curves marked by the solid lines (approximation type B). Fig.6 demonstrates that, within scatter limits among the plotted data represented by the points, the dimensionless Euler number increases with the increase of the mentioned complex term. The first conclusion drawn from the inspection of this graph is that the proposed relationship in a general form of Eq.(25) fits the analysed experimental data very well. One of the advantages of the elaborated relation is the possibility to apply it to generalize the experimental results in the whole range of the fluid flow. It should be noticed that the complex form of the proposed relationship (Eq.(25)) results from the obtained hydrodynamic conditions for the analysed static mixers.

The comparison of the obtained results for various types of the motionless mixers is presented in Fig.6f. This figure shows that the calculated values of the dimensionless Euler number are strongly dependent on the term $\left[\dfrac{Pe}{Br}\left(\dfrac{1}{Re} + \dfrac{Gr\,Fr}{Re^3}\left(\dfrac{d_0}{l_0}\right)^2\right)\right]$. This plot also confirms that the geometrical configuration of the motionless insert has a significant effect on the pressure drop expressed as the mentioned non-dimensional number. It should be noticed that the pressure drop is the most important design criterion for a motionless mixer. It permits to quantify the disspated power in this mixer type. Moreover, the pressure drop expressed as the dimensionless Euler number is also a decisive factor for the estimation of the efficiency of motionless mixers. This number is dependent on several parametrs, such as: flow rates, the type of fluid and the geometrical configuration of a motionless insert of a static mixer (e.g. type, number or arrangements of element of this type of mixer).

In this case study, the pressure drops were measured for various types of static mixers. As follows from Fig.6f the power characteristic for the new type of the mixing device is similar to the power characteristic for practically used motionless devices. As can be seen from this figure, there is a significant difference between the dimensionless Euler numbers for the tested static mixers. It can be seen that the dimensionless Euler number increases slightly with increasing complex $\left[\dfrac{Pe}{Br}\left(\dfrac{1}{Re} + \dfrac{Gr\,Fr}{Re^3}\left(\dfrac{d_0}{l_0}\right)^2\right)\right]$. This figure illustrates an interesting feature that for the new type of the mixing device the enhancement of the dimensionless Euler number is much lower than for different types of static mixers. It is demonstrated that commonly applied mixing devices give a greater enhancement in the pressure drop than the new construction.

In the selection of suitable motionless mixing devices for the heat transfer process it is not sufficient to take into consideration the heat transfer ratio and the hydrodynamic conditions separately. The heat transfer encompasses the thermal homogenization and the heat exchange and the procedures for a heat exchanger design are not included in the estimation for the hydrodynamic conditions in the mixed liquid. In this experimental paper a new criterion (see Eq.(20) or (21)) is defined which takes both the heat transfer enhancement and the hydrodynamic conditions into account. The proposed criterion includes the rates of the heat transfer process and the hydrodynamic conditions. Therefore, the criterion ε is defined as the ratio of the dimensionless Nusselt number to the product of the dimensionless Euler, Brinkman and Reynolds numbers.

This criterion may be used to the comparison of the influence of different types of motionless mixing devices on the heat transfer operations. The calculated value of the obtained criterion varies with the parameters of the mixing process and depends on the operating conditions and the physical properties of the liquid. This relationship can be expressed by the following general equation

$$\varepsilon = f(NTU) \Rightarrow \left[4\frac{Nu}{Eu\,Br\,Re}\left(\frac{l_0}{d_0}\right)\right] = f\left[\frac{\pi d_0 l_0 h_0}{\dot{m}\,c_{p_0}}\right] \tag{26}$$

The results of the measurements and calculations are presented in the form of a plot of the criterion versus NTU indication. The results presented in Fig.7 suggest that the proposed relation can be analytically described by a unique monotonic function. This results in an equation for the whole range of the NTU of the form

$$\varepsilon = p_9 (NTU)^{p_{10}} \tag{27}$$

in which it is necessary to use different values of the exponents p_9 and p_{10} for different types of motionless mixing devices. The constants and exponents are computed employing the principle of least squares.

The experimental results showed that the proposed criterion ε increases when the NTU indication is increased in a wide range. The tested static mixers are highly effective during the heat transfer operation. As follows from the experimental results the heat transfer enhancement of the new construction of the mixing device is high without an excessive increase in the pressure drop. It is evident that the new type of the static mixer tends to be superior to a comparably different type over most of the operating range shown.

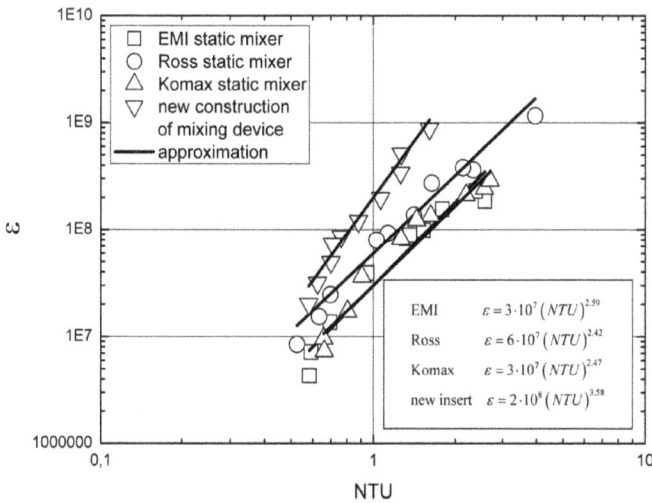

Figure 7. Comparison of the criterion ε values for the different type of motionless mixing devices

5. Conclusion

Salient remarks resulting from the present investigation may be summarized as follows:

1. Various motionless mixing devices have different influence on the heat transfer process and the pressure drop enhancement.
2. The theoretical description of the problem and the equations predicted in the present article is much more attractive because it generalizes the experimental data of the heat transfer process (see Eq.(23)) and the pressure drop database (see Eq.(25)) taking into consideration various parameters in the form of the terms $\left[Pe \dfrac{d_0}{l_0} \left(1 + \dfrac{Gr\,Fr}{Re^2} \left(\dfrac{d_0}{l_0} \right)^2 \right) \right]$ or

$\left[\dfrac{Pe}{Br} \left(\dfrac{1}{Re} + \dfrac{Gr\,Fr}{Re^3} \left(\dfrac{d_0}{l_0} \right)^2 \right) \right]$, respectively. The proposed new-type correlations should be adequate for design purposes when used within the range of experimental verification.

3. With respect to very useful criterions given in the literature which are singly obtained for static mixers it would be interesting to see the new general criterion related to this problem. From the practical point of view, this criterion may be used to the selection of a static mixer for the heat transfer problems.
4. The criterion ε may be used to the comparison of the influence of different types of motionless mixing devices on the heat transfer operations with respect to the hydrodynamic conditions.

Nomenclature

c_p	-	specific heat, $J \cdot kg^{-1} \cdot K^{-1}$
d	-	pipe (mixer) diameter, m
F	-	surface, m^2
g	-	gravitional acceleration, $m \cdot s^{-2}$
h	-	heat transfer coefficient, $J \cdot m^{-2} \cdot K^{-1} \cdot s^{-1}$
k	-	thermal conductivity, $J \cdot m^{-1} \cdot K^{-1} \cdot s^{-1}$
l	-	mixer length, m
$\dot{m}$	-	mass flow rate, $kg \cdot s^{-1}$
N	-	power, $J \cdot s^{-1}$
p	-	pressure, $kg \cdot m^{-1} \cdot s^{-2}$
Δp	-	pressure drop in static mixer, $kg \cdot m^{-1} \cdot s^{-2}$
$\dot{Q}$	-	heat transfer rate, $J \cdot s^{-1}$
r	-	radial cylindrical coordinate, m
R	-	correlation coefficient
t	-	time, s
T	-	temperature, ^0C
T_{in}	-	temperature of flowing and mixed liquid, ^0C
$T_{w,in}$	-	inside wall temperature, ^0C
$\left[T_{in} \right]_{avg}$	-	average value of liquid inside temperature, ^0C
$\overline{w}$	-	velocity vector, $m \cdot s^{-1}$
w_r	-	radial component of liquid velocity, $m \cdot s^{-1}$
w_z	-	axial component of liquid velocity, $m \cdot s^{-1}$
V	-	volume of the static mixer, m^3
z	-	axial cylindrical coordinate, m
$\dot{V}$	-	volumetric flow rate, $m^3 \cdot s^{-1}$

Greek symbols

β	-	coefficient of thermal expansion, K^{-1}
ε	-	thermal-power efficiency criterion
η	-	liquid viscosity, $kg \cdot m^{-1} \cdot s^{-1}$
η_V	-	volumetric viscosity ("second viscosity"), $kg \cdot m^{-1} \cdot s^{-1}$
ρ	-	liquid density, $kg \cdot m^{-3}$
Φ_V	-	viscous dissipation function, $kg \cdot m^{-1} \cdot s^{-3}$

Dimensionless numbers

Br	-	Brinkman number
Eu	-	Euler number
Fr	-	Froude number
Gr	-	Grashof number
Nu	-	Nusselt number
Pe	-	Péclet number
Pr	-	Prandtl number
Re	-	Reynolds number

Superscripts

*	-	dimensionless value

Subscripts

steam	-	heating medium
in	-	Inside
avg	-	average value
0	-	reference value

Abbreviations

MAE	-	mean approximation error
MPE	-	mean percentage error
MRE	-	mean relative error
NTU	-	number of transfer units

Author details

Rafał Rakoczy, Marian Kordas and Stanisław Masiuk
*Institute of Chemical Engineering and Environmental Protection Process, West Pomeranian
University of Technology, Poland*

Acknowledgement

This work was supported by the Polish Ministry of Science and Higher Education from
sources for science in the years 2012-2013 under Inventus Plus project

6. References

Al Taweel, A.M.; Azizi, J.Y.F.; Odedra, D. & Gomma, H.G. (2005). Using in-line static
mixers to intensity gas-liquid mass transfer processes. *Chemical Engineering Science*,
Vol.60, pp. 6378-6390

Engler, M.; Kockmann, N.; Kiefer, T. & Woias, P. (2004). Numerical and experimental investigations on liquid mixing in static micromixers. *Chemical Engineering Journal*, Vol.101, pp. 315-322

Fourcade, E.; Wadely, R.; Hoefsloot, H.C.J.; Green, A. & Iedema, P.D. (2001). CFD calculation of laminar striation thinning in static mixer reactors. *Chemical Engineering Science*, Vol.56, pp. 6729-6741

Hobbs, D.M.; Swanson, P.D. & Muzzio, F.J. (1998). Numerical characterization of low Reynolds number flow in the Kenics static mixer. *Chemical Engineering Science*, Vol.53, pp. 1565-1584

Joshi, P.; Nigam, K.D.P. & Nauman, E.B. (1995). The Kenics static mixer: new data and proposed correlations. *Chemical Engineering Journal*, Vol.59, pp. 265-271

Kumar, V.; Shirke, V. & Nigam, K.D.P. (2008). Performance of Kenics static mixer over a wide range of Reynolds number. *Chemical Engineering Journal*, Vol.139, pp. 284-295

Lemenand, T.; Durandal, C.; Valle, D.D. & Peerhossaini, H. (2010). Turbulent direct-contact heat transfer between two immiscible fluids. *International Journal of Thermal Science*, Vol.49, pp. 1886-1898

Li, H.Z.; Fasol, C. & Choplin, L. (1997). Pressure drop of newtonian and non-newtonian fluids across a Sulzer SMX static mixer. *Trans IChemE*, Vol.75, pp. 792-796

Li, H.Z.; Fasol, Ch. & Choplin, L. (1996). Hydrodynamics and heat transfer of rheologically complex fluids in a Sulzer SMX static mixer. *Chemical Engineering Science*, Vol.51, pp. 1947-1955

Liu, S.; Hrymak, A.N. & Wood, P.E. (2006). Laminar mixing of shearing thinning fluids in a SMX static mixer. *Chemical Engineering Science*, Vol.61, pp. 1753-1759

Masiuk, S. & Rakoczy, R. (2007). Power consumption, mixing time, heat and mass transfer measurements for liquid vessels that are mixed using reciprocating multiplates agitators. *Chemical Engineering and Processing*, Vol.46, pp. 89-98

Masiuk, S. & Szymański, E. (1997). Static mixing device, Polish Patent PL No. 324150

Masiuk, S. (1999). Power consumptions measurements in a liquid vessel that is mixed using a vibratory agitator. *Chemical Engineering Journal*, Vol.75, pp. 161-165

Masiuk, S. (2000). Mixing time for a reciprocating plate agitator with flapping blades. *Chemical Engineering Journal*, Vol.79, pp. 23-30

Rama Rao, N.V.; Baird, M.H.I.; Hrymak, A.M. & Wood, P.E. (2007). Dispersion of high-viscosity liquid-liquid systems by flow through SMX static mixer elements. *Chemical Engineering Science*, Vol.62, pp. 6885-6895

Regner, M.; Östergren, K. & Trägårdh, C. (2005). An improved numerical method of calculating the striation thinning in static mixers. *Computers and Chemical Engineering*, Vol.30, pp. 376-380

Ruivo, R.; Paiva, A. & Simões, P.C. (2006). Hydrodynamics and mass transfer of a static mixer at high pressure conditions. *Chemical Engineering and Processing*, Vol.45, pp. 224-231

Simões, P.C.; Afonso, B.; Fernandes, J. & Mota, J.P.B. (2008). Static mixers as heat exchangers in supercritical fluid extraction processes. *Journal of Supercritical Fluids*, Vol.43, pp. 477-483

Thakur, R.K.; Vial, Ch.; Nigam, K.D.P.; Nauman, E.B. & Djelveh, G. (2003). Static mixers in the process industries – A review. *Trans IChemE*, Vol.81, pp. 787-826

Visser, J.E.; Rozendal, P.F.; Hoogstraten, H.W. & Beenckers, A.A.C.M. (1999). Three-dimensional numerical simulation of flow and heat transfer in the Sulzer SMX static mixer. *Chemical Engineering Science*, Vol.54, pp. 2491-2500

Permissions

The contributors of this book come from diverse backgrounds, making this book a truly international effort. This book will bring forth new frontiers with its revolutionizing research information and detailed analysis of the nascent developments around the world.

We would like to thank Salim N. Kazi, for lending his expertise to make the book truly unique. He has played a crucial role in the development of this book. Without his invaluable contribution this book wouldn't have been possible. He has made vital efforts to compile up to date information on the varied aspects of this subject to make this book a valuable addition to the collection of many professionals and students.

This book was conceptualized with the vision of imparting up-to-date information and advanced data in this field. To ensure the same, a matchless editorial board was set up. Every individual on the board went through rigorous rounds of assessment to prove their worth. After which they invested a large part of their time researching and compiling the most relevant data for our readers. Conferences and sessions were held from time to time between the editorial board and the contributing authors to present the data in the most comprehensible form. The editorial team has worked tirelessly to provide valuable and valid information to help people across the globe.

Every chapter published in this book has been scrutinized by our experts. Their significance has been extensively debated. The topics covered herein carry significant findings which will fuel the growth of the discipline. They may even be implemented as practical applications or may be referred to as a beginning point for another development. Chapters in this book were first published by InTech; hereby published with permission under the Creative Commons Attribution License or equivalent.

The editorial board has been involved in producing this book since its inception. They have spent rigorous hours researching and exploring the diverse topics which have resulted in the successful publishing of this book. They have passed on their knowledge of decades through this book. To expedite this challenging task, the publisher supported the team at every step. A small team of assistant editors was also appointed to further simplify the editing procedure and attain best results for the readers.

Our editorial team has been hand-picked from every corner of the world. Their multi-ethnicity adds dynamic inputs to the discussions which result in innovative outcomes. These outcomes are then further discussed with the researchers and contributors who give their valuable feedback and opinion regarding the same. The feedback is then collaborated with the researches and they are edited in a comprehensive manner to aid the understanding of the subject.

Apart from the editorial board, the designing team has also invested a significant amount of their time in understanding the subject and creating the most relevant covers. They scrutinized every image to scout for the most suitable representation of the subject and create an appropriate cover for the book.

The publishing team has been involved in this book since its early stages. They were actively engaged in every process, be it collecting the data, connecting with the contributors or procuring relevant information. The team has been an ardent support to the editorial, designing and production team. Their endless efforts to recruit the best for this project, has resulted in the accomplishment of this book. They are a veteran in the field of academics and their pool of knowledge is as vast as their experience in printing. Their expertise and guidance has proved useful at every step. Their uncompromising quality standards have made this book an exceptional effort. Their encouragement from time to time has been an inspiration for everyone.

The publisher and the editorial board hope that this book will prove to be a valuable piece of knowledge for researchers, students, practitioners and scholars across the globe.

List of Contributors

Otilia Nedelcu and Corneliu Ioan Sălișteanu
Department of Electronics, Telecommunications and Energy Engineering, Valahia University of Targoviste, Romania

E. Martínez and G. Soto
Universidad Autónoma Metropolitana – Azcapotzalco, Mexico City, Mexico

W. Vicente and M. Salinas
Instituto de Ingeniería, Universidad Nacional Autónoma de México, Ciudad Universitaria, Mexico City, Mexico

A. Campo
Department of Mechanical Engineering, University of Texas at San Antonio, San Antonio, TX, USA

Anwar Ja'afar Mohamed Jawad
Al-Rafidain University College, Baghdad, Iraq

Aleksandra Rashkovska, Roman Trobec, Matjaž Depolli, Gregor Kosec
Jožef Stefan Institute, Slovenia

Lei Zhang
School of Mechanical Engineering, Shandong University, P.R. China

Abdlmanam S. A. Elmaryami and Badrul Omar
University Tun Hussein Onn Malaysia, Mechanical Engineering Department, Batu Pahat, Johor, Malaysia

Rudi Radrigán Ewoldt
Faculty of Agricultural Engineering, Agroindustries Department, Development of Agroindustries Technology Center, University of Concepción, Chile

Zhong-Shan Deng
Technical Institute of Physics and Chemistry, Chinese Academy of Sciences, Beijing, China

Jing Liu
Technical Institute of Physics and Chemistry, Chinese Academy of Sciences, Beijing, China
Department of Biomedical Engineering, School of Medicine, Tsinghua University, Beijing, China

Abdus Sattar
North South University, Bashundhara Baridhara, Dhaka, Bangladesh

Mohammad Ferdows
Department of Mathematics, Dhaka University, Dhaka, Bangladesh

Ryoichi Chiba
Department of Mechanical Systems Engineering, Asahikawa National College of Technology, Asahikawa, Japan

Rangel-Hernandez V. H.
Mechanical Engineering Department, Engineering Division, University of Guanajuato, Palo Blanco S/N, 36885, Salamanca, Mexico

Damian-Ascencio C. E.
Physics Department, Sciences Division, University of Guanajuato, C.P. 37150, León, Mexico

Dragan Ružić and Ferenc Časnji
University of Novi Sad, Faculty of Technical Sciences, Serbia

Bautista Oscar, Bautista Eric and Arcos Carlos
National Polytechnic Institute of Mexico, College of Mechanical and Electrical Engineering (ESIMEAzcapotzalco), Deparment of thermofluids, México D. F., Mexico

Méndez Federico and Sánchez Salvador
National University of Mexico (UNAM), Division of Mechanical and Electrical Engineering, Department of Thermofluids, México D. F., Mexico

Rafał Rakoczy, Marian Kordas and Stanisław Masiuk
Institute of Chemical Engineering and Environmental Protection Process, West Pomeranian University of Technology, Poland